D0607428

**On Size and Life**

# ON SIZE AND LIFE

Thomas A. McMahon    John Tyler Bonner

SCIENTIFIC
AMERICAN
LIBRARY

An imprint of Scientific American Books, Inc.
New York

Leah Ben-David Val, Picture Editor

**Library of Congress Cataloging in Publication Data**

McMahon, Thomas A., 1943-
On size and life.

Bibliography: p.
Includes index.
1. Body size. 2. Life (Biology) 3. Morphology (Animals) 4. Botany–Morphology. I. Bonner, John Tyler. II. Title.
QH351.M34 1983 574.4 83-16472
ISBN 0-7167-5000-7



Printed in the United States of America

Scientific American Library is published by Scientific American Books, Inc., a subsidiary of Scientific American, Inc.

Distributed by W.H. Freeman and Company, 41 Madison Ave., New York, New York 10010.

1 2 3 4 5 6 7 8 9 0 KP 1 0 8 9 8 7 6 5 4 3

*Dedicated to our wives*
*Carol and Ruth*

# Contents

# Preface

Once, we were smaller than we are now. Children often wonder and speculate about their own size, sometimes imagining what it would be like to be smaller or larger than they actually are. If I were very small, I could have a matchbox for my bed. I could ride a cat like a horse. The world would be entirely different for me, merely because I was small. On the other hand, if I were large, I could walk to town in a single step. The earth would appear flat below me, and even mountains would disappear as obstacles. More than one famous story of imagination is built on such speculations. Alice began her adventures in Wonderland only after she had changed her size. She grew small, then large, then small again, and swam in a sea of her own tears. Tom Thumb reached his full height only four minutes after his birth. He was swallowed by a cow, carried away in a raven's mouth, and later driven in a coach drawn by mice. Stuart Little, E.B. White's civilized mouse, cruised in Central Park on a toy schooner, drove a scale model automobile six inches long, and cut down dandelions with an axe. In his travels, Gulliver changed not his absolute size but only his relative size, with nonetheless important consequences. In all of our literature, but more frequently in literature written for children, people are transformed into giants or dwarfs, and this changes everything for them: it confers upon them both extraordinary powers and extraordinary limitations, and it becomes the means of great adventures.

This book is not explicitly about the imaginative effects of size, although it is impossible to forget giants and dwarfs entirely in taking up the subject before us. Instead, this book is about the observable effects of size on animals and plants, seen and evaluated using the tools of science. It will come as no surprise that among those tools are microscopes and cameras. Ever since Antoni Van Leeuwenhoek first observed microorganisms (he called them "animalcules") in a drop of water from Lake Berkel, the reality of miniature life has expanded our concepts of what all life could possibly be. Some other tools we shall use—equally important ones—are mathematical abstractions, including a type of relation we shall call an allometric formula. It turns out that allometric formulas reveal certain beautiful regularities in nature, describing a pattern in the comparisons of animals as different in size as the shrew and the whale, and this can be as delightful in its own way as the view through a microscope.

Size and scale in the living world is an ancient subject, but the person who was most influential in bringing it to our attention in this century was D'Arcy

Wentworth Thompson, a biologist of extraordinary erudition from St. Andrews University in Scotland. The first edition of his famous book *On Growth and Form* was published in 1917, and one of the shining gems in that remarkable volume is the chapter "On Magnitude." In it, Thompson traces the history of the subject back to Archimedes, who argued from elementary mathematics that, in similar solid geometric figures of different sizes, the surface increases as the square of the linear dimensions and the volume increases as the cube. Thompson called this relationship of large and small bodies the principle of similitude, and he added:

> But it was Galileo who, well nigh three hundred years ago, had first laid down this general principle of similitude; and he did so with the utmost possible clearness, and with a great wealth of illustration drawn from structures living and dead. He said that if we tried building ships, palaces or temples of enormous size, yards, beams and bolts would cease to hold together; nor can Nature grow a tree nor construct an animal beyond a certain size, while retaining the proportions and employing the materials which suffice in the case of a smaller structure. The thing will fall to pieces of its own weight unless we either change its relative proportions, which will at length cause it to become clumsy, monstrous and inefficient, or else we must find new material, harder and stronger than was used before. Both processes are familiar to us in Nature and in art, and practical applications, undreamed of by Galileo, meet us at every turn in this modern age of cement and steel.

If we trace the subject in the years after Galileo, all the important advances were made by the physicists and the engineers. The invention of such mechanical devices as the steam engine by the Scot James Watt, and ultimately the industrialization of Britain and other Western countries, generated a great need for understanding how size affects a whole variety of mechanical structures, from motors to bridges. Not only were the problems of similitude raised by Galileo fully appreciated, but it was understood to be important to carefully calculate the role of weight, strength, power, and other physical properties in determining the design. It was no longer sufficient to build by trial and error; it was necessary to find ways of making rules that showed exactly how these properties changed with size, and how they could be accurately predicted to produce safe bridges and boilers or efficient motors.

Flying machines illustrate these concerns particularly well. The first gliders and powered airplanes were built largely by a combination of luck, trial, and much error. In this way, the Wright brothers made the first propeller airplane that flew under its own power, and the immediate further advances were the product of empirical improvements in their design. But it was not long before engineers began to ask what were the properties necessary for flight and how were those properties affected by size.

Out of all this pioneering arose an extraordinarily powerful method called "dimensional analysis" (the main subject of Chapter 3 of this book). As a result, it became possible to analyze the size-related properties of any structure, be it an airplane, a boat, or a bridge. Now there were rules that could be imposed on the design of small or large structures. By the early part of this century, magnitude had become a major component in modern engineering, but the biologists had not advanced much beyond Galileo.

The rise of our understanding of the principle of similitude as it applies to plants and animals came from work in a number of branches of biology. Here we will give a thumbnail sketch of each of these branches, mentioning the names of a few of the pioneers (but, for want of space, neglecting those of many who deserve mention).

Physiologists were the first to be aware of the problem of size. In the 1830s, a French physiologist and a mathematician, J. F. Rameaux and P. F. Sarrus, realized that an animal loses heat through the surface of its body while its capacity for heat production is related to its volume and therefore (following Archimedes) that the larger the animal, the greater its heat production relative to its heat loss. C. Bergmann in Germany further refined the argument a decade later, but it was not until this century that we had careful analyses of the exact relation between heat production (or metabolism, as it is more often called) and body size. (This is a subject first raised in Chapter 2 of this book and examined in detail in Chapter 4). The central figure in this more modern work was Max Kleiber of the University of California, but there are others who made large contributions.

This brings us to a second branch of biology, embryology or developmental biology, which concerns itself in part with the change in size of an individual organism—from a minute egg to an elephant or a giant sequoia, to mention two extreme cases. Growth rates and the concomitant changes in size and shape were understood in a general way, but they were difficult to measure in a useful fashion until Julian S. Huxley, the well known British zoologist, and (independently) the French zoologist G. Teissier developed the method of "relative growth" or allometry (which is the subject of Chapter 2).

The third area of biology in which size considerations have played an important role is in understanding the supporting structure of an organism. Perhaps the credit for pioneering here should go to the British mathematician G. Greenhill, who, in the 1880s, saw clearly why a tall tree was disproportionately thick and was able, furthermore, to define precisely the relation between height and diameter for trees (a subject touched upon in Chapter 4 and discussed in detail in Chapter 5). Since then, there have been many studies of the proportions of limbs, bones, branches, and other structural components of organisms. Perhaps the most interesting trend in very recent times is the growing appreciation of the role of elasticity in biological supporting structures (a topic treated in Chapter 4).

The fourth biological phenomenon that is enormously affected by size is animal locomotion. Important early studies on swimming in fishes were done in the 1930s by Sir James Gray of Cambridge University, who provided a great stimulus to the study of locomotion in general—not only of swimming but also of running and flying. (These subjects, especially as they are influenced by scale, are examined in detail in Chapter 5.)

The locomotion of small animals and microbes involves different physical problems than the ones mentioned in the previous paragraph: It is another world. This had been appreciated for a long time, but recent studies on the swimming of bacteria and sperm cells, begun by H. C. Berg, C. J. Brokaw, and others in the 1970's, have opened the eyes of all of us. We suddenly have a new view into this microworld, in which viscosity, the forces of cohesion, and diffusion become paramount, while the effects of gravity are negligible. This is the fifth area of biology in which size impinges on living organisms (and it is discussed in detail in Chapter 6).

The sixth and final area in which biologists have concerned themselves with the sizes of organisms has been in the study of evolution and ecology. The well-known American paleontologist Edward Drinker Cope, in the second half of the last century, was the first to emphasize that many fossil sequences of vertebrates show consistent trends of size increase over long time spans, raising the reasonable possibility that size may be subject to natural selection. Size also plays a role in ecology, for if one looks at animals and plants in nature in any one habitat, one finds a mixture of the large and the small. The significance of this was appreciated by C. S. Elton in the 1920s, who pointed out that large carnivores ate more numerous, smaller animals, and from this he developed the concept of food pyramids. But there are other ramifications of size in evolutionary biology (and these are touched upon in Chapter 1 and examined in detail in Chapter 7, the last chapter of the book).

* * * *

Since the study of size and life requires the fusing of engineering and of biological principles, it is not coincidence that one of us is an engineer (T. M.) and the other (J. T. B.) is a biologist. Each of us has previously worked and written on the subject of size in biology, and we are indebted to our friends

Robert May of Princeton University and Peter Renz of W. H. Freeman/Scientific American Books for suggesting that we join forces and do a book on the subject for the general reader. It has been a rewarding and exciting experience for us both, and we have learned from each other. The idea of the book is to reveal a part of science that may be new to some but that will be illuminated in a new way for those who are already familiar with it.

* * * *

During the course of preparing this book, we received help from many people. We are grateful to Peter Renz, who played an essential role, first as a catalyst and then as a thoughtful and creative editor. R. McNeill Alexander, Shelly Copley, Nick Humez, Eric Lander, Susan Middleton, Richard O'Hara, Frederick Pearson, Daniel Rubenstein, Steven Strogatz, and Milton Van Dyke, as well as a number of anonymous reviewers, read and criticized the manuscript (or parts of it), and we wish to thank them all for their help. Their advice was invariably good, and despite the fact that occasionally some of it was conflicting, we, of course, followed it all. Special thanks go to Sharon McDevitt, who struggled with two idiosyncratic authors and a word processor and produced the typescript with speed and loving care.

THOMAS A. MCMAHON
Cambridge, Massachusetts

JOHN TYLER BONNER
Princeton, New Jersey

September 1983

**On Size and Life**

# Chapter 1

## The Natural History of Size

"Noah's Ark," by Jan Brueghel the Elder (1568–1625).

When we say that an animal is large or small, we usually mean that it is larger or smaller than ourselves. We think of an elephant as large because it towers over us, and a mouse is small because it does not come up to our ankles. There is also the mechanical limitation of our eyes. A bacterium is so small that we cannot see it without the help of a powerful microscope, and therefore we think of it as belonging to a world that we can observe only by means of technical tricks.

Let us begin this book about the sizes and shapes of animals and plants with a quick panoramic view of the range of sizes of living organisms. To do this, we will borrow from a remarkable book published in 1932—*The Science of Life,* by H. G. Wells, J. S. Huxley, and G. P. Wells. This formidable trio designed a series of figures (shown on pages 2 and 3) that gives an excellent picture of the extremes of size to be found among living things.

The largest animal alive—and the largest that ever existed—is the blue whale. A fully grown individual may be in excess of 22 meters long and may weigh more than 100 tons. The largest dinosaur was *Brachiosaurus,* whose weight is estimated to have been more than 50 tons, and the longest was *Diplodocus,* whose length was almost 29 meters. The largest tree is the giant sequoia, which may achieve a height of well over 100 meters. (The weight of such a tree has little meaning, because the bulk of any tree is mainly dead wood. The living tissue is found only in a layer a few millimeters thick beneath the bark and in the leaves. The amount of "skeleton" in a tree far exceeds anything found in a vertebrate or any other animal.) In the figure on page 2, it is possible to see, by comparison, the record sizes of many other kinds of organisms: the largest jellyfish, clam, worm, crab, fish, reptile, bird, flying vertebrate, and mammal. The record land mammal for all time, the *Baluchitherium,* is also shown.

Extremes of middle-sized organisms are shown in the top figure on page 3. They range from the smallest amphibian, bird, and mammal to the largest protozoan, polyp, land snail, starfish, beetle, and bird's egg. Also shown are a few ordinary animals to give the viewer some points of comparison.

The next step downward is provided by the figure at the lower left, which shows organisms that are very small but that can still be viewed with the naked eye. The smallest vertebrate, a tropical frog, is repeated from the figure above it. Also shown are a flea, a housefly, one of the smallest fishes, and the smallest land snail.

The figure at the lower right brings us down to the lower limit of cells with nuclei—that is, all cells larger than bacteria. Shown here is the largest ciliate protozoan (a single-celled creature covered with hairlike cilia that it uses for

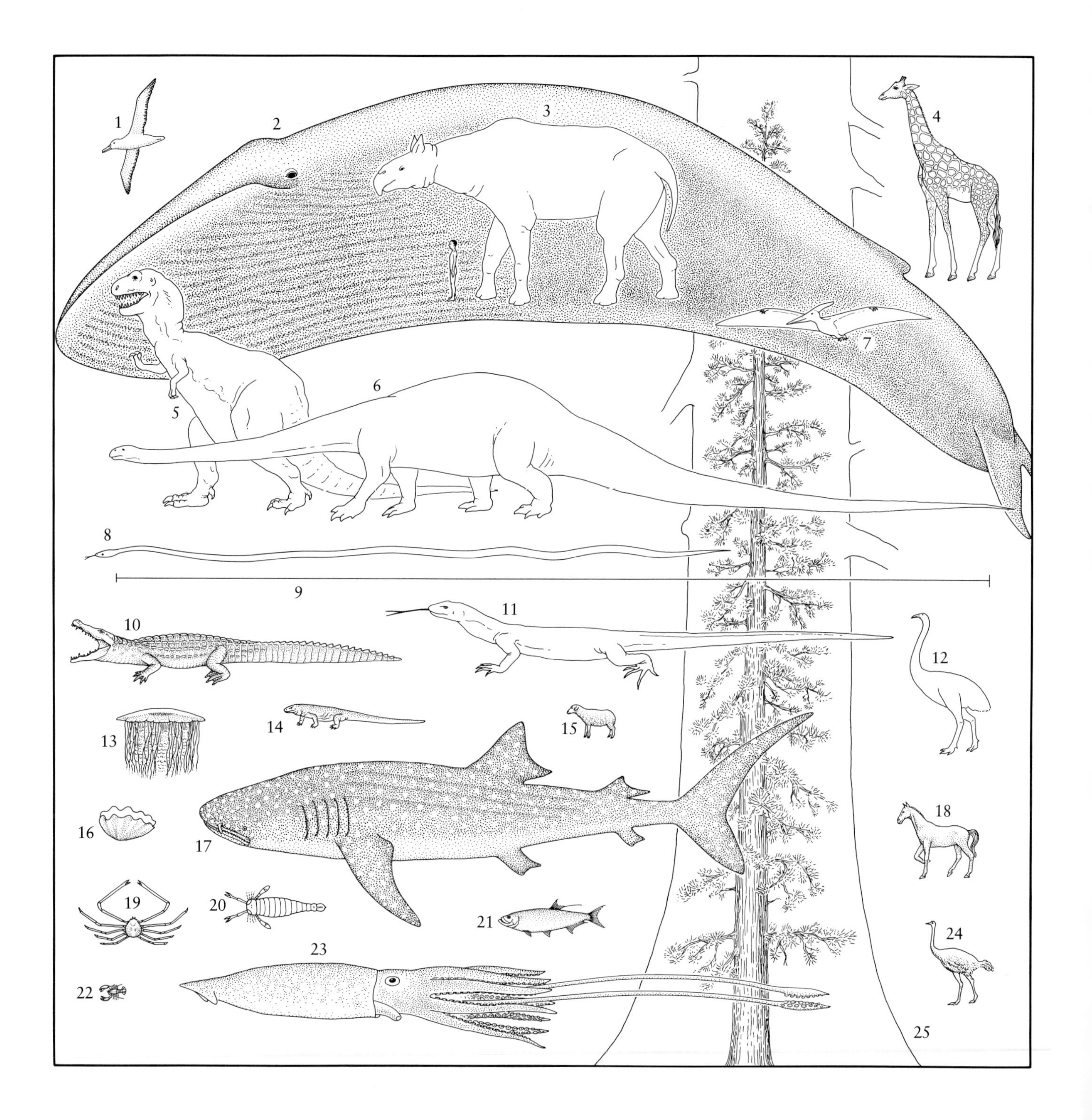
1
2
3
4
5
6
7
8
9
10
11
12
13
14
15
16
17
18
19
20
21
22
23
24
25

The biggest living things (*left*). All the organisms are drawn to the same scale. *1*, The largest flying bird (albatross); *2*, the largest known animal (the blue whale), *3*, the largest extinct land mammal (*Baluchitherium*) with a human figure shown for scale; *4*, the tallest living land animal (giraffe); *5*, *Tyrannosaurus*; *6*, *Diplodocus*; *7*, one of the largest flying reptiles (*Pteranodon*); *8*, the largest extinct snake; *9*, the length of the largest tapeworm found in man; *10*, the largest living reptile (West African crocodile); *11*, the largest extinct lizard; *12*, the largest extinct bird (*Aepyornis*); *13*, the largest jellyfish (*Cyanea*); *14*, the largest living lizard (Komodo dragon); *15*, sheep; *16*, the largest bivalve mollusc (*Tridacna*); *17*; the largest fish (whale shark); *18*, horse; *19*, the largest crustacean (Japanese spider crab); *20*, the largest sea scorpion (Eurypterid); *21*, large tarpon; *22*, the largest lobster; *23*, the largest mollusc (deep-water squid, *Architeuthis*); *24*, ostrich; *25*, the lower 105 feet of the largest organism (giant sequoia), with a 100-foot larch superposed.

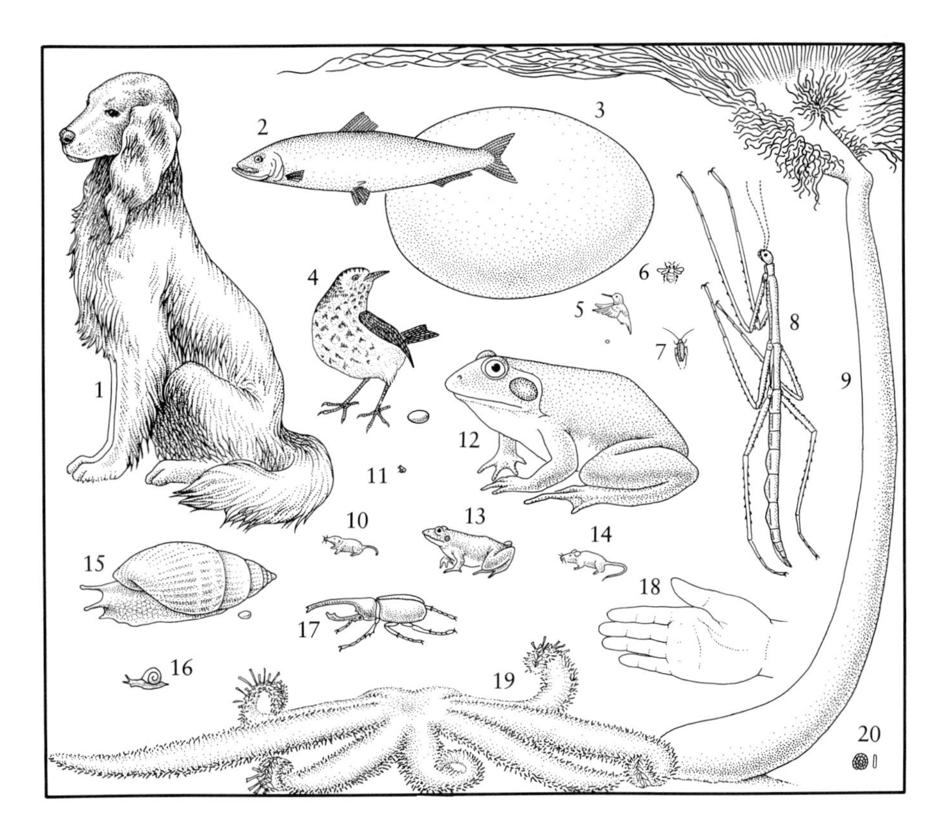


Medium-sized creatures (*above*). *1*, Dog; *2*, common herring; *3*, the largest egg (*Aepyornis*); *4*, song thrush with egg; *5*, the smallest bird (hummingbird) with egg; *6*, queen bee; *7*, common cockroach; *8*, the largest stick insect; *9*, the largest polyp (*Branchiocerianthus*); *10*, the smallest mammal (flying shrew); *11*, the smallest vertebrate (a tropical frog); *12*, the largest frog (goliath frog); *13*, common grass frog; *14*, house mouse; *15*, the largest land snail (*Achatina*) with egg; *16*, common snail; *17*, the largest beetle (goliath beetle); *18*, human hand; *19*, the largest starfish (*Luidia*); *20*, the largest free-moving protozoan (an extinct nummulite).

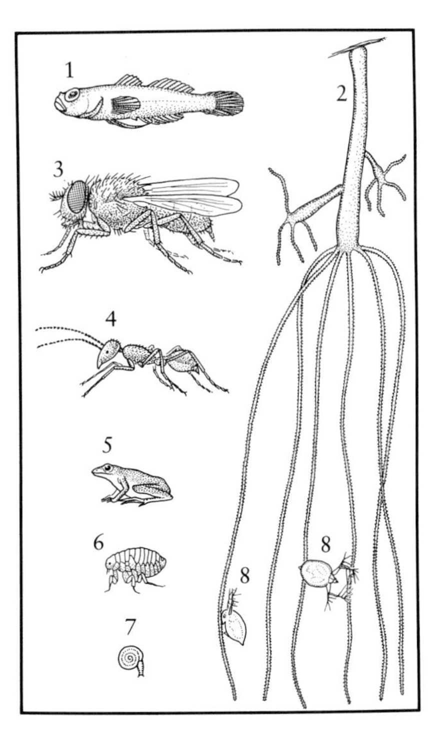


Small, "naked-eye" creatures (*lower left*). *1*, One of the smallest fishes (*Trimmatom nanus*); *2*, common brown hydra, expanded; *3*, housefly; *4*, medium-sized ant; *5*, the smallest vertebrate (a tropical frog, the same as the one numbered *11* in the figure *above*); *6*, flea (*Xenopsylla cheopis*); *7*, the smallest land snail; *8*, common water flea (*Daphnia*).

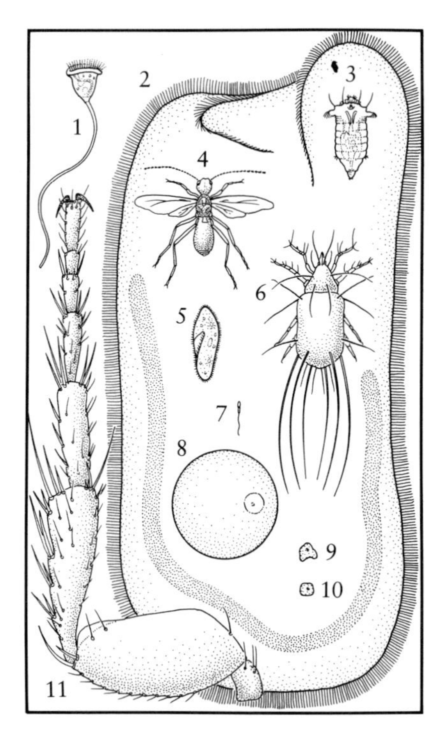


The smallest "naked-eye" creatures and some large microscopic animals and cells (*below right*). *1*, *Vorticella*, a ciliate; *2*, the largest ciliate protozoan (*Bursaria*); *3*, the smallest many-celled animal (a rotifer); *4*, smallest flying insect (*Elaphis*); *5*, another ciliate (*Paramecium*); *6*, cheese mite; *7*, human sperm; *8*, human ovum; *9*, dysentery amoeba; *10*, human liver cell; *11*, the foreleg of the flea (numbered *6* in the figure to the *left*).

propulsion), along with the leg of a flea, a species of rotifer that is the smallest multicellular organism, and the smallest flying insect. Among these extremes are two other species of protozoans of average size, a human egg and sperm, and a human liver cell.

If one pursues small size below these limits, one comes into the world of bacteria and viruses. Viruses do not fully qualify as living organisms because they can propagate themselves only as parasites in the cells of true organisms. They lack most of the cell machinery characteristic of life, and they function by usurping the metabolic machinery of the organisms they parasitize. Bacteria

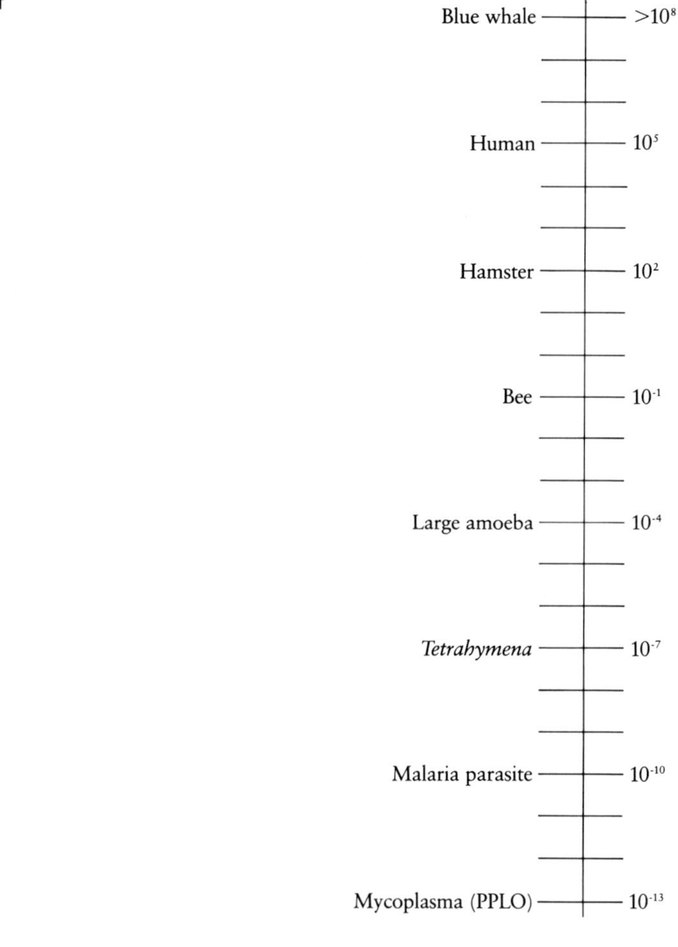


Logarithmic scale showing the spectrum of size in living things.

are roughly a tenth as long as the cells of most organisms, and a virus is minute by comparison: no more than a package of macromolecules equipped to slip through the membranes of cells.

It would be helpful now to put all this information together in one place. The fact that the size range of organisms is so enormous makes it impossible to put everything in one drawing. One way to solve this problem is to put the mass in grams of a whole spectrum of organisms on a logarithmic scale from the lightest to the heaviest as we have done in the figure on the facing page. In this way, it can clearly be shown that there is a range of 21 orders of magnitude—that is, the largest blue whale is more than $10^{21}$ times heavier than the smallest microbe. This factor of $10^{21}$ is very large indeed. The scale on the figure is of the masses of the animals and plants, but one could equally well express this in terms of length. This is an astoundingly large size range, producing a mental image of worlds within worlds. But it is more than an image: it is a reality.

## The Evolution of Size

During the course of evolution, there has been an increase in the upper size limit of animals and plants. Ever since earliest times, small organisms have flourished, but the range of sizes present at any one moment in earth history has progressively increased over the ages.

The fossil evidence for this increase in the range of sizes is scanty but unequivocal. The oldest known fossil-bearing rocks are estimated to be about 3,400 million years old. These rocks contain only fossils of minute bacteria. The next major group of fossils comprises multicellular filaments of simple blue-green algae, which have been found in deposits that are as much as 2,700 million years old. From these two data points and a knowledge of the relative sizes of bacteria and blue-green algae, one may conclude that, in a few hundred million years, the size range of organisms increased in linear dimensions at least 20 times.

A note on the primitive nature of the cells of bacteria and blue-green algae might be helpful here. These cells differ from those of all other organisms in a number of significant respects. Their genetic material (DNA) is not enclosed in a membrane, nor is it packaged in chromosomes as it is in all higher organisms. Thus, they lack a true mitosis, the characteristic chromosome division and separation that occurs when the cells of higher organisms grow and cleave.

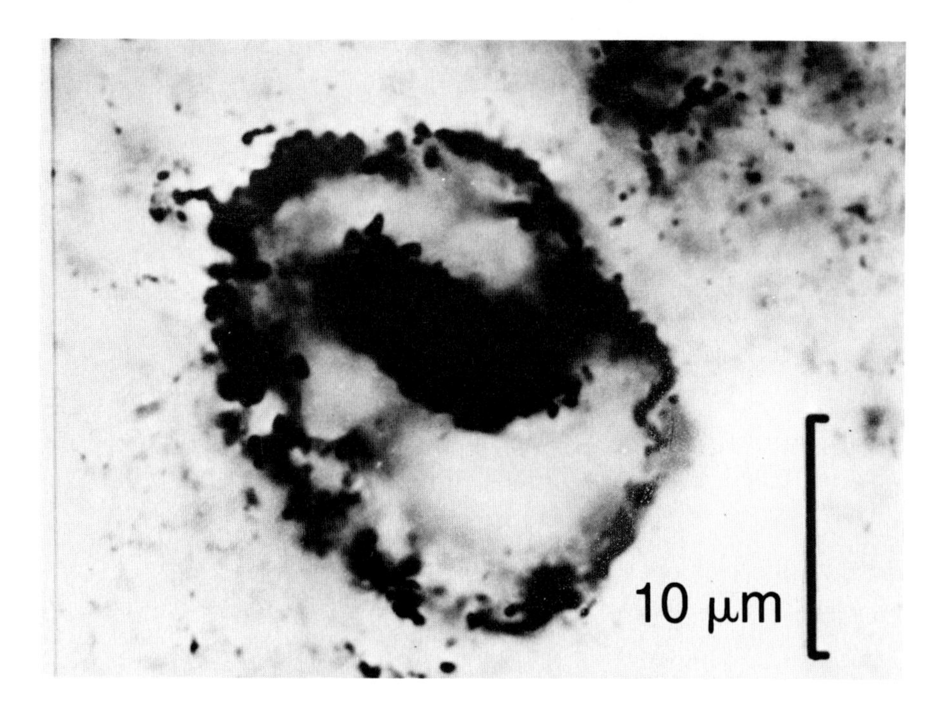


A fossil of one of the oldest known organisms, *Archaeosphaeroides barbertonensis,* a small bacterium-like cell. It was found in the Swartkoppie Formation of the Swaziland sequence, South Africa. It is about 3,400 million years old.

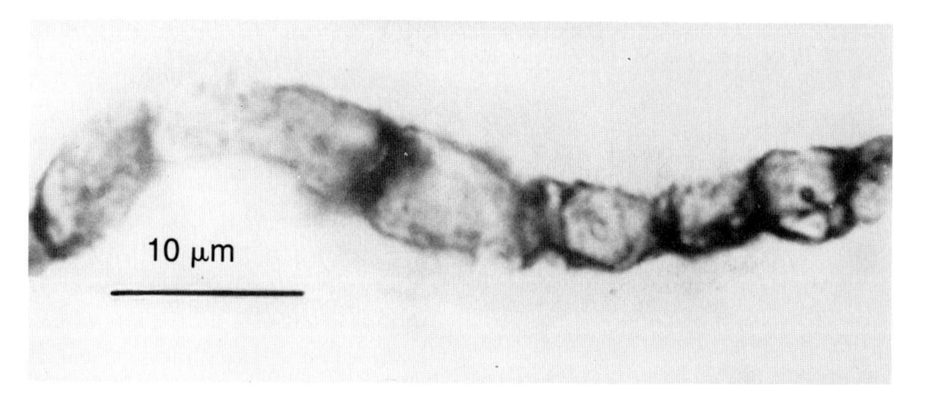


A fossil of the blue-green alga *Gunflintia grandis.* This fossil, which was found in the Gunflint Formation, Lake Superior, Canada, is about 2,700 million years old.

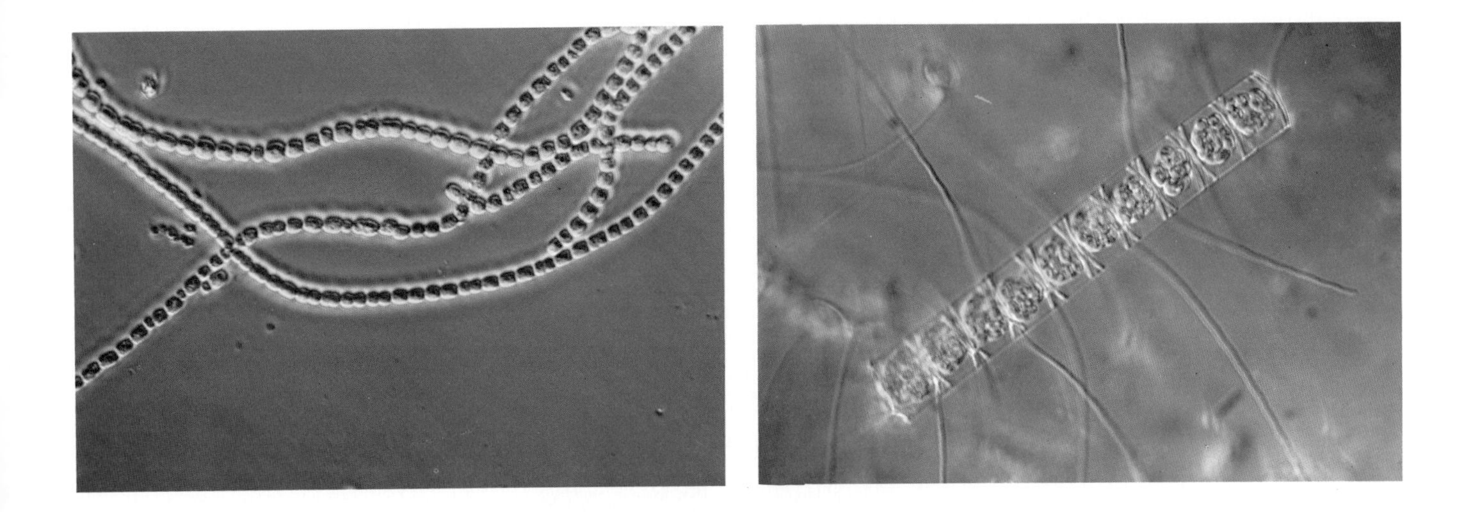

Two types of modern blue-green algae. *Left, Anabena, right, Chaetoceras densus.*

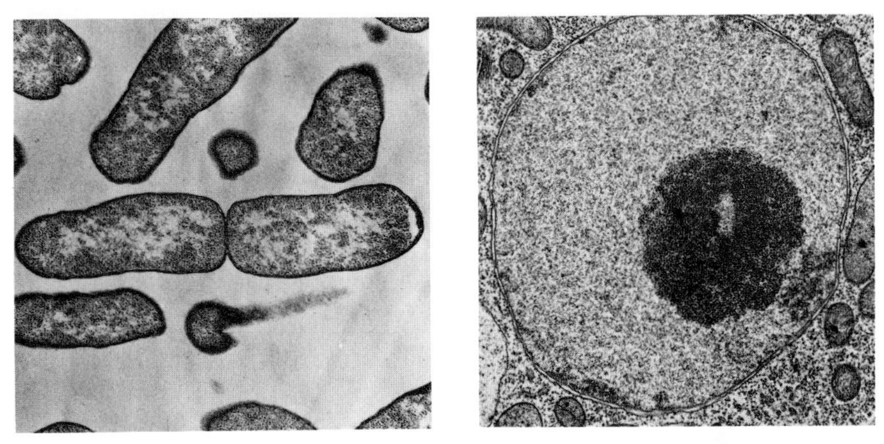

The relative sizes of a prokaryotic cell (the bacterium *Escherichia coli*), and a typical eukaryotic cell. The *Escherichia coli* is about 3.5μm in length. The eukaryotic cell (root tip parenchyma from *Arabidopsis thaliana*) is about 10μm in diameter.

There are many other differences between these ancient cell types (which are called *prokaryotes*) and the cells of all other groups of animals and plants (which are called *eukaryotes*). One of the main differences between bacterial cells and eukaryotic cells is size, for eukaryotic cells are in the size range of the cells of blue-green algae. Prokaryotes (from bacteria to blue-green algae) took about 500 million years to evolve, but the evolution of the first eukaryotic cells took 2,500 million years. Clearly, the development of this more elaborate kind of cell construction has been a crucial step, for early eukaryotic cells began the evolutionary path that led to complex multicellular animals and plants.

Becoming multicellular is an especially significant way of becoming large. The advent of multicellularity opened the floodgates of evolution, making possible the appearance of huge animals with brains, giant trees, and flying birds, bats, and insects. It is extraordinary that, within this enormous variety of large multicellular eukaryotic organisms, the one thing that has remained constant within remarkably strict limits is cell size. Most cells of most organisms, from

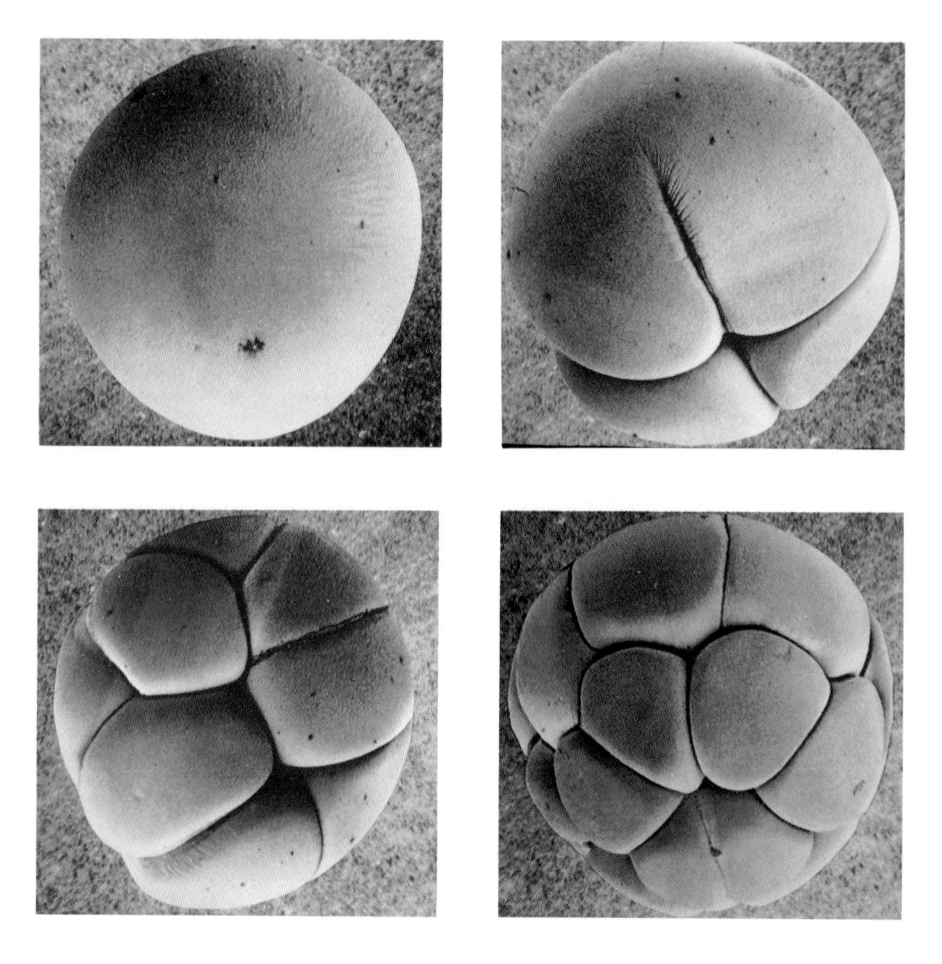

Becoming multicellular is an especially significant way of becoming large. These scanning electron micrographs show a frog's egg first as a single cell and then as its development proceeds to the four-celled, eight-celled, and sixteen-celled stages.

minuscule nematode worms to enormous whales, are roughly 10 micrometers in diameter (a micrometer is $10^{-6}$ meter).

One sometimes hears that there are exceptions to this general rule. For instance, an unfertilized egg of an ostrich is, in a sense, a single cell. But it is a single cell enormously swollen by the addition of yolk and albumen. If one considers only its nucleus and cytoplasm, the cell's diameter is close to 10 micrometers. The cell proper is only a tiny speck that sits on top of the huge yolk. A nerve cell (neuron) may have an extraordinarily long extension, called an axon, that runs, say, the length of a giraffe's neck (or, even more remarkably, the length of the neck of some gigantic dinosaur, such as *Brachiosaurus*). In this case, however, the cell body of the neuron, which contains the nucleus,

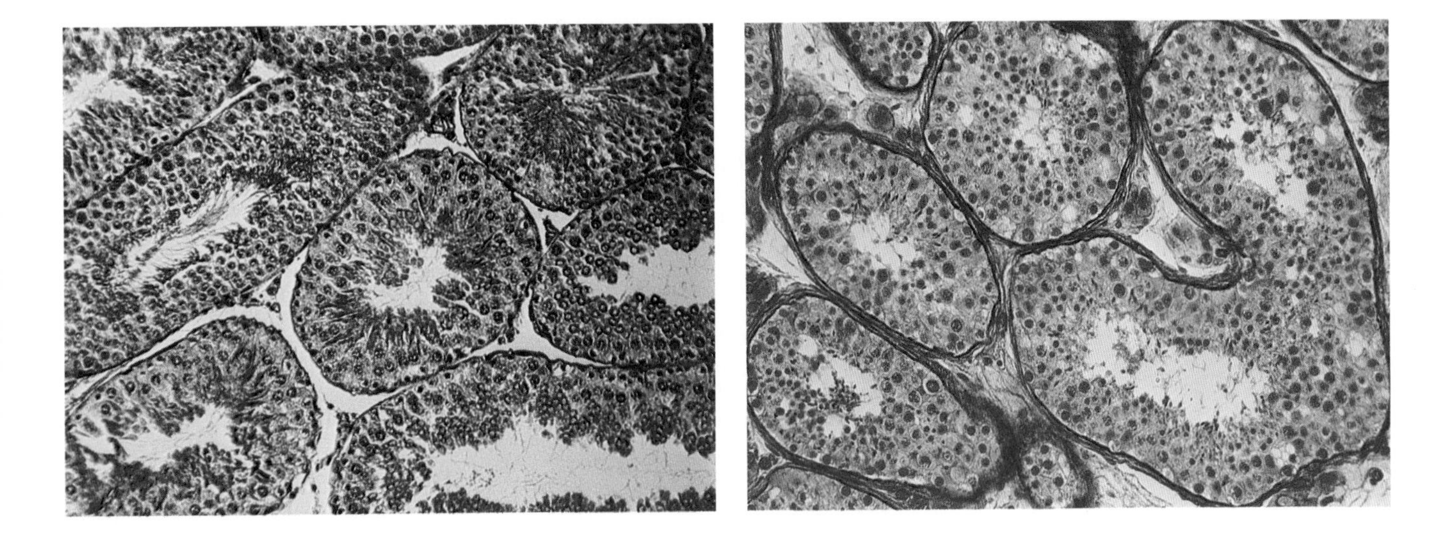

Photomicrographs showing cross sections of the sperm-forming tubes (seminiferous tubules) in the testis of a rat (*left*) and the testis of a human (*right*). Individual sperm cells may be seen forming in the walls of the tubules. Magnification is 181.5:1 for both.

is within the normal size range for cells. A third exception has already been mentioned: some single-celled organisms can be very large, as we saw in the figure on page 3. However, because the large nucleus of such an organism has the genetic material repeated many times, it might be considered an evolutionary experiment in size increase that was successful in producing an organism well adapted to life in ponds and other aquatic environments over the face of the earth but that did not produce a kind of construction well suited to the evolution of larger and more complex forms.

This brings us to the important conclusion that, if the eukaryotic cell remains approximately the same size, large multicellular organisms must have more cells than small ones, and indeed this is the case. A giant sequoia has more cells than a pine tree, a rat has more than a mouse, and a raven has more than a crow. Even if one considers individual organs, the same principle holds true: we are not much larger than chimpanzees; yet our brains are substantially larger, which means that they contain more neurons than the brains of our less endowed cousins.

## Natural Selection and Size

The world contains organisms of many different sizes. The fundamental reason for this is the evolution of new species and their winnowing by means of natural selection.

Three animals with highly developed special features that have arisen by natural selection. A chameleon has a long tongue capable of rapid movement. It is beautifully adapted for catching insects. A giraffe is well adapted to reach food in tall trees. A leopard is splendidly adapted for hunting. It has strong claws and teeth, stealth, and speed.

Let us begin by being absolutely clear what we mean by natural selection. In 1859, Charles Darwin, in his *Origin of Species,* stated the concept unambiguously. All species vary, and many of those variations are inherited; that is, to put it in more modern terms, there are variations in the genetic constitutions of individuals. Depending on the environment, certain of those variant individuals will be favored, and the result will be that they will be the most successful in reproduction and will have more offspring and descendants. Therefore, their genes (again, to use modern terminology) will become more prevalent in the population, and the genes of the less successful variant individuals will become relatively scarce after a number of generations. Given enough time, and with the help of new genes produced by mutation, there will be a marked change in the species. By its very nature, the process of evolution by means of natural selection is slow, and it requires long periods of time to effect major changes.

Selection operates on all of the genes of the whole individual organism, but the significant difference between competing organisms may involve one character or trait. For instance, in a particular set of environmental circumstances, genes for better vision, or a longer neck, or a faster pace may mean survival and reproductive success. Under those circumstances, such genes will be strongly favored by natural selection, and their frequency in the population will increase relative to the frequencies of other genes.

Not all traits of an organism are necessarily adaptive. There are two obvious examples. First of all, many genes have more than one effect (a phenomenon called pleiotropy). Suppose that such a gene produced two traits, one of them very strongly favored by natural selection and the other one mildly deleterious. In such a case, the effect of the deleterious trait would be mostly outweighed by

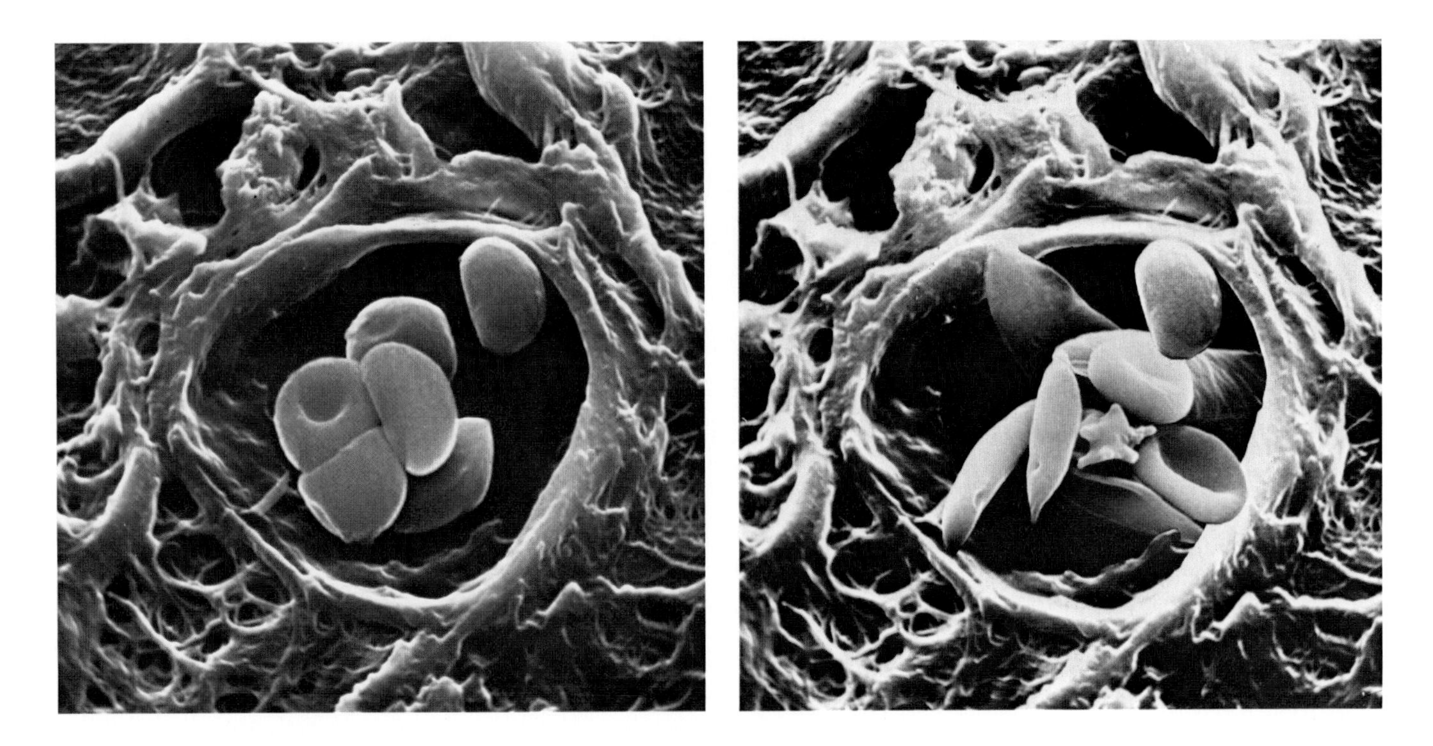

Sickle-shaped erythrocytes (*right*) versus normal erythrocytes (*left*) moving through a small blood vessel. This genetic disease presumably has been retained in human populations in Africa because it provides resistance to malaria. In certain environments, the advantages accompanying the sickle-cell trait outweigh the disadvantages, and therefore the genes for the disease are preserved by natural selection. Both micrographs are at × 4365 magnification.

the effect of the adaptive one, and so the gene would be retained by selection. Another way in which a trait with no adaptive value might remain in a population of a species is by being selectively neutral. If a trait is neither good nor bad, there is no selection for or against it. There remains the question of how such a trait could have come into being. Perhaps it was once adaptive in a previous environment, having become selectively neutral in the present environment. Or perhaps it arose by chance and has always been selectively neutral.

There is yet another aspect of natural selection that occurs within any organism; it is an internal selection. Any new trait that appears as a result of gene mutation or a new gene combination must go through a severe test before the trait ever appears in an adult. The trait must not interfere with the normal development of the organism, so the whole journey from egg to adult can be completed without a hitch. This means that new gene effects that influence early development are less likely to be successful than those whose effects appear very late in development. Development proceeds by an orderly sequence of events, and any change in the order, or any change in one of the steps, could impose such difficulties on what follows that the end result might never be achieved. To make an analogy to building houses, if there is a defect in

A group of male narwhals. The extraordinary tusk of the male apparently arose as the result of sexual selection. Individuals with large tusks are more likely to mate successfully. At the same time, the tusk would appear to be awkward and disadvantageous for locomotion and feeding. If this is so, the fact that the tusk has been retained means that the advantage of having a tusk at mating time outweighs the disadvantages of pushing a tusk around at other times.

the initial frame, the building might well collapse when the walls and the floors are added.

Let us return to the matter of size. One thing is obvious: the larger an animal, the more complex it will be—and the more complex its development will be, and the more time its development will take. It takes much longer to build a skyscraper than a hut. There are a number of reasons for this, which we will examine presently, but let us first point out a consequence of the process of development (or building). The larger the organism (or structure), the more difficult it becomes to make significant changes in the beginning. The legacy of all the many steps of development of a large organism (or of construction of a large building) is that the earlier steps become harder to change. Any tinkering becomes increasingly likely to be lethal. Some of the difficulties have nothing to do with size, but others do, and presently we shall examine the nature of those size-dependent constraints.

There is no doubt that size can be adaptive in itself. Sometimes there are advantages to being relatively large, and other conditions will favor being

Male bighorn sheep (*top*) and pheasants (*bottom*) clashing to win females of their species.

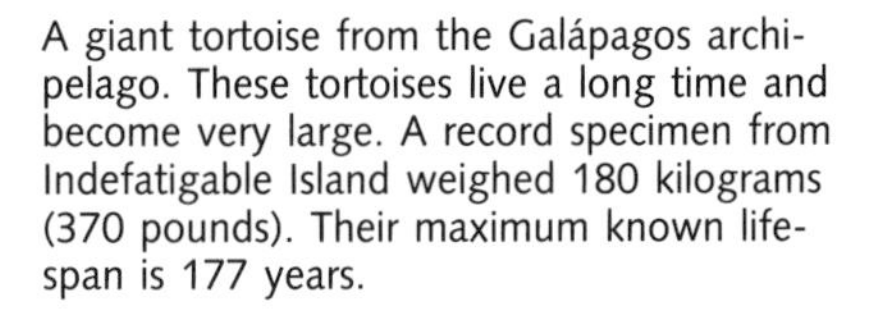

A giant tortoise from the Galápagos archipelago. These tortoises live a long time and become very large. A record specimen from Indefatigable Island weighed 180 kilograms (370 pounds). Their maximum known lifespan is 177 years.

relatively small. For instance, if aggressive fighting is important to reproductive success, then large individuals will be favored. The fact that island species often differ in size from closely related mainland forms is evidence that increases or decreases are common. There are some examples in which the island form is larger, such as the giant tortoises of the Galápagos Islands, the big brown bears of Kodiak Island and other Alaskan islands, and the giant lizards known as Komodo dragons, which are native to small islands in Indonesia. In other cases, the island forms are smaller than their mainland relatives. Good examples are the whitetail deer of the Florida Keys, the tigers of Sri Lanka, and the now extinct elephants of Malta.

A splendid example of the importance of size in predation comes from one of the pioneer modern ecologists, C. S. Elton. He described a toad in Lake Victoria in Africa that would normally be easy prey for a certain species of snake. However, if the toad sees the snake in time, it can swallow enormous amounts of air and puff up like a ball. The sight of such an enormous toad immediately ruins the snake's appetite. One wonders if this might have been the source of Aesop's and La Fontaine's fable in which an unfortunate toad, using the same technique, tries to achieve the size of an ox (page 16).

Of course, it does not follow that all size changes in evolution are due to natural selection. It is quite possible, for example, that some of the island

A Komodo dragon. This is the largest living lizard. It is found on the Lesser Sunda Islands in Indonesia. It grows to a length of 3 meters (10 feet) and attains a mass of 135 kilograms (a weight of more than 300 pounds).

Puffer fishes. When disturbed, these fishes can inflate themselves with water, thereby greatly increasing their size and dismaying their attackers.

The ox and the toad. Illustration from *A Free Translation of Aesop's Fables into Flemish Verse* published in 1579.

*The Ox and the Toad*

An ox while drinking by the road
Stepped on the offspring of a toad.
His mother came; he was not there;
She asked her sons if they knew where
He was. "O Mother, he is dead.
We saw a ponderous quadruped
An hour ago who came this way,
Whose hoof quite squashed him where he lay."
The toad then puffed herself all out,
And asked if it had been about
This size. They told their mother, "Stop,
Lest you be torn apart! You'll pop
To bits before you imitate
A creature so immensely great."

—*Aesop*

forms may have changed in size owing to a random drift of the genes affecting size in small populations. In any one particular instance, it is very difficult or more often quite impossible to be certain of the causes of a size change.

The fossil record is particularly rich in size-change trends. For some reason, almost all of those that are well documented by fossil evidence show a size increase. The most famous is the size change from the early dawn horse, *Hyracotherium*, to full-sized modern horses of the genus *Equus*, all of which occurred over a span of about 60 million years. Similar trends can be found among the elephants, the camels, the ceratopsian dinosaurs, and other large vertebrates. In fact, these trends are so conspicuous among vertebrates that the phenomenon is known as Cope's rule, named for the American paleontologist Edward Drinker Cope (1885), who was the first to point it out. A similar set of trends was shown to be true for a variety of invertebrates by Norman D. Newell (1949) of the American Museum of Natural History. Again, both for invertebrates and for vertebrates, we do not know what selective forces produced these repeated trends, although it is assumed nowadays that they are the direct result of natural selection. We simply do not know enough of the details of the ecology of those early times in earth history to understand how selection affected any one kind of animal. In Cope's day, it was fashionable to interpret this phenomenon in terms of some sort of progressive internal evolutionary drive (known as *orthogenesis*), but since the definitive writings of George Gaylord Simpson (1967), also of the American Museum and later of Harvard University, this notion has been totally discredited as anti-Darwinian and mystical.

What is especially puzzling is that almost all well-documented evolutionary size-change sequences show an increase in size, and few show a size decrease. This is all the more strange because existing evidence from island populations, as pointed out above, shows trends in size decrease as well as size increase. Of course, for island populations, we are considering evolutionary changes in size that occur in a few thousand years. For the fossils that obey Cope's rule, however, such changes span millions of years. Thus, the island differences could be seen as small irregularities in the major trends.

We do know that certain species of organisms that live today must have descended from larger ancestors. Good examples of modern animals that are smaller than their ancestors are hummingbirds, among the vertebrates, and rotifers, among the invertebrates. So there are certainly major trends of size decrease as well as size increase, but why do they seem to be less common? At this point, there is no satisfying answer. This whole area of study is a fascinating one, and perhaps in the years to come we will gain a better understanding of it.

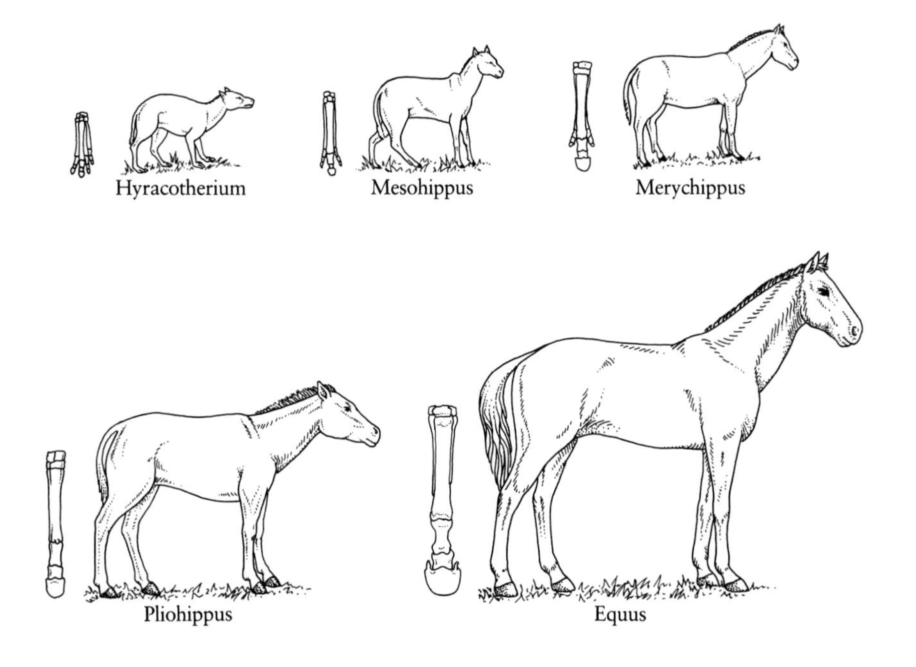


The evolution of the horse required about 60 million years. There has been a regular increase in the size of horses over time. As horses evolved, the structure of the foot changed. In the dawn horse or "Eohippus" (*Hyracotherium*), a tiny creature that walked on the moist floors of tropical forests, there were four distinct toes. Later, however, all of the toes but one disappeared as the horse became a plains-dwelling animal.

## Size and Shape

A central objective of this book is to examine the relation between the size and shape of organisms. We began with a discussion of size and how it is affected by natural selection, but this clearly is only half of the story. If we look at the figures in the beginning of this book showing animals and plants of different sizes, we cannot help being struck by the fact that, not only is there a great range of sizes among living things, but there is an even greater range of shapes. Following the arguments of the previous section, one could show that, in many instances, these differences in shape are also due to natural selection. In Darwinism, we expect to find explanations for the long neck of the giraffe and the tapering wings of a bird. But here this is not our main concern.

Instead, we are interested in those instances in which size appears to impose specific constraints on shapes, in which there is a necessary link or relation between the two. The leg of a 5-kilogram dik-dik differs in shape from the leg of a 500-kilogram buffalo for reasons almost entirely due to the constraints imposed by their differing sizes. And, as we shall see, the same is true of the difference in proportions between the delicate stalk of a wheat plant and the thick trunk of an oak tree. If an animal or plant is large, the physical constraints that act upon it are different from those that would act upon it if it were small, and these constraints may have a profound effect on the shape of the organism. Or, to put it another way, there are certain shapes that are physically impossible for large organisms, and there are others that are equally impossible for small ones. (One must remember that, if a particular shape is physically disadvantageous, or even simply less efficient, it will be eliminated by natural selection.)

It is also true that shape imposes certain restrictions on size; the argument works both ways. If a bird is to fly, its weight, which is directly related to its size, must remain below a strict limit. It is not surprising, therefore, that all the very large birds, such as ostriches, great auks, and dodos, have lost the art of flying.

The whole subject of size limits will be a central theme in this book. Even without a careful examination of the subject, most of us are aware that such limits exist. We all know that the largest animals, the blue whales, are aquatic and that they are able to attain their great size only because they are supported by the water in which they live. We presume that dinosaurs could not have been much larger than they were because they surely would have run into mechanical problems, and it is just such problems that we will examine later in some detail. Why are there flying squirrels but no animals the size of horses

A nocturnal Southern flying squirrel (*Glaucomys volans*) glides through the air using lateral flaps of skin stretched between its forelimbs and its hindlimbs. Flying squirrels often sail 45 meters or more in this way. The flying lemurs of the Malay Archipelago are also capable of long glides. Why don't animals as large as horses do this? In Chapter 5 we discuss why gliding speed increases with size. Using the results of that analysis, one can estimate that a horse would glide at about four times the speed of a squirrel. This means that the kinetic energy per unit volume would be 16 times as great, and, therefore, the horse would be 16 times as likely to break its bones on impact as it crashed into a tree at the end of its flight.

A small lizard (*Draco taeniopterus*) that glides by spreading flaps of skin at its sides like outstretched wings.

Hummingbird, albatross, ostrich. Relation between wing shape and wing function. The size of a bird affects the shape of its wings. A minute hummingbird has small wings for hovering. A large albatross has long wings for soaring. An ostrich is too large to fly, so its wings have become useless and vestigal.

that dare to climb trees and glide about on webs of skin stretched between their legs? How is it that an ant can lift 10 times its own weight while a human being must struggle to pick up and carry a weight greater than his own? Why are the smallest mammal (the dwarf shrew, about 4 centimeters long) and the smallest bird (the Cuban bee hummingbird, about 5 centimeters long) both about the same weight (approximately 3 grams)? Why are the smallest insects and arachnids much smaller still? (Several species of fairy flies and beetle mites are less than 0.25 millimeter long.)

Every student in a high-school biology course learns that the external skeleton of an arthropod imposes certain size limits. Those size limits are different for flying arthropods (such as dragonflies), for terrestrial arthropods (such as scorpions and tarantulas), and for aquatic arthropods (such as lobsters). The textbooks explain that exoskeletons have mechanical problems that make them unsuitable for organisms larger than a certain size (science fiction notwithstanding) and that the vertebrate endoskeleton has mechanical properties that permit the existence of such enormous creatures as dinosaurs, elephants, and whales. A part of our task in the chapters to come will be to see whether the evidence available today supports or contradicts those textbook explanations.

There are size limits also at the lower end of the scale. We have already discussed the fact that bacterial and eukaryotic cells have certain narrow size ranges, and it seems reasonable to presume that this is the result of the mechanical limitations imposed by diffusion and other physical forces involved in cell metabolism. Later, we will discuss the effects of minuteness on locomotion and on the shape of the locomotory organs. Here it will suffice to give an example

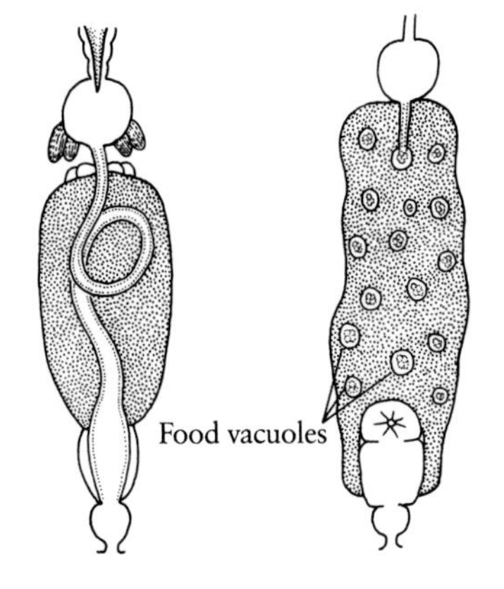


The conventional gut system of a typical rotifer (*left*) and the single-celled gut of an unusally small rotifer (*right*), showing food vacuoles passing through a continuous cytoplasm.

involving digestion. In small protozoans, one finds uniformly that a particle of food is engulfed in a small vacuole, which floats free in the cytoplasm during the digestion process. Once digestion is completed, this food vacuole moves to the surface membrane and opens to the exterior (rather like a gas bubble popping at the surface of a liquid), thus expelling the undigested wastes. In multicellular forms, even in relatively small rotifers, there is usually an alimentary canal with a mouth at one end and an anus at the other. The canal is characteristically lined with cells, some of which secrete the digestive enzymes. As mentioned earlier, rotifers in general show a trend toward decrease in size and are presumed to have descended from larger ancestors, being an exception to Cope's rule. There are some species of rotifers that are especially small, and in them we find a peculiar change in the digestive system. Instead of having an alimentary canal, these minute forms have one "cell" connecting the mouth and the anus, and this cell engulfs food in a vacuole in the same manner as a protozoan and passes the vacuole to the anal end, where the wastes are excreted. It would appear in this case that, once the rotifer evolved to such a small size, it could revert to the protozoan type of digestion. The assumption is that this radical change in construction has been made possible by simple mechanical considerations due to small size. In this case, unfortunately, we do not fully know the details, but in many cases we do, and they will be emphasized in the pages to follow.

## Size and Complexity

There is one further very important consequence of size. An organism is not a statue; it is alive, and it performs all sorts of vital activities. It consumes energy and converts it into mechanical movements; it coordinates the movements; it even coordinates its own growth and development. A large organism is an organized collection of cells that functions as a unit. Only as a unit can it respire efficiently, move materials from one part of its body to another, and have all the other properties of life finely tuned. If it is deficient in any of these respects, it will be eliminated by natural selection.

These activities—keeping the life motor running and performing all the functions associated with life—are as essential to the smallest animals and plants as they are to the largest. But there are big size-related differences among the ways in which the functions are carried out. An increase in size imposes certain restrictions that require an increase in the division of labor among the parts. This size-related increase in the division of labor can be called

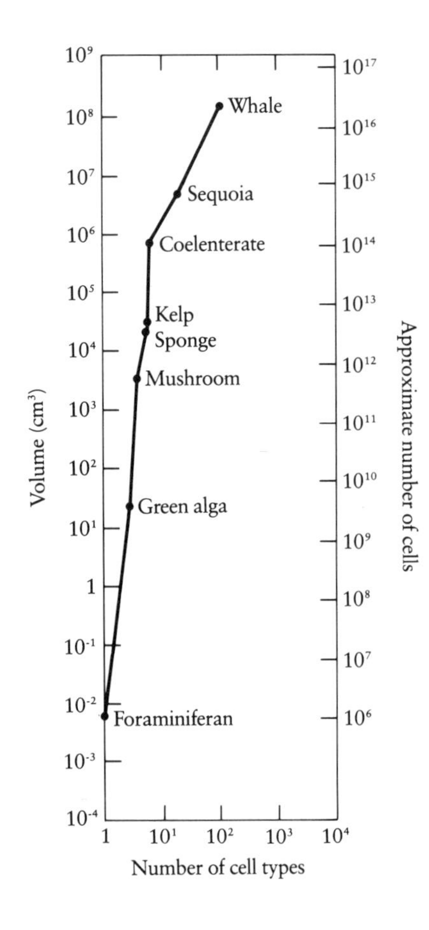


Generally speaking, the larger an organism, the greater its complexity (as measured by counting the number of different types of cells present). This figure shows an estimate of the maximum number of cells (or the largest volume) in various organisms plotted against the number of types of cells found in those organisms.

an increase in *complexity*, and we can say that the complexity of an organism increases with its size. This "rule" has nothing directly to do with natural selection; it is a purely physical constraint imposed by an increase in size. By this we mean that, if such an increase in the division of labor did not occur, the larger organisms would simply not work properly and would die. It would be impossible to build a successful elephant with the kind of division of labor found in an amoeba. An elephant built on such a plan would survive only a few seconds. We (as motile, sensitive, and intelligent mammals) are built with a complex interior division of labor because of our size, and anything less carefully designed would fail mechanically and therefore cause us to be culled by natural selection.

It is difficult to find a good measure of the complexity of an organism. One simple method would be to count the number of types of cells, such as muscle, nerve, cartilage, and so forth. A biologist, however, would immediately recognize a technical difficulty with this method, because each tissue of a complex multicellular organism is made up of more than one type of cell, and it is hard to know where to draw the line in making fine distinctions. Thus, any estimate of the number of types of cells can be only rough at best. As can be seen from the figure on the left, however, there is a good correlation between size and the approximate number of types of cells in an organism. In this figure, the maximum size for an organism is plotted against number of types of cells. The largest living organism, the blue whale, has approximately 120 types of cells, whereas the foraminiferan has cells of only one type.

The principle that complexity increases with size is true for many other things besides living organisms: The larger the university, the army, the business, or the government, the larger the numbers of individuals doing different, specialized jobs. It is a very general principle, but it applies with special force in the construction of living forms.

## On Being the Right Size

Because the capabilities of organisms of different sizes are so vastly different, many people—including J. B. S. Haldane (1928), from whom we have borrowed the title of this section, and Frits W. Went (1968)—have asked questions about the capabilities of organisms of our own size. In particular, Went made a delightful comparison of social human beings and social ants. We differ from ants in many ways, but size is certainly an important difference, important especially because of its consequences. For instance, ants could not use

An ant can lift a load many times its own weight, but an animal the size of a horse cannot carry even one of its own kind on its back.

fire, for even the smallest possible stable campfire flame is larger than an ant. Keeping a wood fire burning would be quite beyond their capacity, because ants are too small to get near enough to add fuel (which, in any event, they would be unable to carry). Ants cannot use tools. A miniature hammer has too little kinetic energy to drive even a miniature nail. Spears, arrows, and clubs, which depend on a suitable ratio of a kinetic energy to a characteristic surface area to do their work, would be ineffective at ant size. Ant-sized books would be impossible to manufacture, or even to open, because the thin pages would stick together owing to intermolecular forces that are relatively powerful at that scale. In any event, reading would likely have few charms for ants because, with their small size, they have very few brain cells. Of course, ants have enough neurons to do all of the remarkable things that ants normally do, but we modestly presume that, in order for an animal to appreciate the joys of literature, it needs to be at least the size of a human being.

And finally, to emphasize our great superiority to ants, Went pointed out that they cannot wash themselves with water. The water droplets of a shower stream come in a certain minimum size. Droplets of even this minimum size would strike an ant like heavy missiles. Even if an ant tried to take a bath in a single drop, surface tension would interfere because the chitin of the ant's body is water repellent. If the ant did somehow manage to get into a drop of water, surface tension would make it difficult to get out again. (It is not uncommon to see a fly struggling to extricate a leg from a drop of liquid on a flat surface.) The answer for the ant is to dry-clean itself by rubbing particles of dry substances over its body and then scraping the particles off.

Went continued with this comparison between ants and people. An ant cannot pour a liquid; the best it could manage would be to squeeze a liquid droplet out of a compliant pouch. (We have seen the astronauts do this in space.) There are certain advantages to being an ant, however. An ant can lift 10 times its own weight. It can fall large distances without injury. At a certain time in the lives of some ants, flying is possible by an awkward mechanism that would never serve to get a human being off the ground. In the chapters to follow, we shall look into the details of all these and more of the biological consequences of physical size. In many places, it will be profitable to compare biological observations with principles from engineering, and we shall not hesitate to do this. One of our first concerns will be the matter of proportions and how they are influenced by body size. This is the subject of the next chapter.

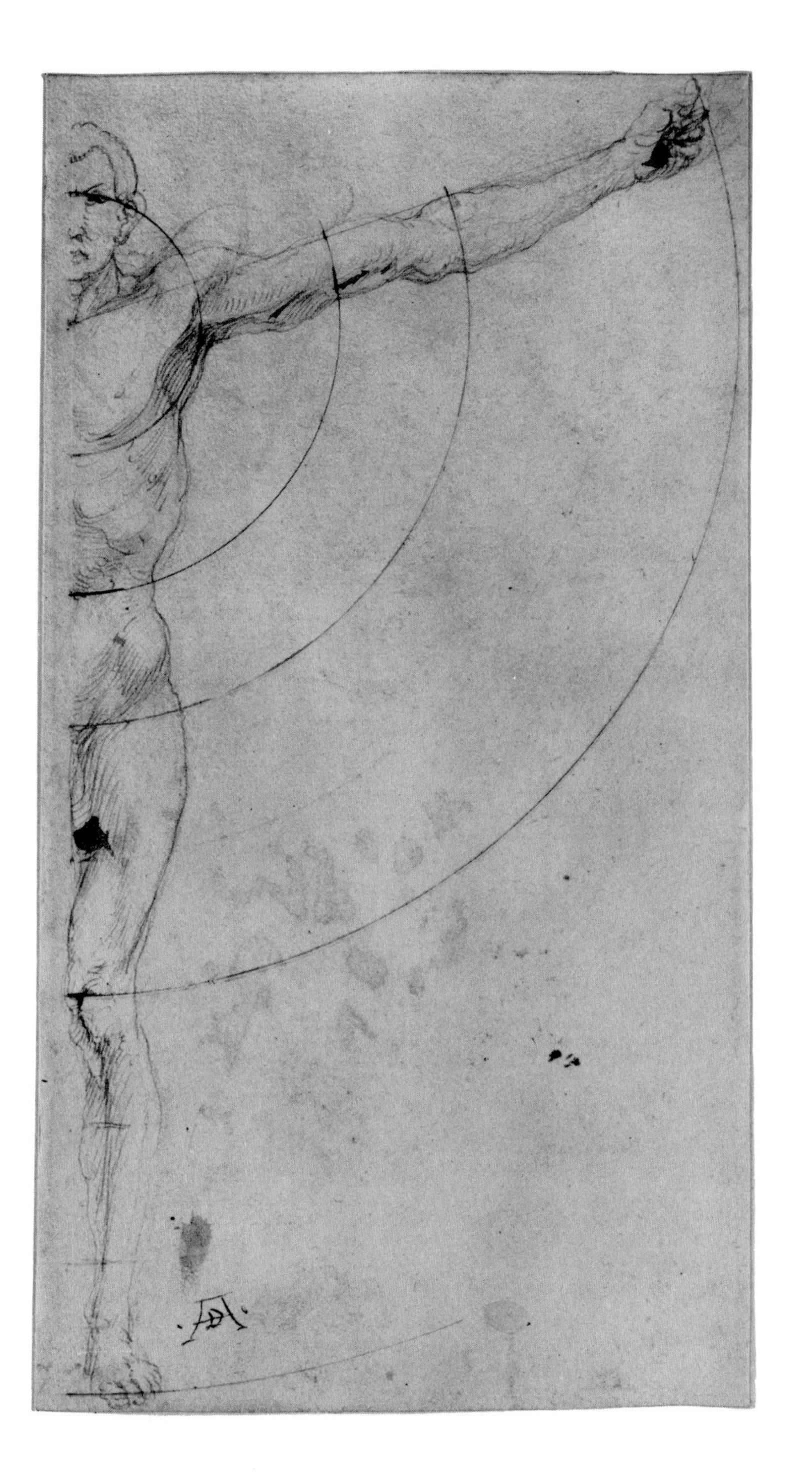

# Chapter 2

# Proportions and Size

If a change in size is accompanied by a change in shape, we should have a way of measuring it. Perhaps the first step would be to find a way of visualizing the changes in proportions, and this, then, might lead us to a method of measurement.

## How to Measure Changes in Proportions

**Cartesian Transformations.** In his famous book *On Growth and Form* (1917), D'Arcy Thompson suggested an ingenious method of showing how shapes may be related. He called it a "Cartesian transformation" and gave credit to others before him, especially the fifteenth-century German artist Albrecht Dürer, for first using this technique. But the main credit goes to Thompson himself, who used it specifically to measure differences in proportion between closely related species.

The procedure is simple in principle. A Cartesian or rectilinear grid is placed on a two-dimensional drawing of an animal or plant. One can then ask how this organism is different from a closely related species of somewhat different proportions. To determine this, a comparable grid is placed over a drawing of an individual of the second species, and the lines are made to pass through homologous points. When this has been done, one can easily see how the grid has been warped or "transformed" in going from one organism to the other. In the figures on the following page one can see that the distortions of the grid provide a clear indication of which parts of one organism are expanded or contracted with respect to those of the other.

Because of the obvious appeal of this method, it has become widely known, and D'Arcy Thompson's figures have been reproduced in an enormous number of places. Despite its popularity, the promise of this method is as yet largely unfulfilled because, to use the words of P. B. Medawar (1958), it is "analytically unwieldy": it is difficult to use in a precise way to follow quantitative changes in form. One difficulty is that it depends upon two-dimensional representations of three-dimensional objects. But, the main problem grows out of its principal virtue: it is a way of visualizing many changes at once, and it becomes exceedingly difficult and impractical to assign numerical values to those continuously changing grid lines.

Albrecht Dürer's sketch of half a man with arcs.

**Allometric Formulas.** In place of the mathematical complexities of the Cartesian transformations of D'Arcy Thompson, a far simpler approach has turned out to be very useful. It was largely developed by J. S. Huxley (1932) with

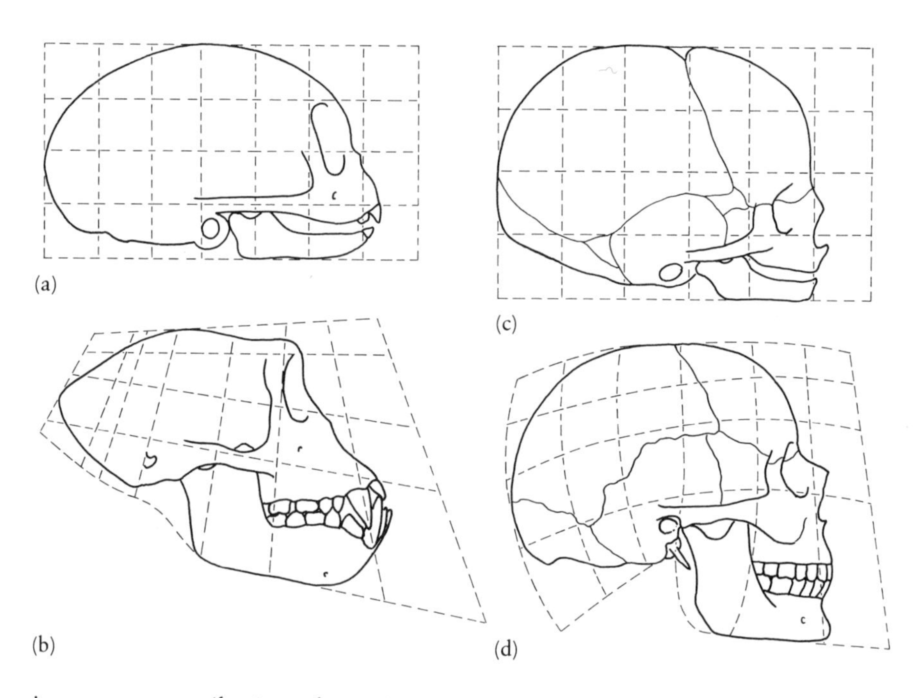


Illustration of Cartesian transformations: *a*, the skull shape of a rhesus monkey a few days after birth; *b*, an old male rhesus monkey; *c*, a newborn human; *d*, an adult human.

important contributions from Georges Teissier (1931). It consists simply of comparing the relation of two measurements, ignoring all the complexities and subtleties of the detailed changes in form. First we will describe this relation by means of a simple formula, and then we will discuss the principle.

If one wishes to compare the relative sizes of two parts, $x$ and $y$, of an organism, then one way to describe this comparison is to write

$$y = bx^a \tag{2.1}$$

in which $a$ and $b$ are constants. Notice that, if $a$ is equal to zero, $y$ is always equal to the constant $b$, no matter how large or small $x$ is. One of the observations of the previous chapter was that the cells found in large and small animals are not very different in size. If we let $y$ be the cell diameter and $x$ be the overall length of the animal's body, making $a$ equal to zero would describe the relation between $y$ and $x$ reasonably well, because $y$ would be equal to the constant $b$, whatever the animal's body length.

Suppose, instead, that we let $y$ be the full armspread and $x$ be the height of adult human beings. In this case, the equation would fit the formula best if $a$ were made to equal to 1, since, in that case, armspread would be directly proportional to height, a fact about human anatomy that has been widely appreciated since Leonardo da Vinci first pointed it out.

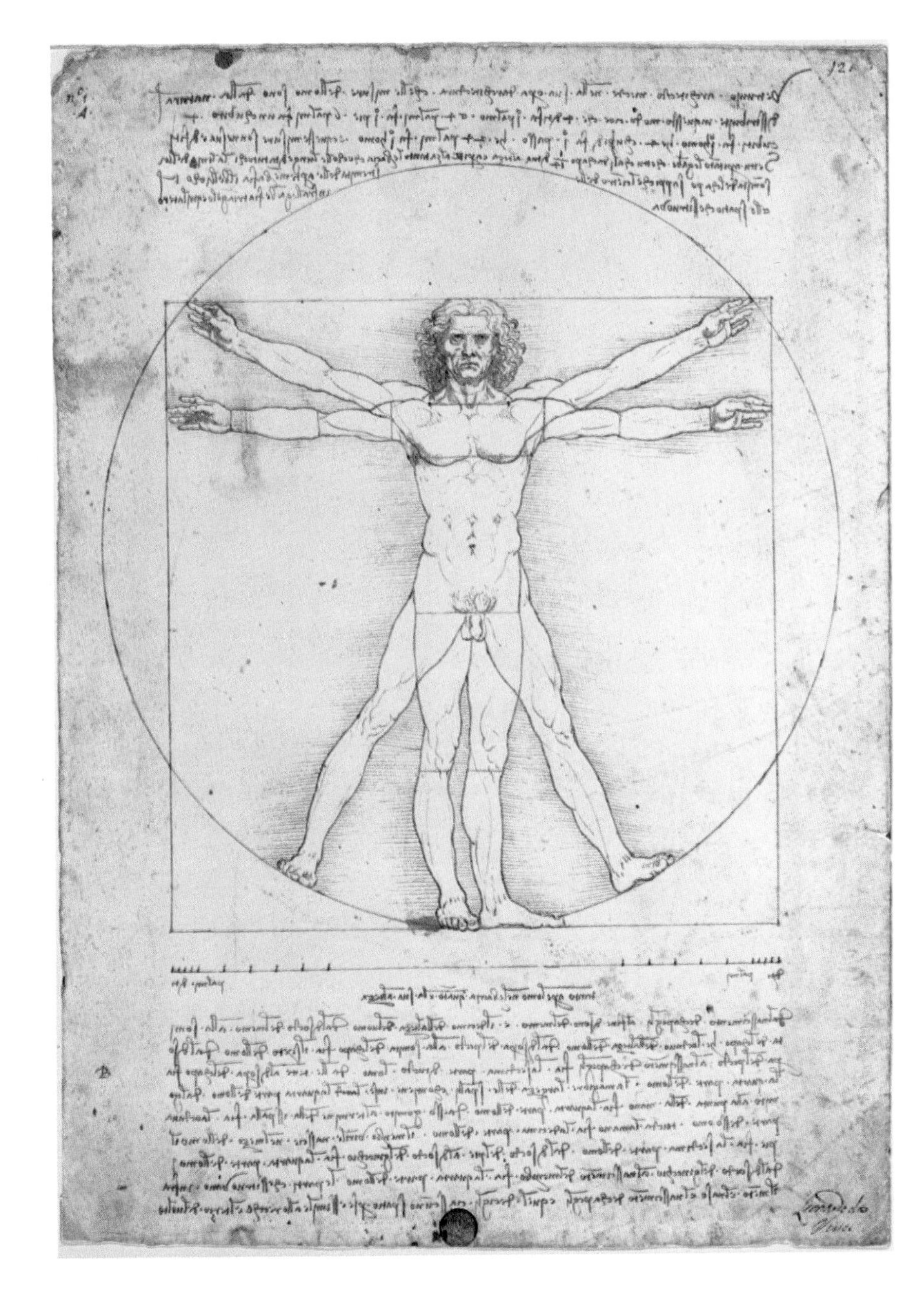

"Vitruvian Man," a drawing from Leonardo da Vinci's notebooks.

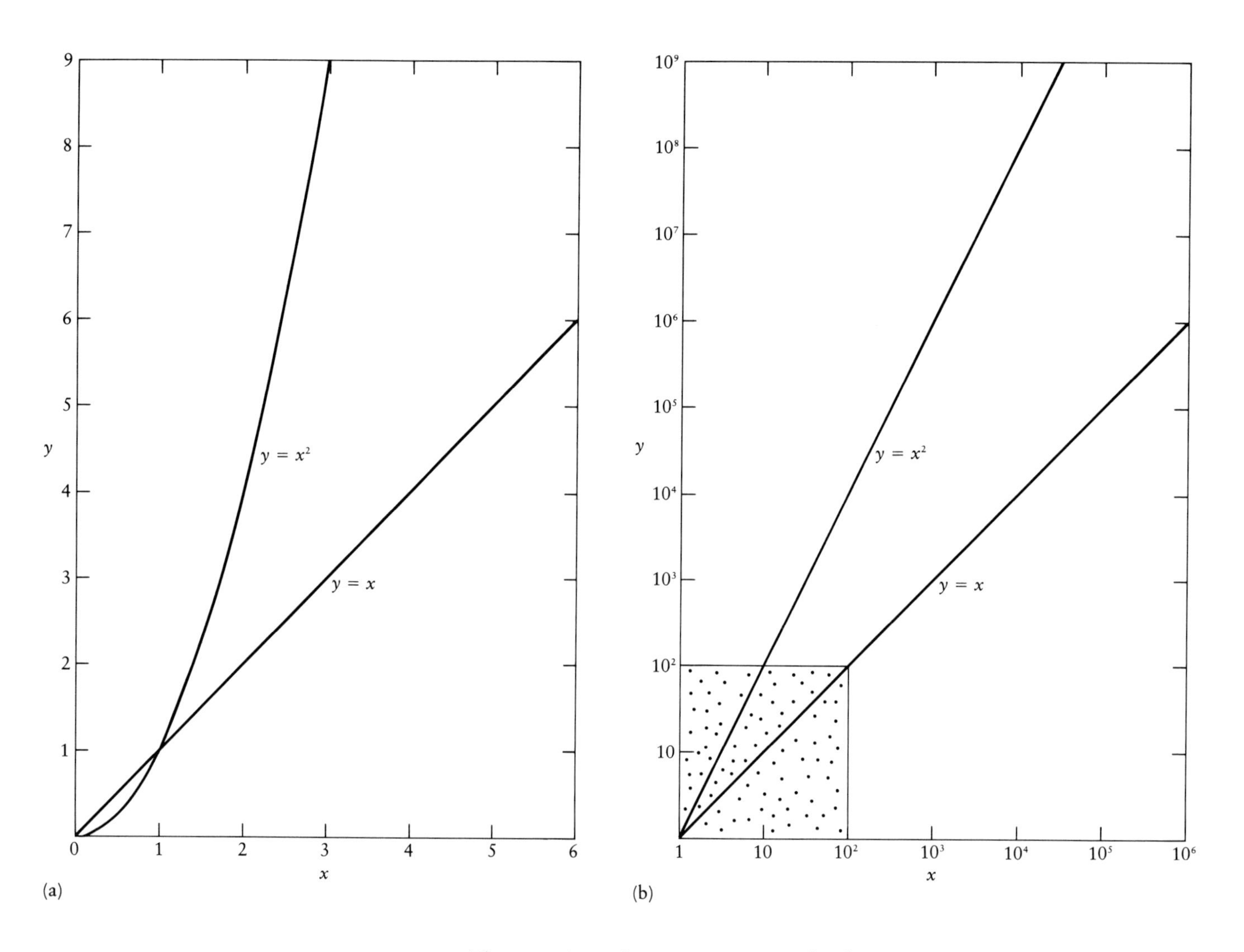


The equation relating $y$ to $x$ can also be written in logarithmic form,

$$\log y = \log b + a \log x \tag{2.2}$$

in which "log" means "logarithm to the base 10." Logarithms are exponents; in this case, they are exponents of the number 10. They have many worthwhile properties—the reader may remember that, if two numbers are to be multiplied, this may be done by looking up the logarithms of both numbers, adding the logarithms, and then looking up the antilog of the sum. Using logarithms, a multiplication operation can be replaced by an addition operation, which sometimes is convenient.

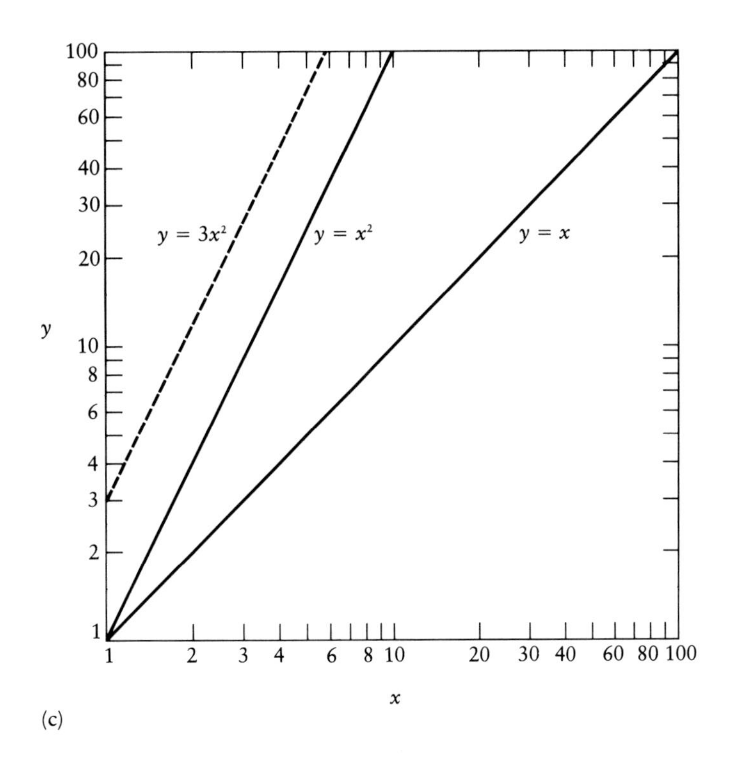


Illustration of the use of log–log paper. In part *a*, the equations $y = x$ and $y = x^2$ have been plotted on ordinary Cartesian coordinates, in which there is an equal spacing between adjacent numbers on both the *x* and *y* axes. In part *b*, there is an equal spacing between adjacent powers of 10. In part *c*, the first two decades (shown stippled in *b*) have been drawn in more detail to illustrate how the spacing between adjacent numbers gets smaller as the numbers get larger on each of the axes of log–log paper.

Often, in the arguments to follow in this book, we shall wish to determine the relation between two size variables, *y* and *x*. To do this, we shall plot log *y* against log *x* on ordinary graph paper or, equivalently, we shall plot *y* against *x* on log–log paper, a special graph paper on which the spacing between adjacent numbers on the axes varies logarithmically. An illustration of the use of log–log paper is given in the three figures shown here. In part *a* of the figure, *y* is plotted against *x* on ordinary Cartesian coordinates. In this graph, a plot of the equation $y = x$ is a straight line of slope 1.0 starting from the origin at $x = 0$, $y = 0$. A plot of the equation $y = x^2$ yields a curve that starts at the origin and rises ever more quickly as *x* increases.

In part *b* of the figure, the two equations have been replotted on log–log scales. Now, instead of there being an equal spacing between the numbers 1, 2, 3, and so on, the *x* and *y* axes have been labeled so that there is an equal spacing between adjacent powers of 10. Thus there is an equal distance between 1 and 10, between 10 and 100, and so on. The interval between 1 and 100 for *x* and *y* has been shown in more detail in part *c* of the figure, and additional numbers have been put in between the powers of 10 to make the scales more useful.

There are a number of points worthy of notice. First, as may be seen by examining the axes of part *c*, there is an equal spacing not only between 1 and 10 but also between 2 and 20 and between 5 and 50. On logarithmic scales, any pair of numbers different by a factor of 10 is separated by the same distance as any other such pair. Second, log–log plots may be used to get a huge range of numbers on the same graph. All three parts of the figure show the two functions $y = x$ and $y = x^2$, but, whereas $x$ cuts off at 6 and $y$ cuts off at 9 on the Cartesian graph of part *a*, $x$ ranges from 1 to $10^6$ and $y$ ranges from 1 to $10^9$ in the log–log graph of part *b*. In order to accommodate this tremendous range, the log–log graph collapses the spacing between numbers when the numbers are large and blows up the spacing when the numbers are small. One could argue that this is a very natural, even a "biological" way of dealing with information. We do something similar ourselves when we hold something small up close to our eyes and when we step back to get a whole view of something big. Similarly, our ears are more sensitive to small variations in faint sounds than to small variations in loud sounds.

Finally, there is the point that power-law formulas of the type $y = bx^a$ give straight-line plots on log–log paper, and the slope of the log–log plot is the exponent $a$ of the power law. Thus, the slope of the line $y = x$ is 1 in parts *b* and *c* of the figure, and the slope of the line $y = x^2$ is 2. The value of the multiplicative factor $b$ merely raises or lowers the line $y = bx^a$ in a log–log plot. For example, the line specifying $y = 3x^2$ (shown broken in part *c* of the figure) is parallel to the line $y = x^2$ but passes through the point $y = 3$, $x = 1$, and therefore lies above the line $y = x^2$ for all values of $x$.

If the data points describing a given experiment or set of observations fall along a straight line when plotted on log–log paper, we can be fairly confident that $y = bx^a$ describes the relation between $y$ and $x$. Furthermore, the exponent $a$ and the constant $b$ can be read directly from the slope of the log–log plot and the value of $y$ when $x = 1$, as we shall illustrate.

Suppose that many measurements are collected relating one dimension of a skull, $y$, to another dimension, $x$. As we examine skulls of different sizes, we write down the measurements in the table shown on the facing page. Then we plot $y$ against $x$ on a log–log plot. The slope of the line may be obtained directly from the graph. It is found by measuring, with an ordinary ruler, the increase in vertical height between any two points on the line and dividing by the increase in horizontal distance between the points, measured with the same ruler. Using this procedure, we can find that the slope of the line in this case is 1.75. The value of $y$ when $x = 1$ is 3.5. Thus, an equation describing the table of data is $y = 3.5x^{1.75}$.

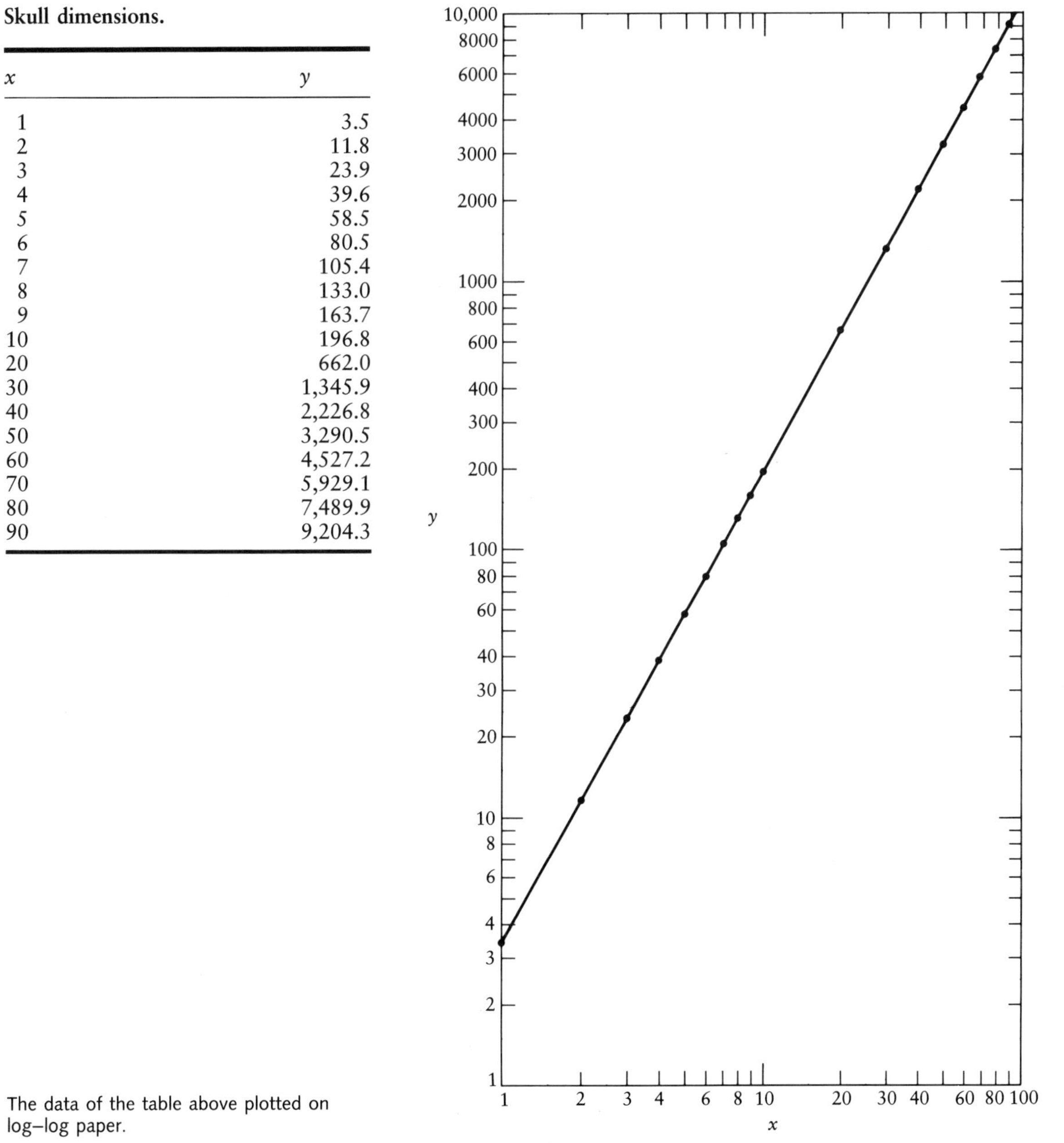

**Skull dimensions.**

| $x$ | $y$ |
|---|---|
| 1 | 3.5 |
| 2 | 11.8 |
| 3 | 23.9 |
| 4 | 39.6 |
| 5 | 58.5 |
| 6 | 80.5 |
| 7 | 105.4 |
| 8 | 133.0 |
| 9 | 163.7 |
| 10 | 196.8 |
| 20 | 662.0 |
| 30 | 1,345.9 |
| 40 | 2,226.8 |
| 50 | 3,290.5 |
| 60 | 4,527.2 |
| 70 | 5,929.1 |
| 80 | 7,489.9 |
| 90 | 9,204.3 |



The data of the table above plotted on log–log paper.

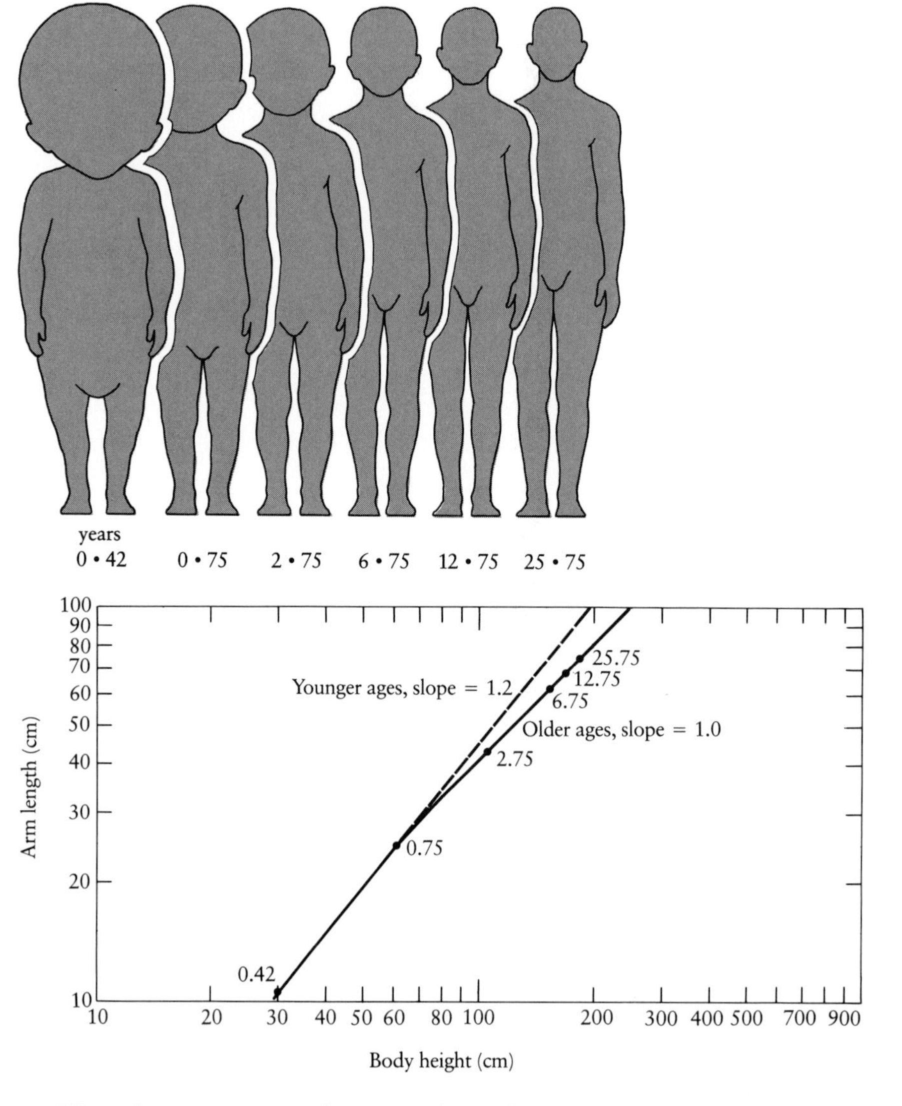


Human development, showing change in body shape with increasing age. The last three figures, from age 6.75 years onwards, are roughly the same shape.

Arm's length against body height on an allometric plot. The points are taken from the six stages of human development shown in the figure above. A straight line of slope 1.2 fits the early stages of development, but the slope changes to 1.0 in the later stages. The numbers shown beside the points refer to the age in years.

When the equation $y = bx^a$ is used to relate one dimension, $x$, to another dimension, $y$, in a range of organisms of different sizes, it is called an *allometric formula*. (The term *allometric* means literally "by a different measure," from the Greek *alloios*, different. By contrast, *isometric* means "by the same measure." A good example of isometry (exponent $a = 1$) was given when we noted that the armspread of adult human beings is proportional to their height. Consider now not the comparison of adults of different sizes but the growth of a

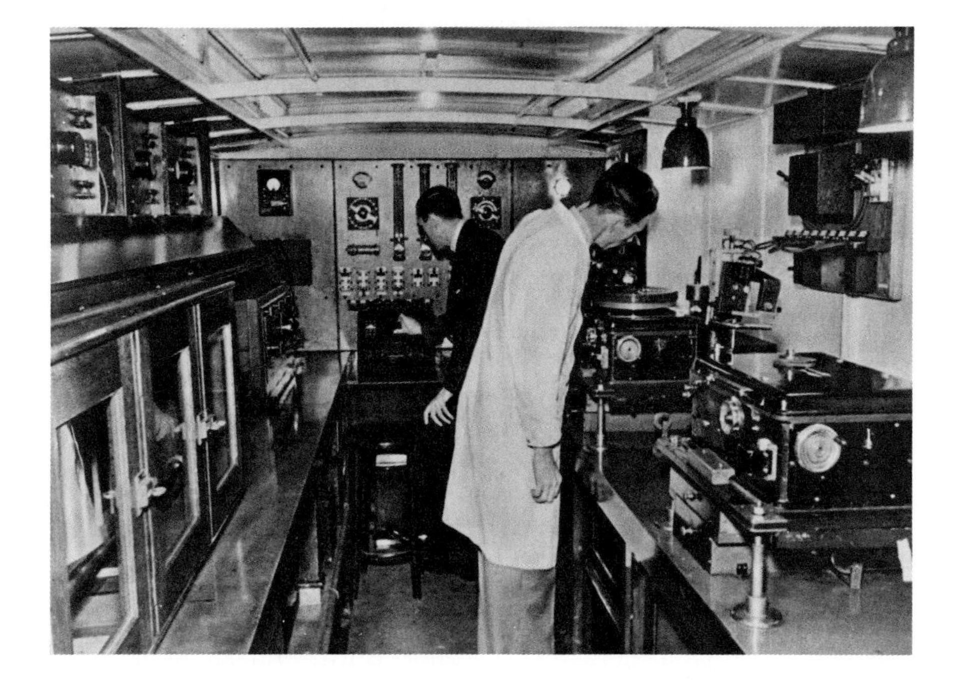

This photo, taken in 1938, is from a study of animal language carried out by Julian Huxley and Ludwig Koch. Here we see Huxley (foreground) and Koch inside the recording van checking the equipment.

single human being. The length of any structure (such as the arms) varies relative to that of any other structure (such as the total body height) at different stages of development. For example, the arms of a baby boy are shorter relative to the rest of the body than are the arms of a man as can be seen in the figure at the top of page 32. On a log–log graph, one could plot arm length ($y$) against body height ($x$) at the early stages of development, and the result would give a straight line with a slope greater than 1 (bottom figure on page 32). In the later stages of development, the slope is rather close to 1.

The slope $a$ of the line gives the ratio of the exponential rate constant for the growth of $y$ divided by that for the growth of $x$. Thus, $a$ has the same meaning as the ratio of the interest rates for two different types of savings account in a bank. In this example, the fact that $a$ is greater than 1 in early development means that the arms are growing faster than the body as a whole, something we know to be true.

The constant $b$ is also of significance because, as we have said above, it fixes the value of $y$ when $x$ is equal to 1. Consider a hypothetical case in which two pairs of structures have the same value of $a$ but differ in their values of $b$. One can see in the graph at top of page 34 that the slopes of the two lines will be identical and will produce two parallel lines intersecting the $x = 1$ line at two

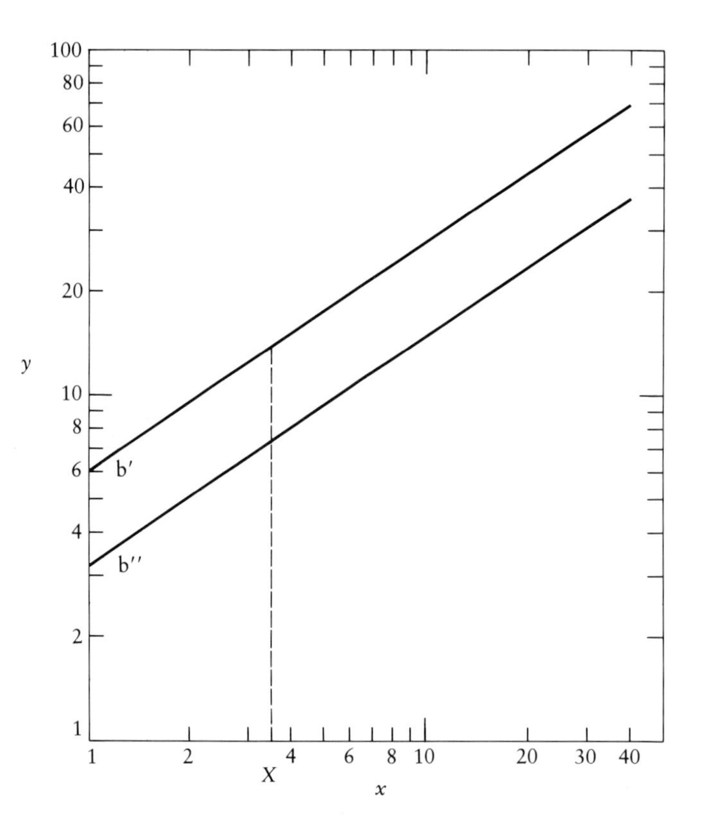


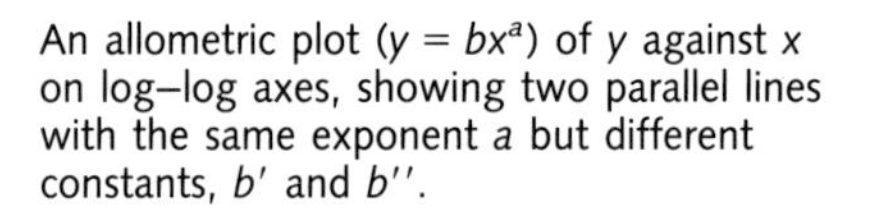
An allometric plot ($y = bx^a$) of $y$ against $x$ on log–log axes, showing two parallel lines with the same exponent $a$ but different constants, $b'$ and $b''$.

different points ($b'$ and $b''$). Let us now assume that these are the arm lengths of two different species of hypothetical primates. At any given body size ($X$), the $b'$ animal will have much longer arms than the $b''$ animal, yet the ratio of the exponential growth-rate constants of arm length and body height is the same for both species. This means that the starting time of the $b'$ arm primordium must have been earlier than that of the $b''$ arm primordium: the difference is due simply to a head start. By the time that the $b''$ arm primordium began, the $b'$ arm had already undergone a considerable amount of growth.

**The Uses and Limitations of Allometry.** The reader should be cautioned that this allometric relation is descriptive; it is not an explanation of proportions. Many parts of many animals and plants were measured by J. S. Huxley and many authors who followed him, and a surprising number of these organisms showed a straight line when the data were plotted on log–log graph paper. But not all such measurements give perfectly straight lines. In some instances, when the measurements have been plotted, they will clearly indicate two or more straight lines. This means a sudden change in the value of $a$, which may reflect a transition from one stage of development to another, such as before and after the attainment of sexual maturity. The graph on page 32 is an example—arm length is an increasing fraction of body height in the early stages of growth, but it becomes an approximately constant fraction of body height in the later stages. If there is a shift in $a$ (a break in the slope of the curve), this means that the ratio of the exponential rate constants for $x$ and $y$ has become altered by some major developmental event, such as something associated with the increased production of a hormone by newly mature gonads. There are other cases in which the value of $a$ is changing in a continuous manner, giving a curved line throughout the entire log–log plot. In these cases, the use of allometry is not very helpful.

Thus far, most of the examples that have been given are those of changes in proportion that occur during development. But allometry can be used equally effectively in comparing the adults of different species, as D'Arcy Thompson did with his Cartesian transformations. As a good example of the comparison of species, consider the log–log graph on page 36, which shows the relation between horn length and skull length in different species of fossil titanotheres. Note that, when we compare full-grown adults of different species, the horn length increases more rapidly than the length of the skull (the slope, $a$, is much greater than 1), which means that large titanotheres had very much larger horns.

Drawing an allometric plot and putting the exact numbers in an allometric formula do not in themselves explain the change in proportions; they merely show precisely that there is a change in proportions with a change in size. One of the main purposes of this book will be to try to shed some light on why there are such changes in proportions (or a lack of them). As we shall see, one cannot dogmatically answer the question of "why," but one can set up interesting and reasonable hypotheses.

Before continuing in our exploration of allometric relations, we will make one final caution. In many allometric plots, there is a wide scattering of the data points, even though the scatter is deceptively minimized in log–log graphs. The result is that one often cannot be certain of either the exact slope or the value of $b$ for the line (the titanotheres are a good example) and, even more seriously, one cannot be sure whether it is a straight line, a series of straight lines with well-defined breaks, or a continuous curve. If one is simply aware of these dangers, the chance of being misled to unlikely conclusions is minimized.

Those cases that do show clear, straight lines, even with some approximation, are those that we now wish to try to understand. What is the explanation of such allometric relations? In the first place, it is obvious that genes control the intrinsic rates of growth and therefore that they are responsible ultimately for the value of $a$ that reflects a ratio of two exponential rate constants. The idea that genes control rates of biological processes is an old one, and there are innumerable examples and much evidence to support it. One then might ask how the particular genes that control the rates of growth of particular struc-

The length of the horn is disproportionately larger, by comparison with the length of the skull, as larger titanotheres are compared with smaller ones. Titanotheres were large land animals that lived in the Eocene epoch (between 53 million and 37 million years ago).

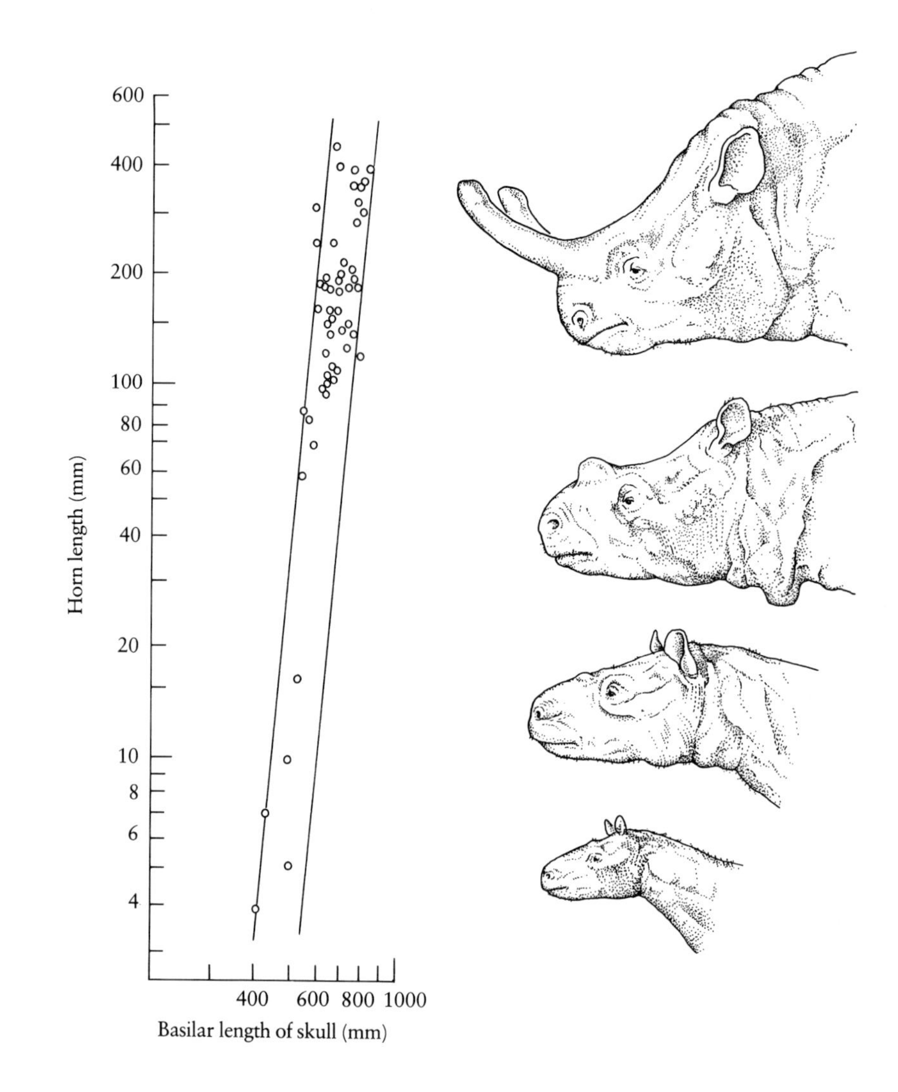


tures actually arose. Here the biologist has an unequivocal answer: by natural selection. A particular set of "rate genes" must produce a set of exponential rates for different parts of the growing body that produced a successful individual, successful in the sense of producing relatively long-lived offspring and thereby spreading those particular genes in the population.

But the possibilities for mutation of the genes that affect the rates are within very strict limits: There is not an infinite range of rates that can be imposed on any one body part, and here we return to the notion of constraints. Again, there are constraints at all levels. In the first place, the rates are limited by the size of the job. If a full-length arm of a human being takes 18 to 20 years to

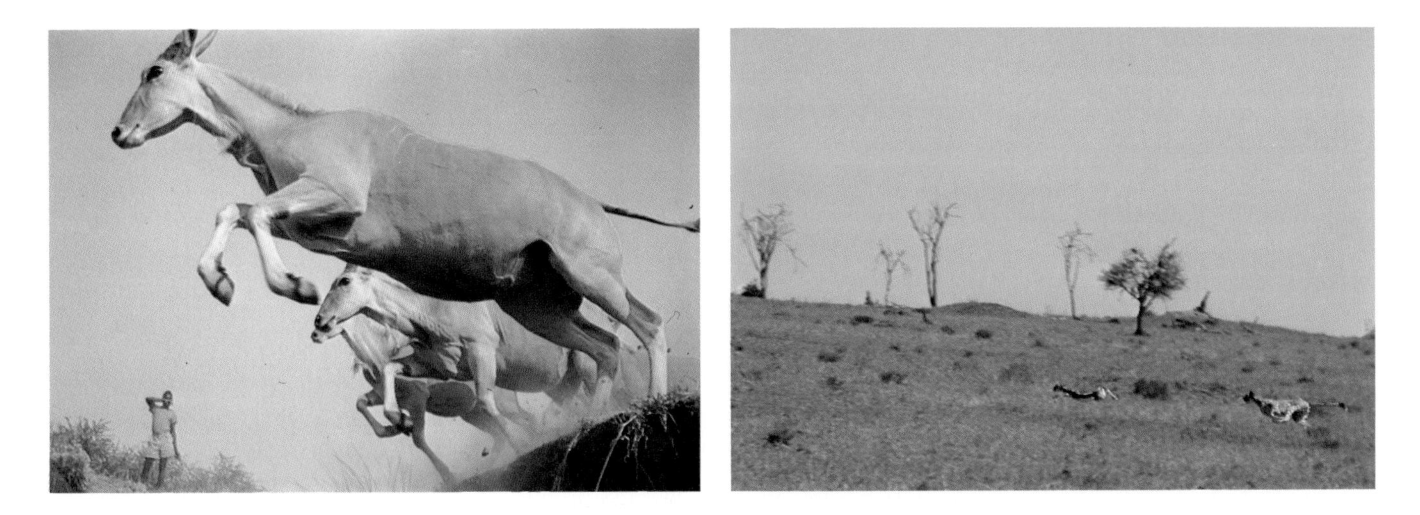

Speed is an excellent adaptive feature, both in predators and prey. An eland (*left*) jumps a ditch, relying on its speed to launch it into flight. A cheetah (*right*) pursues a gazelle.

build, one can hardly expect to find gene changes that will make the same process occur in five minutes. Food energy simply cannot be converted into cells that rapidly, nor can cells divide or differentiate that rapidly. This is the same point we made earlier: It would be just as impossible to build the World Trade Center in five minutes.

Beyond this rather obvious developmental constraint, there are much more subtle physical forces that also play a part. For instance, suppose that there exists a species of antelope in which some individuals have slightly shorter legs than others of similar body size. We discover that, even though this mutation appears frequently, those genes never seem to settle in the population: they are quickly eliminated by natural selection. Because the shorter-legged individuals cannot run as fast as the others, they are more likely to be caught by predators before they reach the age of reproduction.

Thus far, we have laid out the basic principles of allometry and have given several cautions about its use. Now it is time to illustrate these principles with a number of specific examples, for it is only through examples that we can gain a deep appreciation of the principles.

## Isometry

Let us begin with examples in which the exponent $a$ of the allometric equation equals 1. This special case is known as *isometry* or *geometric similarity*. If $x$ doubles in length, then so does $y$. Or, to put it another way, the proportions

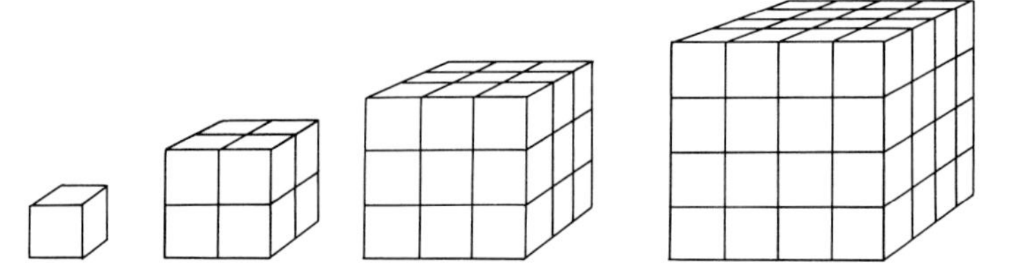

Cubes of different sizes. The edges of the cubes in this series are 1, 2, 3, and 4 centimeters long.

remain constant; they do not change with a change in size. An easy way to picture this in one's mind is to think of a photographic enlargement. If it is a photograph of a horse, then the horse's proportions remain the same whether one looks at a tiny contact print or at a huge enlargement.

**Length, Area, and Volume.** A very simple example of isometry is provided by the set of four cubes of different sizes shown above. The edges of these cubes are 1, 2, 3, and 4 centimeters long. The areas of the faces of these cubes are therefore 1, 4, 9, and 16 square centimeters, and the volumes of the cubes are 1, 8, 27, and 64 cubic centimeters.

Suppose we let the edge length of a cube be $l$. The perimeter of one face of the cube, therefore, is directly proportional to this length (the constant of proportionality is 4). The area of one face is proportional to the square of the length (with a proportionality constant of 1), and the volume is proportional to the cube of the length (again with a proportionality constant of 1).

For a sphere, the length dimension chosen could be either the radius or the diameter. If we choose the diameter to be the measure of length, the perimeter of the equator is proportional to $l$, and the constant of proportionality is $\pi$. The surface area $S$ of a sphere is easily calculated:

$$S = \pi d^2 \qquad \text{or} \qquad S \propto l^2 \tag{2.3}$$

in which the symbol $\propto$ means "is proportional to."

The volume $V$ of a sphere is also easily calculated:

$$V = \frac{\pi}{6} d^3 \qquad \text{or} \qquad V \propto l^3. \tag{2.4}$$

This means that, with a change in the size of a sphere (for example, from a marble to a basketball), if the diameter is increased 10 times, the surface will have increased $10^2$ or 100 times, while the volume will have increased $10^3$ or 1,000 times. The same obviously applies to the horse mentioned earlier. If

A water strider depends on the forces of surface tension to support its weight.

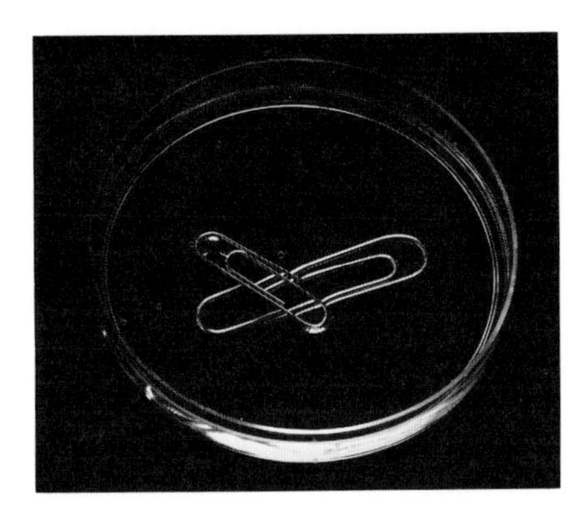

Surface tension is powerful enough to keep a small paper clip floating at the surface of a dish of water, but a larger paper clip sinks.

those geometrically similar horses were real horses and not simply photographic images, a tenfold increase in the horse's height would mean a hundredfold increase in its surface area and a thousandfold increase in its volume (which is proportional to its mass). As we shall see, in these elementary and simple-minded examples lies a principle of profound biological significance.

Size change in geometrically similar living bodies has numerous effects, and we will make a brief list of them here. Some of these effects are clearly dependent on the length dimension ($l$), but most are dependent on the length dimension squared ($l^2$), and a few are dependent on the length cubed ($l^3$).

**Length ($l$).** The most obvious things that vary linearly are height or length—for example, height in trees or leg length in running animals. Suppose $l$ is designated as the length of one of a water strider's legs. Then the perimeter of its foot and leg also varies as $l$ in a series of geometrically similar insects. Because the force due to surface tension depends on the perimeter but the weight to be supported depends on $l^3$, small insects of a suitable design can walk on water, although a large insect of exactly the same design could not. For the same reason, the small wax-coated steel paper clip in the photograph on the left floats on the surface of the water, whereas a large wax-coated steel paper clip of the same shape immediately sinks.

**Length Squared ($l^2$).** The exchange of gases, especially oxygen and carbon dioxide, occurs through surfaces such as those of the lungs or the gills (or, in plants, the leaves). In animals, food is assimilated through the surface of the gut. Heat loss takes place through the skin, which covers the body surface, and across the surfaces of the tongue and pharynx. The strength of a bone, a

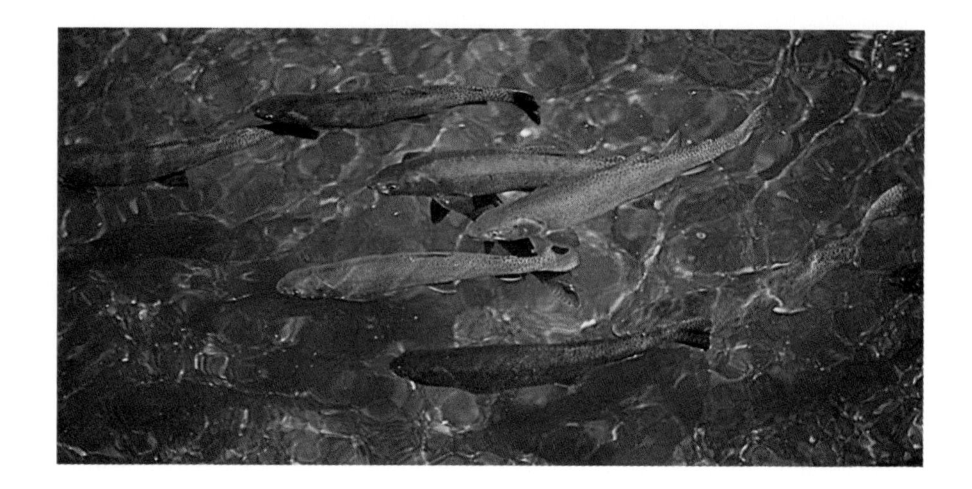

The drag force acting on a swimming animal depends, among other things, on its wetted surface area. This picture, taken at Yellowstone National Park, shows Cutthroat Trout swimming.

muscle, or even the inert structure of a tree is a function of its cross-sectional area. The metabolic power consumed by a muscle is related to the cross-sectional area of the muscle. In locomotion, the drag of a swimming animal is, under certain conditions, proportional to its wetted surface area, and the capacity of a bird to generate lift when in the air is related to the area of its wings. Throughout this book (particularly in Chapters 5 and 6), we will see more examples of how properties related to the surface and cross-sectional areas of organisms of different sizes play a role in performance and survival.

**Length Cubed** ($l^3$). Here the most significant factor is mass. We will return again and again to this theme: The larger the animal or plant, the greater the significance of gravity and the more important the part played by any support mechanism that helps it to compensate for the effects of gravity and inertia.

Incidentally, the fact that buoyancy also scales as $l^3$ means that a scaled-up floating body floats just as well as a smaller one. Neither of the steel paper clips shown in the photograph on page 39 was buoyant; the smaller one was kept afloat by surface-tension forces. On the other hand, all ice is less dense than water, so a massive iceberg floats as well as a small ice cube because both mass and displacement scale the same way. Most mammals, like most ice cubes, float with about the same fraction of their bulk out of the water, illustrating the fact that mammals all have approximately the same mass density.

At this juncture, we are concerned with *isometric* size changes in which geometric similarity is maintained and there are no compensatory changes in proportions. One might ask, for example, why all globular, more or less spher-

Most mammals float with about the same fraction of their bulk out of the water. At the right is a floating sea otter, left are hippos.

ical organisms are less than a millimeter in diameter. The answer is quite obvious. These organisms must exchange oxygen and carbon dioxide and assimilate food, and their capacity for these activities varies as the linear dimensions squared ($l^2$). The only way in which there could possibly be large spherical organisms a meter in diameter would be if they were to change their proportions and vastly increase their surface area by means of great convolutions so that the surface/volume ratio could remain constant. In that way, each internal portion of the huge sphere would have some bit of folded surface near it for gas and food exchange. From this we can conclude that strict adherence to geometric similarity generally imposes limits on the range of sizes available to organisms of a particular design.

In many instances, there may be a considerable range of possible sizes with perfect isometry, but such size differences are not without consequences. Before considering a biological example, we shall consider two nonliving examples. A solid material like wood is difficult to ignite in bulk, but it is quite easy to ignite in the form of wood shavings. Further, if wood is ground up into minute dust particles and mixed with air, not only will it ignite easily, but the combustion will be so rapid that it will explode. A more homely example may be found in the kitchen. Suppose that it is Christmas Day and that you have lost your cookbook. You must figure out how long to roast the turkey. All you can remember is that, when you last roasted a small chicken, it took one hour. If you estimate that your turkey is one and a half times as long as the chicken and you increase your roasting time by that amount, your dinner will be a failure; but, if you estimate the time difference on the basis of the square of the

length, your Christmas will be merry. (The reason the cooking time increases as the square of the length is discussed in Chapter 3.)

The biological example involves song in animals. In some groups of mammals, the length of the vocal cords or other sound-producing structures is roughly proportional to the length of the animal. Therefore, the longer the animal, the deeper the voice. Beautiful recordings have been made of the songs of humpback whales and, as one would expect for such enormous animals, the songs are both deep and slow. Again for size-related reasons, the ears of whales are tuned to low frequencies. Partly because of their low pitch, whale songs travel enormous distances in the ocean. It would be fair to say that the animal's sound-producing apparatus, its sound-receiving apparatus, and the distance the sound travels are all "tuned" by its size to satisfy its needs. In general, animals vocalize and hear on frequency bands established by their size, and they may be comparatively deaf to the communications of those much larger or smaller than themselves. There is a mouse-repelling device for sale that emits a high-pitched noise. It cannot be heard by humans, but apparently it sounds like the end of the world to mice, and they stay away. Merely by listening to the sound of an animal's voice, it should be possible to determine whether it is larger or smaller than you. For this reason, the miniature lion in Mandrake's crater-world shown on page 43 should have squeaked; a growl would really have been below its vocal range.

There is an interesting case reported in the recent literature in which Fowler's toad, which is native to the eastern United States, makes use of these principles (Fairchild, 1981). Apparently, the largest males are most successful in attracting females during the mating season, and their appeal comes from their deep voices. As one would expect, the size of the animal and the pitch of the song are related, and the females seek out the deeper voices in the swamp. There is the interesting possibility, not yet proved, that some males employ a trick to improve their chances. They remain in the coldest places and, by this means, deepen their voices, for the cold directly affects the muscular function of the sound-producing structures. If this turns out to be true, it would show that, even for toads, all is fair in love and war.

**Rowing Shells.** Another example of an application of isometry is given by a comparison of the rowing shells used in college competitions. The boats are rowed by one, two, four, or eight rowers. If one examines the proportions of the different-sized boats, they are very close to being geometrically similar, at least over their wetted surfaces (see figures on pages 44 and 45). This is estimated by measuring the length of the boat, $l$, and the width (called the

An episode from *Mandrake the Magician.* (King Features Syndicate, Inc.)

Baboon, barking treefrog; red-tailed hawk; dolphin. A large frog has a deeper voice than a smaller frog and the same is true of comparisons between large and small individuals for the other animals shown.

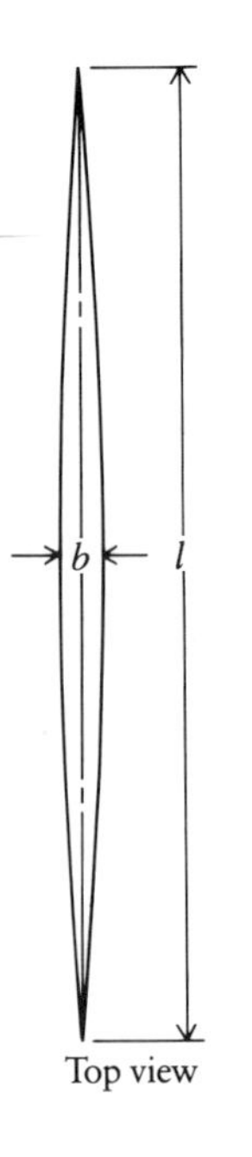

The shape of a rowing shell. This same shape describes boats seating one, two, four, and eight oarsmen.

Rowing shells: clockwise from left, single scull seating one oarsman; pair-oared shell; four-oared shell; eight-oared shell.

**Shell dimensions and performances.**

| No. of oarsmen | Modifying description | Length, $l$ (m) | Beam, $b$ (m) | $l/b$ | Boat mass per oarsman (kg) | Time for 2000 m (min) | | | |
|---|---|---|---|---|---|---|---|---|---|
| | | | | | | I | II | III | IV |
| 8 | Heavyweight | 18.28 | 0.610 | 30.0 | 14.7 | 5.87 | 5.92 | 5.82 | 5.73 |
| 8 | Lightweight | 18.28 | 0.598 | 30.6 | 14.7 | | | | |
| 4 | With coxswain | 12.80 | 0.574 | 22.3 | 18.1 | | | | |
| 4 | Without coxswain | 11.75 | 0.574 | 21.0 | 18.1 | 6.33 | 6.42 | 6.48 | 6.13 |
| 2 | Double scull | 9.76 | 0.381 | 25.6 | 13.6 | | | | |
| 2 | Pair-oared shell | 9.76 | 0.356 | 27.4 | 13.6 | 6.87 | 6.92 | 6.95 | 6.77 |
| 1 | Single scull | 7.93 | 0.293 | 27.0 | 16.3 | 7.16 | 7.25 | 7.28 | 7.17 |

"beam") and making a ratio of these two values, which, as can be seen in the table above, remains fairly constant even though the length almost triples going from the smallest shell to the largest. Even the boat mass per oarsman is similar within very narrow limits for all of the shells. Armed with this basic information, we will now explain what at first appears to be a puzzling fact: that the larger boats will go faster than the smaller ones.

The explanation comes directly from the geometric similarity of the shells. The total mass of the oarsmen plus that of the boat equals the mass of water displaced by the boat, and, if the boats are geometrically similar, this displacement is proportional to $l^3$, which in turn is proportional to the number of oarsmen.

The drag produced by the water as it moves past the boat is due to two effects, skin friction and wave-making. A more complete discussion of the drag on boats and ships is presented in Chapter 3, but, for the moment, it will be sufficient to note that the long, thin shapes of rowing shells serve to minimize that part of the drag that is due to wave-making. Tests in full-scale shells show that the drag that is due to the eddying wake and wave-making amounts to less than 8 percent of the total drag at 20 kilometers per hour, a speed close to the usual racing speed. This means that the drag that is due to skin friction is the major component, and it is known that skin-friction drag is proportional to the product of the wetted area and the square of the speed. Using this fact, and making the additional assumption that the power available is proportional to the number of oarsmen, one may calculate that the speeds of a range of rowing

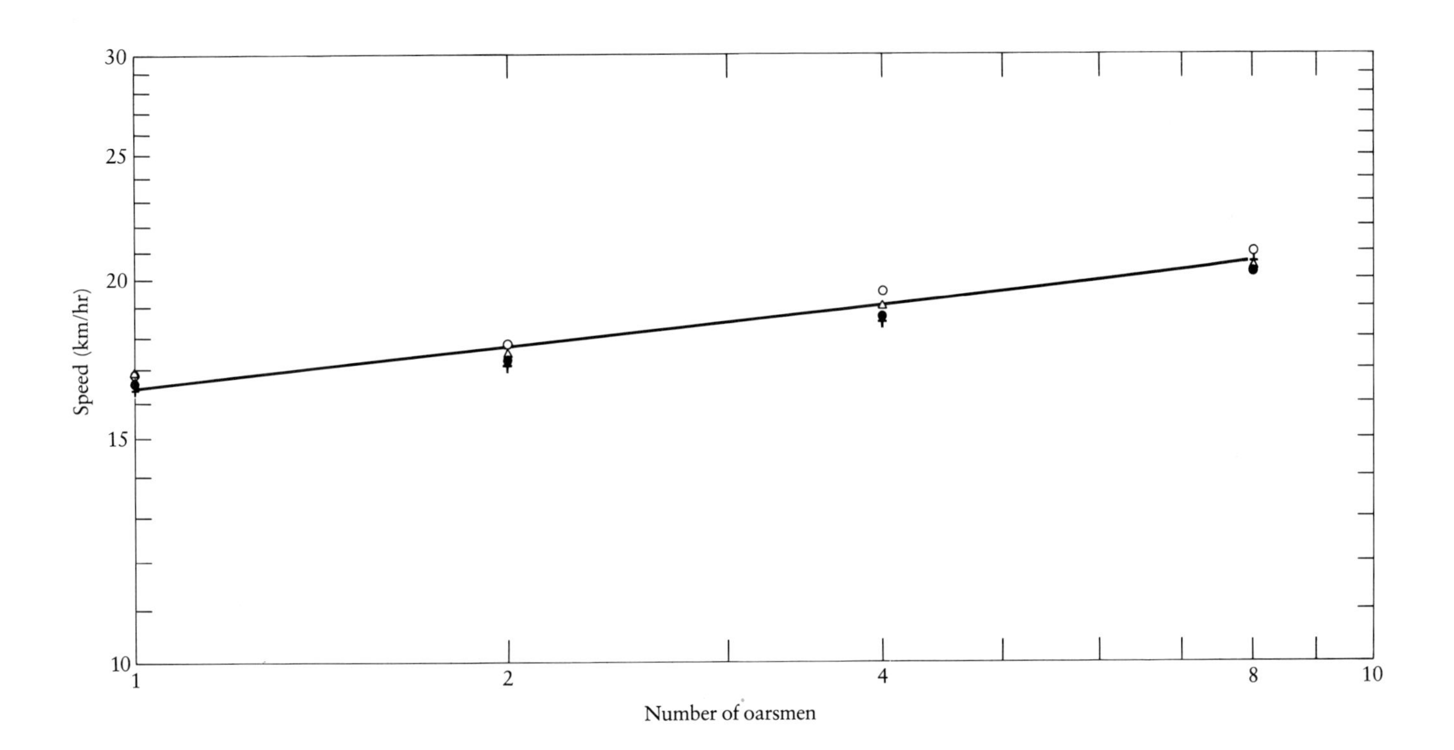


Average speeds over a 2000-meter course for rowing shells seating one, two, four, and eight oarsmen. *Triangles* show the 1964 Olympics, Tokyo; *solid circles* are the 1968 Olympics, Mexico City; *crosses*, the 1970 World Rowing Championships, Ontario; *open circles*, the 1970 Lucerne International Championships. The races were rowed in calm or near-calm conditions. The solid line shows the theoretical result.

shells of different sizes should be proportional to the number of oarsmen rowing raised to the one-ninth power. (The details of this calculation are given in the Appendix at the end of this chapter.)

This prediction is checked in the figure above, which shows the speeds for 2000-meter races in two Olympic competitions and two world championships. The theoretical line with a slope of $1/9$ fits through the data points on this log–log plot quite neatly.

In summary, because rowing shells seating one, two, four, and eight oarsmen happen to be built with the same proportions—that is, as isometric models of one another—a particularly simple example is available from which it can be predicted that speed is a function of boat size, and this theoretical prediction agrees with the data from actual races.

**Isometry in Spirals.** In speaking of isometry so far, we have made comparisons between separate shapes—that is, separate spheres or separate rowing shells. Isometry may also be the organizing principle underlying single forms of spiral shape.

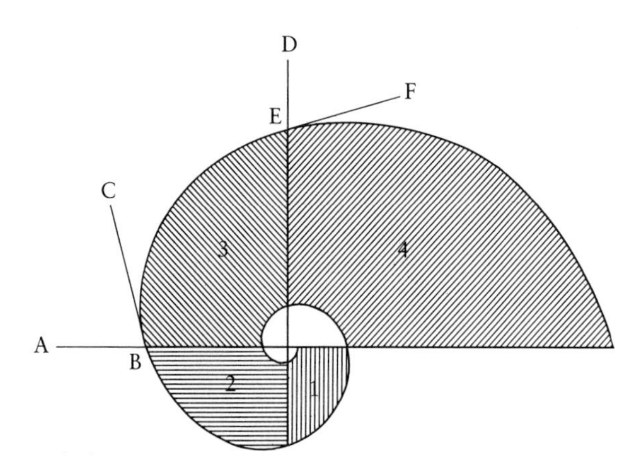


An equiangular spiral (*above*). The angles made by the intersection of any radial line with a line drawn tangent to the spiral are equal. Thus, angle *ABC* is equal to angle *DEF*. The chambers marked 1, 2, 3, and 4 are of exactly the same shape.

The chambered nautilus. *Top*, each of the chambers is geometrically similar. *Bottom*, a shell with an added chamber, shown by a broken curve (*left*), is almost identical to the same shell magnified and turned by the angle occupied by one chamber in the spiral (*right*).

The chambered nautilus is a cephalopod mollusc, a relative of the octopuses and the squids. It differs from them in a number of ways, perhaps the most conspicuous of which is the possession of an external shell that is coiled in a spiral. The nautilus builds its shell in such a way that the shape of the living chamber (and, therefore, the shape of the animal) remains the same as it grows larger. As the nautilus increases in size, it adds to its shell, and periodically it partitions off a part of its living chamber that it has just outgrown. These partitions form a series of compartments in the old shell.

The shape of the nautilus shell is an equiangular spiral, a shape first recognized and described by the French philosopher and mathematician René Descartes in 1638. The equiangular spiral shown in the diagram (*top left*) on page 48 demonstrates that the angle between a tangent to the spiral and any radial line drawn from the center is a constant (for example, angle *ABC* is equal to angle *DEF*). Descartes deduced that, in a spiral having this property, the chambers cut off by successive radii drawn at equal increases in radial angle have exactly the same shape. Thus, chamber 4 is a perfect scale model of chamber 3, only dilated in every dimension and rotated by 90 degrees.

Each time the nautilus outgrows part of its old living chamber and partitions it off, it creates a new empty chamber about 6.3 percent larger than its predecessor (*top right*). The result is that, in the course of building the 18 chambers needed to bring this spiral full circle, the size of the chambers triples. As we can see by comparing a shell with an added chamber (*bottom left*) to a shell turned one-eighteenth of a circle and magnified by 6.3 percent (*bottom right*), these are essentially identical. Thus, the graceful spiral form of nautilus shells follows from the constant shape (isometry) of the animal's living space as it curls its new rooms around the old ones.

The nautilus can descend to great depths by using its rigid shell as a buoyancy tank (Currey, 1977). It leaves the closed chambers filled with gas at slightly less than atmospheric pressure. At depths at which the hydrostatic pressure squeezing the shell is as much as 10 atmospheres, the shell is liable to fail either by buckling or by crushing. Calculations show that in a series of isometrically scaled pressure chambers made of the same material, the failure pressures will be the same if the wall thickness scales isometrically with the size of the chamber. Careful examination of the photograph at the top (*right*) of page 50 shows that the wall thickness increases directly with the radius of the nautilus shell, exactly as would be required to keep the failure pressure constant as the animal grows.

The chambered nautilus *Nautilus macromphalus* (*top left*); detail of the interior (*top right*). The moundlike perforation in each chamber wall allows the passage of a tube by which the animal can adjust the pressure of the gas in the compartments. Buoyancy can be changed by changing the relative quantities of gas and fluid in the chambers.

Isometry also dictates the compound spiral patterns observed in the arrangement of the parts of many flowers and fruits. The spiral patterns found in chrysanthemums, for example, are produced as new tiny flowers (florets) are added near the center of the growing composite flower head. Space appears at the center of the flower head as the florets at the periphery grow.

The development of a flower may be simulated by a photographic trick shown below. The left-hand image is a photograph of a chrysanthemum flower, showing two sets of equiangular spirals, one set spiraling out to the left and the other to the right. The middle image is the same photograph, except

Chrysanthemum: *a*, two sets of spirals, one set turning to the left and the other to the right; *b*, expanded version of *a* with a hole left in the center; *c*, same as *b*, except that the center hole has been filled by a patch from the central region of *a*.

that it has been enlarged and a hole has been cut out of the center. In the right-hand image, the hole has been filled in with the central part of the original photograph, which has been rotated slightly to ensure that the spiral patterns of the expanded flower are continuous with those of the original.

The undifferentiated cells toward the center of a flower are organized into the spiral pattern in an analogous way. New florets are added in the midst of existing florets, and then the whole pattern expands to make room for another course of florets at the center. The florets farthest from the center are growing at the greatest rate, which guarantees that the angle subtended by two radial lines drawn to the edges of a floret will stay the same as the floret expands. The individual florets maintain their shape, and therefore the spirals maintain their shape, throughout the growth of the flower.

In both the nautilus and the chrysanthemum, the overall spiral pattern is determined by adding isometric elements—that is, elements of the same shape as those already present but of a different size. In the nautilus, these elements may be thought of as the chambers added at the periphery, while in the chrysanthemum the isometric elements are the new florets added at the center. There is a special pleasure in discovering that all this spatial harmony follows from one simple principle, that the shape of the elements and their position in the pattern should not change as the pattern gets larger.

**Isometry Within a Species.** Although comparisons between species often show a regular change in shape with increasing size, comparisons among different-sized individuals of the same species generally reveal a reasonably faithful isometry. For example, the three horned lizards shown on page 52 are all of the same species, and, even though they are quite different in body mass, they are almost exactly the same shape.

The same is true of individual cockroaches of the same species but of different sizes. Precise measurements of any part of the cockroach anatomy (for example, the tarsus of the second leg) show that the body mass increases with the leg-segment length raised to a power very close to 3, as would be required by isometric scaling (see figure on page 53). Measurements of body width, body length, and leg-segment width are all proportional to leg-segment length (Prange, 1977). Isometric scaling was also found in a range of individual wolf spiders of various sizes.

The tables compiled by life-insurance companies showing the relation between height and weight in humans show that body weight is generally proportional to height raised to a power of about 2.9. Colin Harris of Columbia University has found a description of one human giant whose height and

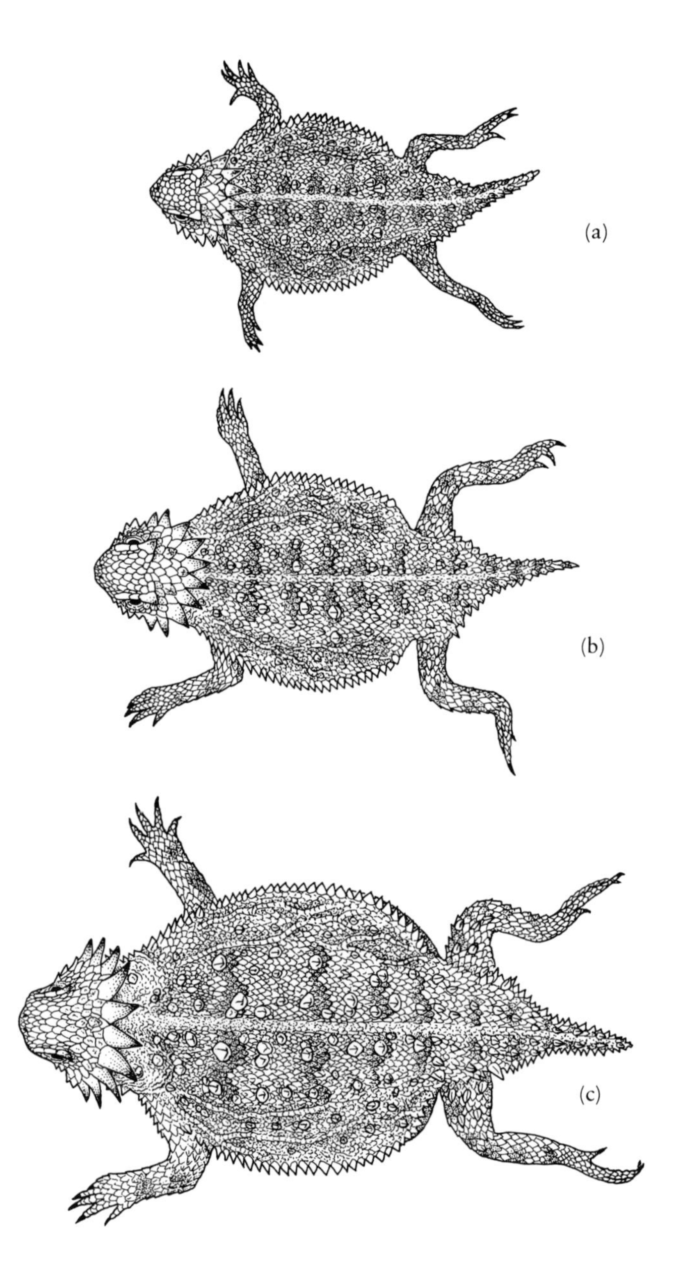


Regal horned lizards (*Phrynosoma solare*) showing a range of adult sizes. Individual *a* weighs 12.4 grams, *b* weighs 29.4 grams, and *c* weighs 85.5 grams. They are very nearly the same shape, although the smallest and largest differ in body weight by about a factor of 7.

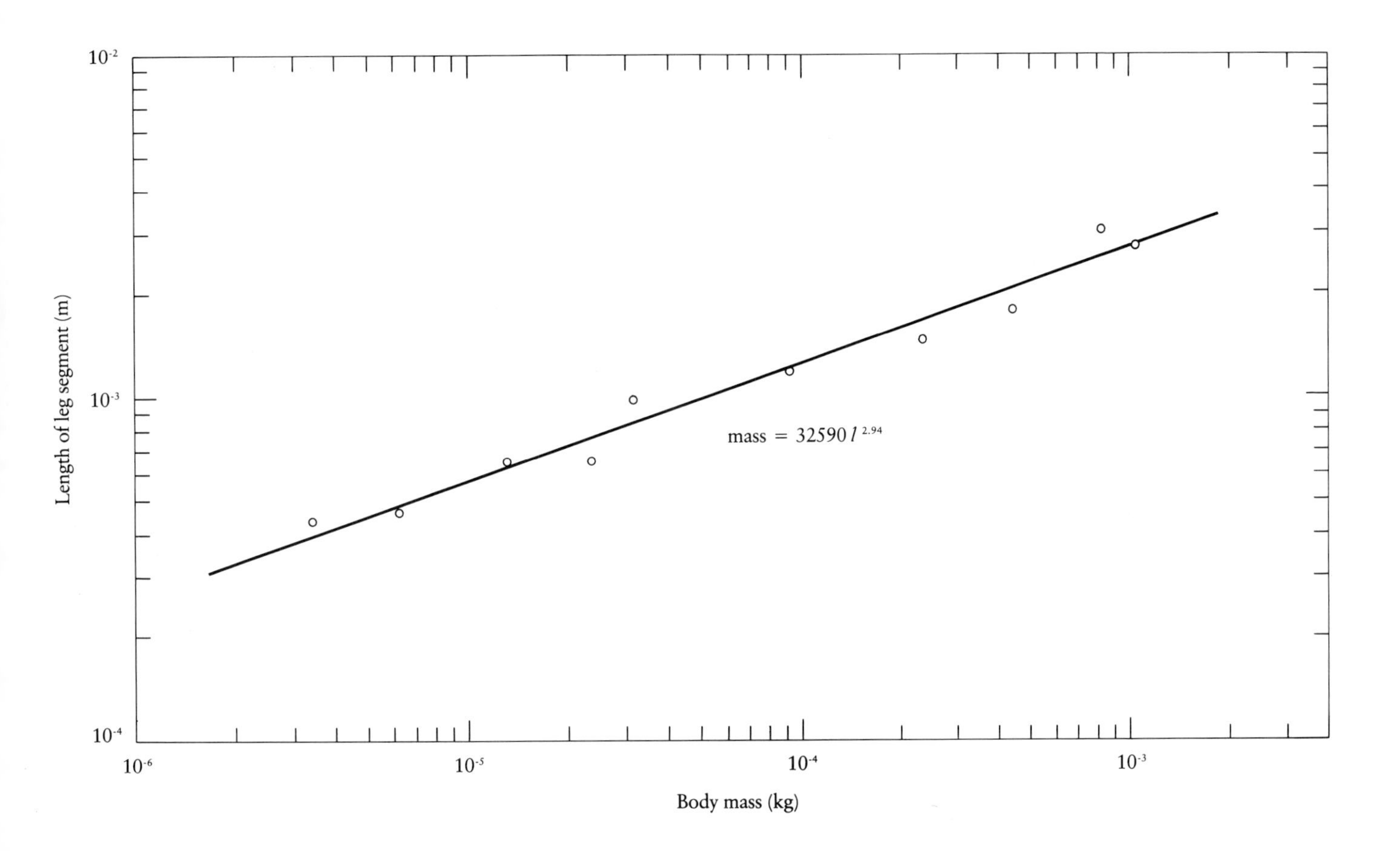


Length of a leg segment as a function of body mass in a single cockroach species, *Periplaneta americana*. Body mass is proportional to leg-segment length raised to a power of close to 3.

weight were well documented from the age of 8 years (6 feet 0 inches, 169 pounds) to the age of 21 (8 feet 8¼ inches, 491 pounds). He died at age 22 following a minor injury. The general trend is that very tall and heavy individuals suffer an increased frequency of foot, leg, and spinal problems, and they tend to lead shorter lives.

According to the rules for isometry, area should scale as the square of length. For geometrically similar (isometric) animals of the same mass density, body mass $m$ is proportional to the cube of body length, $l^3$, so $l$ is proportional to the cube root of $m$. It follows that areas should scale as body mass $m$ raised to the ⅔ power. (If you are not used to dealing with fractional exponents, you may be pleased to discover that many pocket calculators can evaluate fractional powers as well as integral ones through the use of the $y^x$ key.)

In a number of species, body-surface areas proportional to the ⅔ power of body mass are just what is found. In the figure on page 55, the total body-surface area for a species of salamander is shown against body mass on a

This man, Robert Wadlow, was 20 years old and 8 feet 10¾ inches tall when this photograph was taken in 1939. His father, standing next to him, was 5 feet 11 inches tall. An abnormal functioning of the pituitary can result in such a human giant. The excessive growth is most apparent in the head and legs.

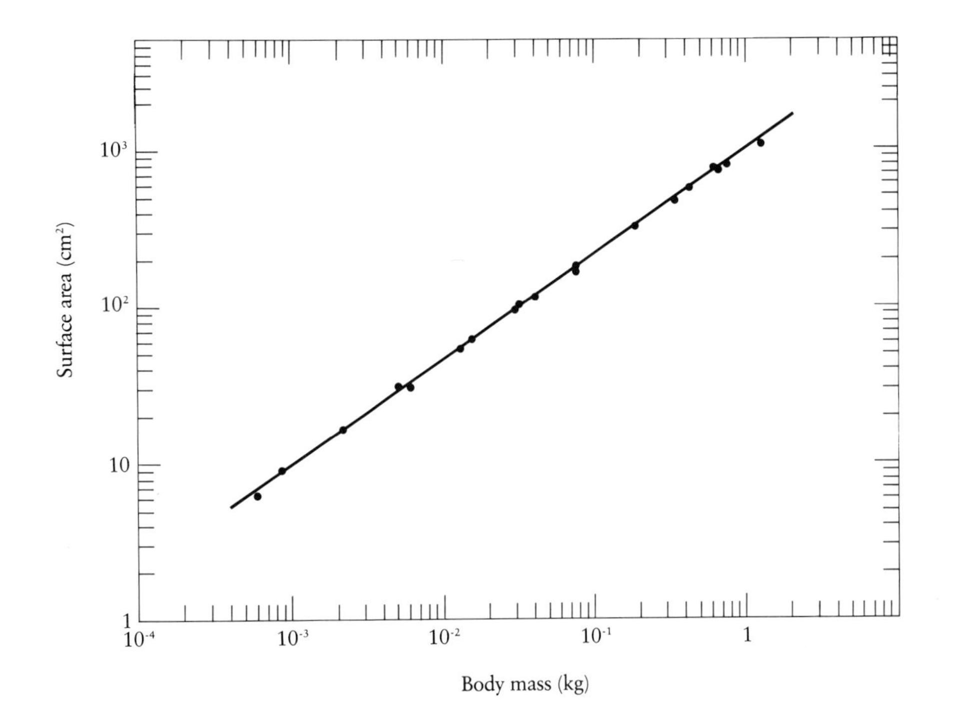


Surface area vs. body mass for a species of salamander, the dwarf siren (*Pseudobranchus striatus*). The surface area is proportional to body mass raised to the 0.67 power.

log–log plot. The mass exponent 0.67 fits the isometric prediction precisely.* The fact that salamanders havc oxygen-permeable skin makes this correlation particularly important.

The graph at the top (*left*) of page 56 demonstrates the relation between basal (resting) metabolic rate and body mass in one species of guinea pig. The rate of oxygen consumption is proportional to $m^{0.67}$, just as predicted by isometric arguments, supposing that strict isometry is preserved in the lung and that the interior surface area of the lung limits the basal rate of oxygen consumption.

The last piece of evidence concerning the maintenance of isometry within a species comes from the record books of competitive weight lifting. In the figure at the bottom (*left*) of page 56, the total weight lifted in three lifts—the press, the snatch, and the clean-and-jerk— is shown to be an increasing function of the body weight of the lifter. Weight-lifting championships are usually decided on the basis of such totals. The figure shows that the weight lifted in each of the body-weight classes up to 198 pounds is quite precisely proportional to the

---

**Note*: Here and throughout this book, we shall use whole fractions, such as ⅔, for the exponents in theoretical results and decimal fractions, such as 0.67, for experimental results.

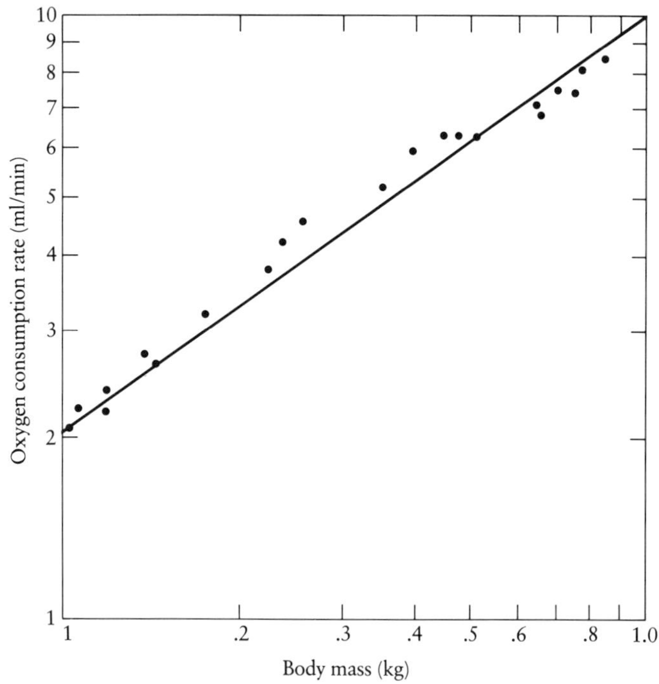


Rate of oxygen consumption (basal metabolic rate) for resting guinea pigs of various sizes. This rate is proportional to body mass raised to the power 0.67.

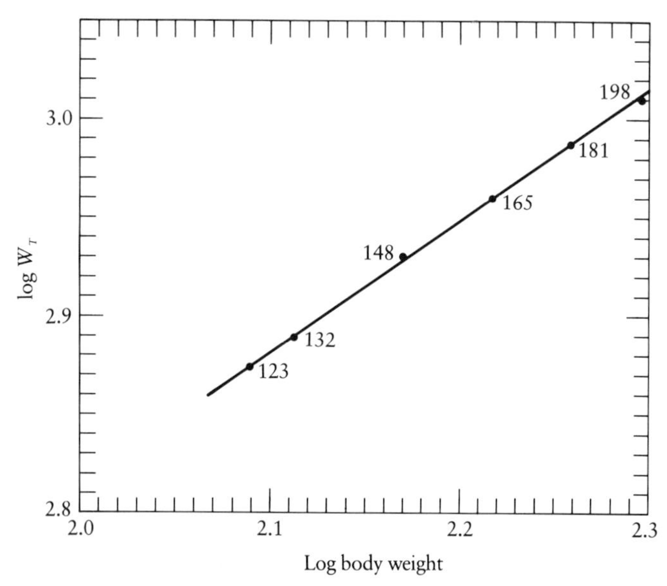


World weight-lifting records, represented by log $W_T$, plotted against log body weight. Here, $W_T$ is the total weight lifted in three lifts: the press, the snatch, and the clean-and-jerk. The numbers beside each point indicate the body weight class, given in pounds.

Weight lifted in body weight classes up to 198 pounds is precisely proportional to the 0.67 power of body weight, that is, the ⅔ power of body weight in animals scaled by isometry.

0.67 power of body weight, as would be predicted by an argument that muscle stress is invariant of body size, so that muscle force, and therefore total weight-lifting ability, is proportional to the cross-sectional area of the body (that is, the ⅔ power of body weight in animals scaled by isometry).

M. H. Lietzke (1956), who plotted these data, observed that the point that indicates the record for the 148-pound class falls slightly above the line. Therefore, he concluded, the Soviet athlete Kostilev, who held that record, was the world's best weight lifter as of the date these world records were current (when proper consideration was given to his small size).

## Allometry

As stressed earlier, it is almost invariably true that, in a comparison of terrestrial animals or plants of greatly differing sizes, proportions change with size and are allometric. Isometry is the rule only for comparisons made over a limited size range, within a species, or under other special circumstances. Furthermore, we pointed out that these allometric changes in proportions were compensations for competing requirements dependent on surface area (diffusion, friction, and heat loss), cross-sectional area (tensile strength and convective transport), and volume (body mass, volume of metabolites, and buoyancy). Here we shall give a number of examples of allometric variations, introducing some that will receive further attention in later chapters and emphasizing the parallels and contrasts between allometry in the engineering and biological worlds.

**Nails.** The standard sizes of common nails are illustrated in the figure at the top of page 58. Common nails range from 1 to 6 inches long, and they are very different in weight—2-penny nails, the smallest, average 847 to the pound, whereas 60-penny nails, the largest, are only 11 to the pound. Although it is not readily apparent by inspection of this figure, an allometric plot (top of page 59) shows that the longer nails are relatively thinner than the shorter ones. The data plotted in the figure include the 16 standard sizes of common nails plus the four standard sizes of round wire spikes and the two nonstandard long sizes. The points fit reasonably well on a straight line whose allometric formula is $d = 0.07\ l^{2/3}$, with $l$ and $d$ both given in inches. Other varieties—including casing nails, finishing nails, and box nails—obey the same allometric formula $d = bl^a$, with the same exponent $a$ but somewhat different values for the constant $b$. A line of slope 1, showing how things would look if nails were made according to isometry, is also drawn on the same graph.

Common nails arranged by size from 60 penny (6 inches) to 2 penny (1 inch).

What could be the explanation for this regular distortion of shape with increasing size? The actual loads to which a nail is subjected when it is struck by a hammer can be quite complicated to calculate. These loads are determined by a number of things, including the energy of the hammer, the speed with which the nail parts the fibers of the wood, and the coefficient of friction between the nail and the wood. In addition to these complex matters, there is also the possibility of dynamic (as opposed to static) buckling of the nail, in which the inertia of the nail plays a role. A plausible approach to explaining the regular distortion of nail shape might be to neglect most of this complexity and to apply only the principles of static buckling.

First, let us imagine that we do a number of experiments in which we measure the force necessary to pull a nail straight out of a piece of wood. The empirical result is that the pull-out force is proportional to the area of the imbedded part of the nail—that is, the circumference of the nail times the depth of penetration. Next, we do a number of nail-driving experiments, from which we find that, when a nail buckles, it almost always does so when the point has been driven the same short distance into the wood (about a quarter of an inch in pine), no matter whether the nail is large or small. Taking a giant leap, we make an assumption (which may not be realistic at all) that the force applied to the head of the nail when the nail is most likely to buckle is roughly measured by the pull-out force for that depth and is, therefore, proportional to the diameter of the nail. A standard result from engineering says that columns made of the same material buckle under a critical load that depends on the

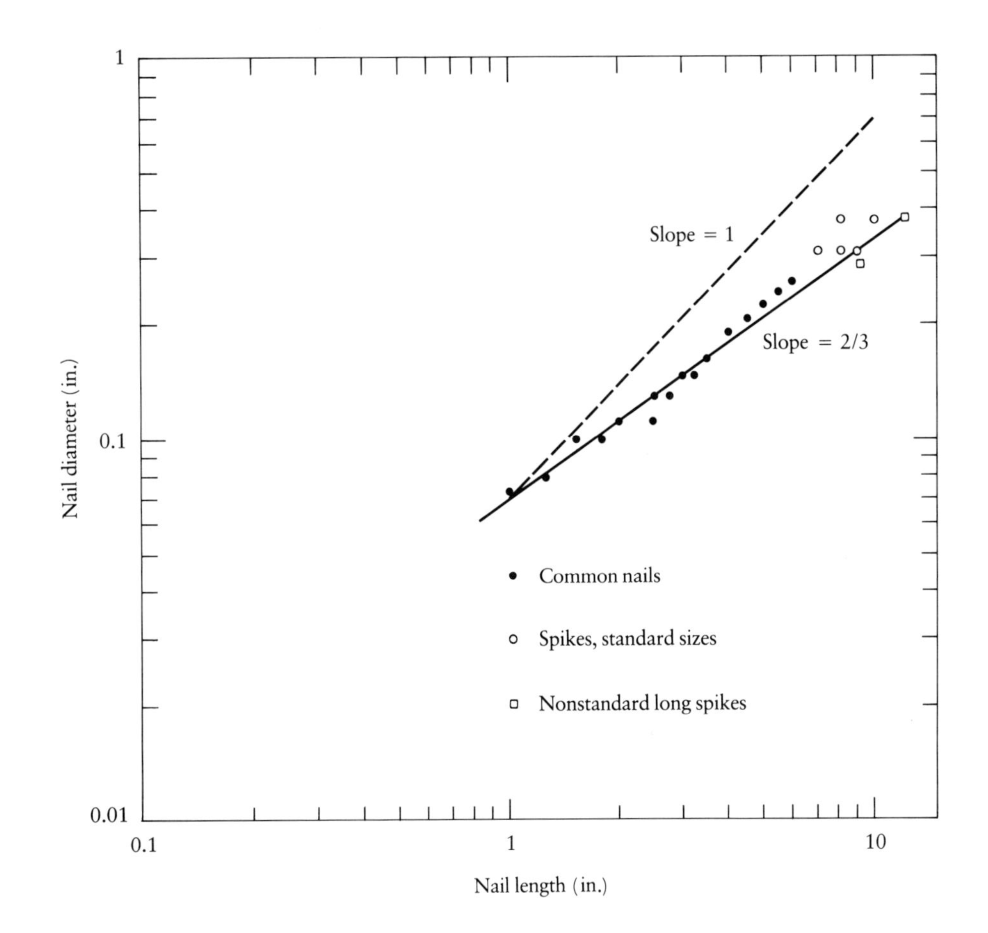


The ratio of nail length to nail diameter on a log–log plot, showing the allometric formula $d = 0.07l^{2/3}$. A broken line of slope 1.0, representing strict isometry, is also shown.

ratio $d^4/l^2$. Therefore, in order for the ratio of the buckling force to the diameter to be the same in each of the nails, we require that the ratio $(d^4/l^2)/d$, or $d^3/l^2$, be the same for all nails. Another way of stating this is to say that the diameter of a nail should be made proportional to the two-thirds power of its length, just as we observed.

It is interesting to notice that, if this is actually a reasonable way to look at the design of nails, U.S. Steel is missing an opportunity to economize in the manufacture of common nails in the size range from 20 penny to 60 penny (4 to 6 inches in the figure on page 58) and in the manufacture of some of the standard spikes. According to the preceding simplified analysis, these nails and spikes could all be made somewhat thinner without jeopardizing their ability to be driven.

**Engines.** Internal-combustion engines come in a great variety of sizes and shapes. Most small engines burn gasoline, have two-stroke cycles, and function without benefit of valves. Middle-sized engines generally have four-stroke cycles, and, as they become larger, more cylinders are added. Large reciprocating aircraft engines in use just before the widespread adoption of jet engines had as many as 24 cylinders. The largest engines, such as those used in railroad locomotives and stationary powerplants, burn diesel fuel rather than gasoline and may have as many as 16 cylinders, as shown in the table below. With such fundamental matters as the type of fuel, the type of ignition, and the number of cylinders changing with size, it must be clear that rigorous isometry goes out the window.

Even so, there are some interesting comparisons to be made when various parameters are plotted against engine mass on a log–log graph. As revealed in the graph on page 63, both the bore (diameter) and the stroke (distance the

| Engine | Mass (kg) | Number of Cylinders | Strokes per Cycle | Displacement ($cm^3$) | Bore (mm) | Stroke (mm) | Maximum Brake Horsepower, BHP | Revolutions per Minute, RPM |
|---|---|---|---|---|---|---|---|---|
| 1. Webra Speedy | 0.135 | 1 | 2 | 1.8 | 13.5 | 12.5 | 0.45 | 22,000 |
| 2. Motori Cipolla | 0.15 | 1 | 2 | 2.5 | 15 | 14 | 1.0 | 26,000 |
| 3. Webra Speed 20 | 0.25 | 1 | 2 | 3.4 | 16.5 | 16.0 | 0.78 | 22,000 |
| 4. Webra 40 | 0.27 | 1 | 2 | 6.5 | 21.0 | 19.0 | 0.96 | 15,500 |
| 5. Webra 61 Blackhead | 0.43 | 1 | 2 | 10 | 24.0 | 22.0 | 1.55 | 14,000 |
| 6. Webra 61WR | 0.49 | 1 | 2 | 10 | 24.0 | 22.0 | 2.76 | 19,000 |
| 7. Enya 60-4C | 0.61 | 1 | 4 | 10 | 24 | 22 | 0.84 | 11,800 |
| 8. Webra 91RC | 0.67 | 1 | 2 | 14.3 | 27.0 | 25.0 | 2.85 | 13,000 |
| 9. Technopower radial | 0.913 | 7 | 4 | 20 | 15.9 | 14.0 | 0.82 | 8,700 |
| 10. Webra T4 | 0.93 | 1 | 4 | 14.7 | 27.0 | 25.0 | 1.43 | 9,300 |
| 11. Kavan FK50 | 2.45 | 2 | 4 | 50 | 34 | 28 | 4.1 | 8,000 |
| 12. McCulloch M2-10 | 3.5 | 1 | 2 | 55 | 44 | 35 | 6.5 | 11,000 |
| 13. Honda 450 | 34 | 2 | 4 | 450 | 70 | 58 | 43 | 8,500 |
| 14. KFM 104 | 68 | 4 | 4 | 916 | 90 | 72 | 79 | 3,900 |
| 15. Lycoming 0-145A | 75 | 4 | 4 | 2,370 | 92 | 89 | 65 | 2,550 |
| 16. Continental A65 | 79 | 4 | 4 | 2,800 | 98 | 92 | 65 | 2,300 |
| 17. Franklin 4AC-176 | 82 | 4 | 4 | 2,800 | 100 | 87 | 75 | 2,500 |
| 18. Franklin 6AC-296 | 118 | 6 | 4 | 4,700 | 106 | 87 | 130 | 2,600 |
| 19. Continental C115 | 119 | 6 | 4 | 3,000 | 102 | 92 | 115 | 2,350 |
| 20. Continental C140 | 135 | 6 | 4 | 3,000 | 102 | 92 | 140 | 3,000 |
| 21. Lycoming GO-290A | 150 | 4 | 4 | 4,750 | 124 | 98 | 145 | 3,000 |
| 22. Ranger 6-440C | 171 | 6 | 4 | 7,200 | 105 | 129 | 200 | 2,450 |
| 23. Lycoming GO-435B | 182 | 6 | 4 | 7,100 | 124 | 98 | 220 | 3,000 |
| 24. Jacobs R-775 | 229 | 7 | 4 | 12,400 | 133 | 127 | 225 | 2,000 |
| 25. Chrysler 340 | 245 | 8 | 4 | 5,569 | 101 | 84 | 275 | 5,000 |

| Engine | Mass (kg) | Number of Cylinders | Strokes per Cycle | Displacement ($cm^3$) | Bore (mm) | Stroke (mm) | Maximum Brake Horsepower, BHP | Revolutions per Minute, RPM |
|---|---|---|---|---|---|---|---|---|
| 26. Ranger SGV-770C | 331 | 12 | 4 | 12,600 | 102 | 120 | 450 | 3,000 |
| 27. Allison V-1710 | 735 | 12 | 4 | 28,000 | 194 | 152 | 1,000 | 2,600 |
| 28. Wright GR-2600 | 885 | 14 | 4 | 42,700 | 155 | 160 | 1,350 | 2,300 |
| 29. Allison V-3420 | 1,180 | 24 | 4 | 56,000 | 140 | 152 | 2,100 | 2,600 |
| 30. Daimler-Benz 609 | 1,400 | 16 | 4 | 61,800 | 165 | 180 | 2,450 | 2,800 |
| 31. GMC Electromatic 645 | 1,775 | 16 | 4 | 183,000 | 239 | 254 | 3,300 | 900 |
| 32. Daimler-Benz 613 | 1,960 | 24 | 4 | 89,000 | 162 | 180 | 3,120 | 2,700 |
| 33. Sultzer 16-LVA | 2,000 | 16 | 4 | 220,000 | 250 | 280 | 4,000 | 1,100 |
| 34. Nordberg | 5,260 | 16 | 4 | 488,000 | 356 | 407 | 3,000 | 400 |
| 35. Ingersoll-Rand PKVT | 10,850 | 16 | 4 | 1,165,000 | 406 | 559 | 4,400 | 360 |
| 36. Fairbanks-Morse | 12,860 | 12 | 4 | 1,327,000 | 508 | 546 | 12,000 | 400 |
| 37. Cooper-Bessemer V-250 | 13,500 | 16 | 4 | 1,334,000 | 457 | 508 | 7,250 | 330 |
| 38. Sultzer RD-90 | 97,200 | 12 | 4 | 11,850,000 | 900 | 1,550 | 27,600 | 119 |
| 39. Burmeister and Wain | 102,300 | 12 | 4 | 11,900,000 | 840 | 1,800 | 27,800 | 110 |

piston moves) in a single cylinder increase almost isometrically with the size of the engine. A line with slope ⅓ passes through most of the points representing both bore and stroke in engines ranging in size from tiny two-stroke model-airplane engines through the largest diesels. Furthermore, as shown in the graph on page 62, the engine rotational speed, measured in revolutions per minute (RPM), fits a line with slope −⅓ in this log–log graph. From these two observations, we can conclude that the peak speed of the pistons, which is proportional to the product of RPM and piston stroke, is nearly a constant in engines over a huge range of different sizes.

It would be fair to say that this last observation amounts to a principle of design. If the speed of the pistons midway through their stroke is independent of engine size, so is the kinetic energy per unit volume, which is calculated as one-half the mass density of the pistons times the velocity of the pistons squared. As a consequence, the strain energy per unit volume transiently stored by the pistons, connecting rod, and crankshaft when each piston is at the top and the bottom of its stroke is independent of the size of the engine. Because materials like steel are prone to failure when the strain energy per unit volume exceeds a certain critical level, most of the engines compared here appear to have been designed using the same factor of safety.

This model airplane engine (no. 9 in the table) is a quite faithful scale model of the Jacobs R-775 aircraft engine (no. 24 in the table). Both model and prototype are four-stroke, seven-cylinder radial engines. The ratio of power to weight is somewhat higher in the large engine than in the model.

Another important matter in engine design is the ratio of the power to the mass. The line of slope 1 in the graph on page 62 indicates the ideal condition in which the ratio of the maximum brake horsepower (BHP) to the engine mass is a constant, independent of engine size. Engines along this line may be said to be equivalently well designed; no one is better than another. The large high-performance aircraft engines numbered 27 through 32 in the figure were used on fast military aircraft. They were designed to optimize the ratio of power to mass, even when that resulted in great complexity and cost. In general, increasing the size of an engine makes it difficult to maintain the ratio of

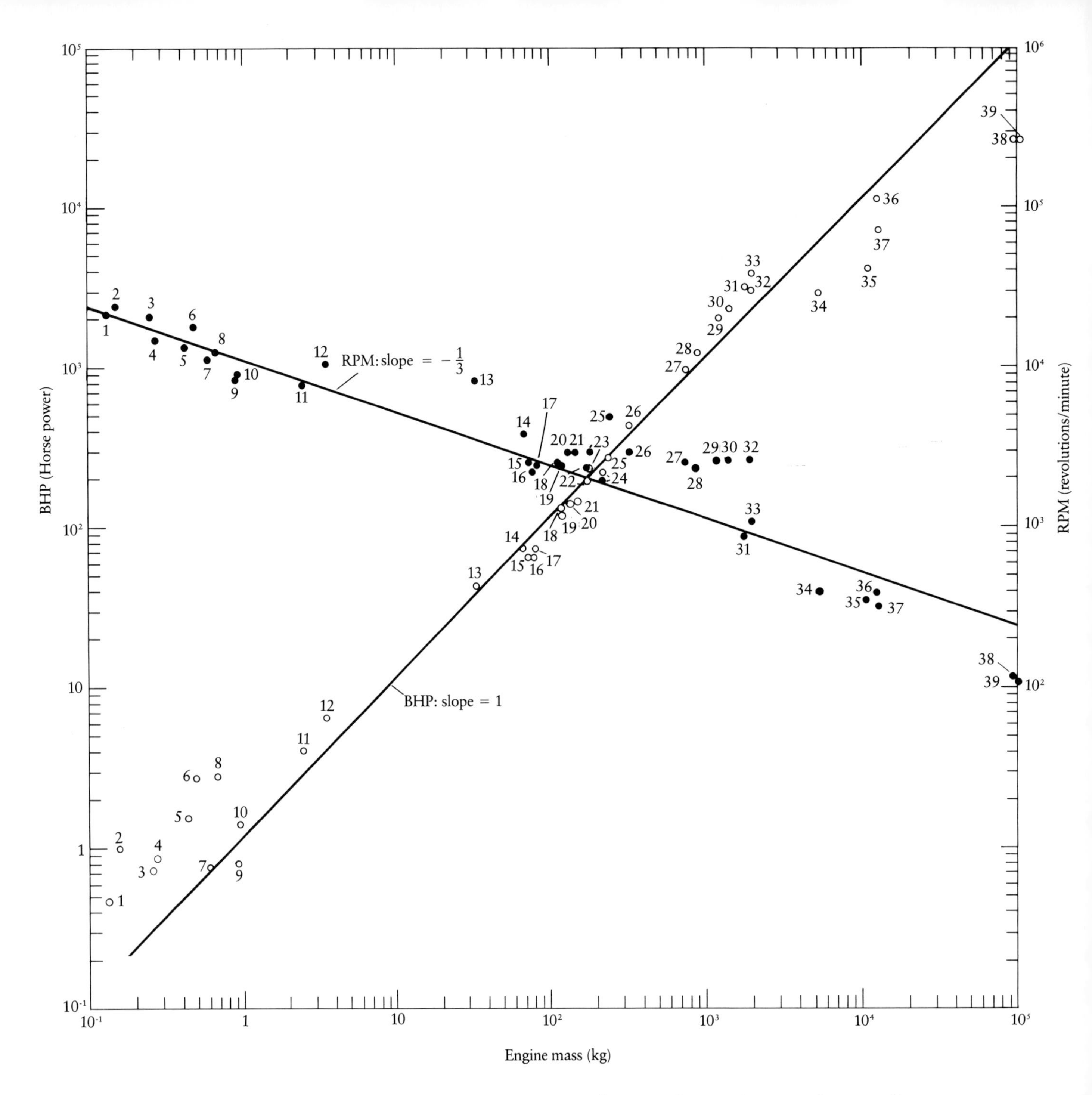


The allometry of internal-combustion engines. Although the number of cylinders, type of fuel, and type of ignition change with size, there are allometric relations between RPM, power, displacement, and engine mass.

power to mass constant because the requirements for a cooling apparatus are relatively more expensive, both in subtracted power and added mass, the larger the engine becomes. It is clear from the figure that nearly all the small model-airplane engines lie above the line and all the largest diesels lie below it. Part of the reason large airplanes and ships have several engines instead of one is that

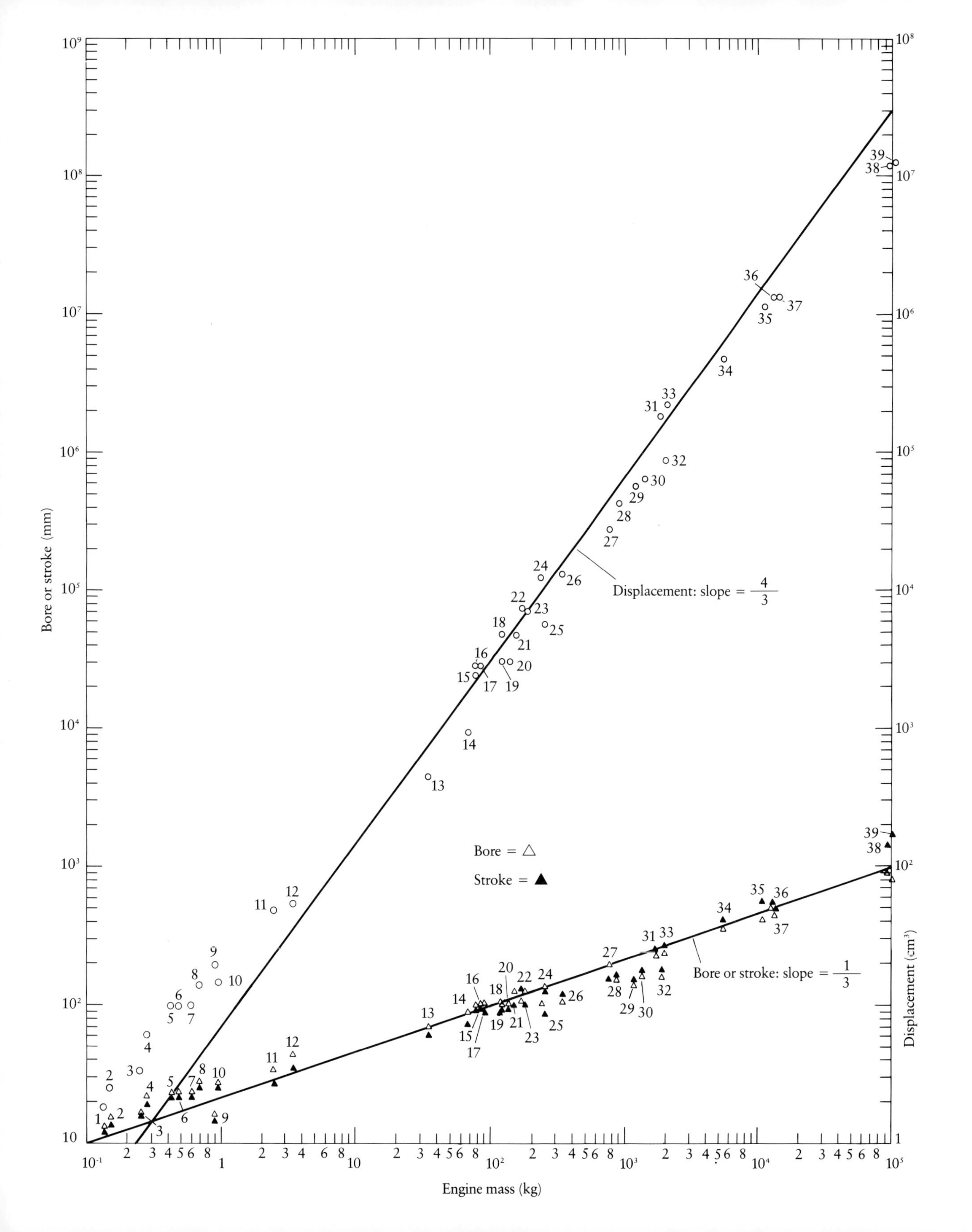
Bore or stroke (mm)
Displacement (cm³)
Engine mass (kg)
Displacement: slope = 4/3
Bore or stroke: slope = 1/3
Bore = △
Stroke = ▲

it is safer; you can lose one or two engines out of four and still have enough power to continue flying or steaming. Another part of the reason is that the same total engine mass yields more power if it is divided up among several small engines rather than concentrated in one large one. A total of 757,778 Webra Speedys add up to the same mass as one Burmeister and Wain, but at 341,000 horsepower collectively, the smaller engines have more than 12 times the power. As a final point, one may see from the figure that the volume of air that moves through an engine in one cycle (displacement) is not proportional to the mass of the engine but more nearly to the mass raised to the power $4/3$. This is a consequence of the two competing design goals mentioned earlier, a constant ratio of power to mass and a constant peak speed of the pistons. For the engines considered here, the power depends roughly on the volume rate at which air and fuel moves through the engine, and this volume rate is proportional to the displacement multiplied by the RPM. If RPM is proportional to mass raised to the power $-1/3$, displacement must be made proportional to mass raised to the power $4/3$ in order for the volume rate of flow of air and fuel to be proportional to engine mass to the power 1. While most of the data points for displacement fall along the ideal line of slope $4/3$ in the graph on page 63, again there is a trend for the smallest engines to lie above the line and the largest engines to lie below it.

**Kleiber's Law.** The metabolic machinery of animals shows a fascinating and important allometric variation with body size. The mass-specific (intrinsic) basal rate of heat production, comparable to the power-to-mass ratio for engines, is plotted in the top figure on page 65 on semilogarithmic coordinates so that a large range of body sizes can be shown. The smallest animals have the highest intrinsic metabolic rates, and it is often remarked that this explains why they eat so often and appear to have such relatively high rates of activity.

We have already seen how the basal rate of oxygen consumption within a species is roughly proportional to $m^{0.67}$ (top graph on page 56). This makes some intuitive sense when we remember that both the surface area and the cross-sectional area of body forms scaled by isometry are proportional to the 0.67 power of the body volume. A remarkable observation, first made by the American veterinary scientist Max Kleiber in 1932, is that the basal metabolic rate, whether measured by the rate of food consumption, oxygen consumption, or heat production, varies in an even more regular and repeatable way when the comparison is made not *within* a species but *between* species of mammals. The lower figure on page 65 shows the relation that has come to be known as Kleiber's law, which says that metabolic rate in a comparison between species does not show isometry and is proportional to $m^{0.75}$ over a

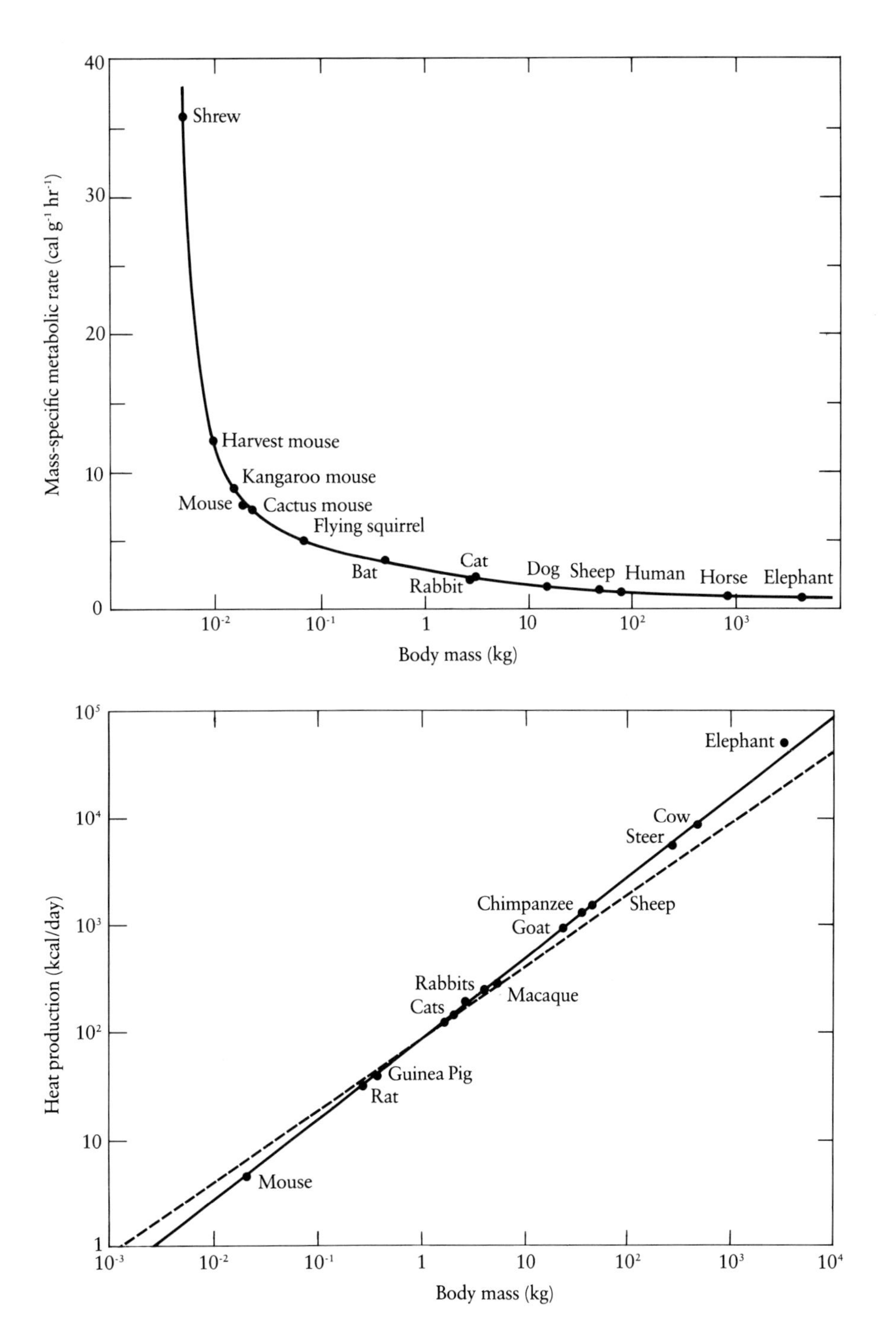


Mass-specific rate of heat production vs. body mass, plotted on semilogarithmic coordinates (only the body-mass axis is logarithmic). The intrinsic (mass-specific) metabolic rate of small animals is greater than that of large animals.

Metabolic heat production vs. body mass in an allometric plot. The solid line has a slope of 0.75, as required by Kleiber's law. The broken line, which shows a slope of 0.67, has been included for comparison.

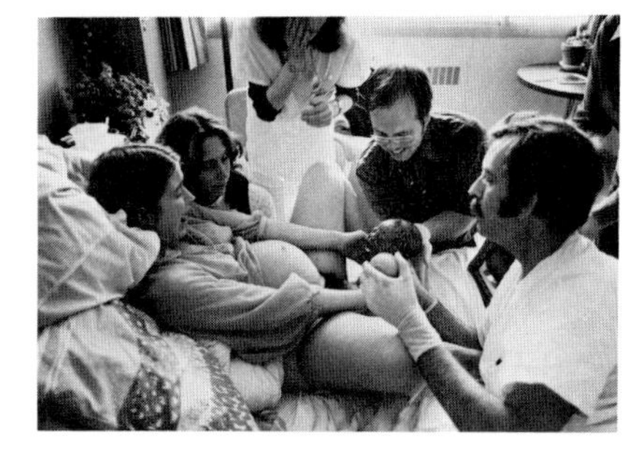

An infant's first breath.

range of animals from mice to elephants. Recently, it has been shown through statistical analysis—by Heusner (1982) and Feldman and McMahon (1983)—that the best fit to all the data available indicates that a mouse-to-elephant curve like the one at the bottom of page 65 is a series of short straight lines of slope 0.67 (one for each individual species) arranged such that their means fall on the Kleiber line (the line that compares different species), which has a slope of 0.75. Certain groups of animals, dogs in particular, span a sufficiently wide range of body size that their individual species slope is also close to 0.75.

Since Kleiber's discovery, many investigations have confirmed the 0.75 exponent to a high degree of confidence. A broken line of slope 0.67 is shown in the figure for comparison. There are clear differences between this line and the Kleiber relation over the mouse-to-elephant range.

The Kleiber relation between size and metabolic rate is also relevant to human development. The English physiologist Douglas Wilkie has called attention to an interesting transition in metabolic rate that occurs in a human infant shortly after birth (Wilkie, 1977). Measurements show that the (intrinsic) basal rate of oxygen consumption in the newborn rises in about 36 hours from a rate of 3.5 milliliters per minute per kilogram (appropriate for an animal the size of the mother) to approximately 7 milliliters per minute per kilogram (appropriate for the body mass of the baby). While the fetus is within its mother, it behaves metabolically as if it were simply one of her organs, ticking over at the relatively low metabolic intensity determined by her weight. Within 36 hours after birth, the rates at which all of its enzymes act and the rate of activity of its mitochondria have changed, speeding up its cellular processes sufficiently to bring it onto that point on the human segment of the Kleiber line determined by its own weight as a separate small individual.

Biology is extremely short on mathematical laws as clear and as general as this one. In the remainder of this book, but particularly in Chapters 4 and 5, we shall be seeking an understanding of Kleiber's law based on simple principles. This will lead us to consider how gravity introduces regular distortions into the shapes of terrestrial animals and plants of large size. But before we take these steps, it will be worthwhile to review the uses to which engineers and physicists put dimensional analysis and the theory of models, a powerful tool for coming to grips with both the isometric and the allometric scaling discussed in the chapters to follow.

## Appendix: The Relation Between Speed and Size in Rowing Shells.

As explained in this chapter, we assume that the power required to move a boat through the water comes mainly from the skin-friction drag, which is proportional to the product of the wetted area and the square of the speed. Because the wetted area is proportional to $l^2$, in which $l$ is the length of the boat, the drag force $D$ is proportional to $V^2l^2$. The power required to move the boat is equal to the drag force multiplied by the speed:

$$\text{power required} = \text{drag force} \times \text{speed} = D \times V$$

so that

$$\text{power required} = (k_1V^2l^2) \times V = k_1V^3l^2 \qquad (1)$$

in which $k_1$ is a constant.

Because the power is provided by the oarsmen, it is therefore proportional to the number of oarsmen, $n$. This may be stated

$$\text{power available} = k_2n \qquad (2)$$

in which $k_2$ is another constant. If the power required is equal to the power available, we obtain the following equation from equations 1 and 2:

$$k_1V^3l^2 = k_2n \qquad (3)$$

Because $l^3$ is proportional to $n$, according to the argument (given in this chapter) that says that the total weight of the oarsmen plus that of the boat equals the weight of the water displaced by the boat,

$$l^3 = k_3n$$

and, therefore,

$$l^2 = (k_3)^{2/3}n^{2/3} \qquad (4)$$

Substituting equation 4 in equation 3 and solving for $V$ gives

$$V = (k_2k_3^{-2/3}k_1^{-1})^{1/3}n^{1/9}$$

or, using the proportional sign,

$$V \propto n^{1/9}.$$

# Chapter 3

# The Physics of Dimensions

Turning vanes in the 16-foot wind tunnel at Langley Field, Hampton, Virginia.

Sometimes we see amazing things at the movies. The World Trade Center is on fire, or a giant cockroach has taken over downtown Schenectady. When these special effects are done well, they can almost make a bad movie worth the price of admission.

Doing special effects properly with small models that are supposed to look large to the theatre audience depends on a knowledge of the science of *dimensional analysis* and the related *theory of models*. We shall see that dimensional analysis matters to more people than movie moguls—engineers and physicists often do experiments on small-scale models when experiments on the full-size airplane, ship, or other structure (hereafter called the *prototype*) would be prohibitively expensive or impossible. Using scale models, the wind currents around a skyscraper may be studied and the likelihood that it will twist and sway in the wind may be investigated before it is built. The stresses in a dam, the deposition of silt in a harbor, the dispersion of the plume from a smokestack—all these may be studied in laboratory models whose scale is conveniently smaller than the actual systems they represent.

Biologists also use dimensional analysis from time to time. The mechanics of red blood cells moving through the capillaries, the flow of seawater around filter-feeding organisms, and the aerodynamics of insects have all been studied in scaled-up, rather than scaled-down, models.

Dimensional analysis even has a dignity of its own, quite apart from its applications. Using the techniques of dimensional analysis, one may get quite a long way toward the answer to a complex physical problem without explicitly using any of the laws of physics. In this way, dimensional analysis is a kind of Physics Made Easy—and it really works, up to a point.

## Dimensional Analysis: The Basic Idea

The word "relatively" can conceal many pitfalls, and it is often misused. People will say, "I guess he's honest, relatively speaking," or "What are you complaining about? You're relatively well off." All too frequently, we are left wondering, "Relative to what?"

The same difficulty comes up in physical problems. It doesn't mean much to say that a stove is relatively hot or that an airplane is going relatively fast until you specify *relative to what*. This may be done properly by giving a relative temperature as the ratio of two temperatures or a relative speed as the ratio of two speeds. An iron stove glows cherry red at a temperature of about 800 degrees Celsius and melts when the temperature reaches 1,535 degrees Celsius.

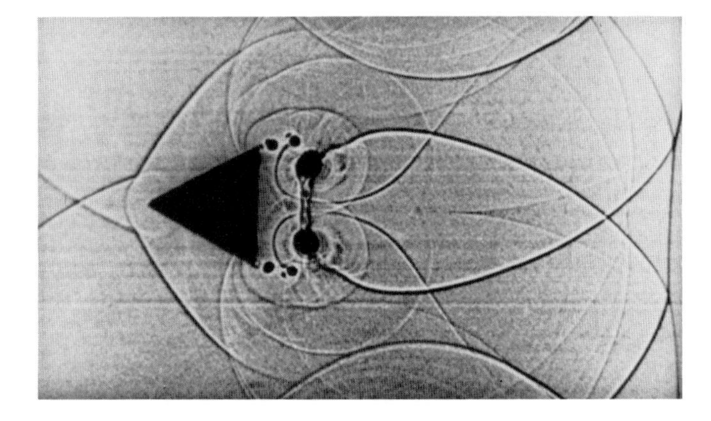

(*Above*) The shock-wave pattern produced by unsteady air flow over a wedge. The air flows from left to right in this shadowgraph visualization. The shock waves are the dark lines.

Laurel and Hardy in *Brats.*

A useful relative temperature for safety considerations might be defined as the temperature of the stove divided by its melting temperature. The flow of air around an airplane in flight changes radically when the airplane begins to move faster than the speed of sound. When the plane is moving faster than sound, it radiates energy in the form of shock waves—these can be heard and felt as sonic booms on the ground—and the energy required costs extra power from the engine. The Mach number is defined as the ratio of the plane's speed to the speed of sound. Since the speed of sound changes with altitude, whether there are shock waves depends not on the plane's absolute speed but on its speed relative to the speed of sound in the air it is flying through. What matters is not a dimensional parameter (the plane's speed) but a dimensionless number (the Mach number).

In dimensional analysis, the object is to substitute a set of dimensionless numbers for the dimensional physical variables that describe a problem—pressure, velocity, density, and so on. Because the dimensionless numbers are products or ratios of the physical variables, this process always succeeds in reducing the number of variables in a problem. For example, when two variables are multiplied or divided, we are left with only one product or ratio. This reduction in the number of variables is worth doing. It frequently saves a great deal of experimental effort, as we shall see. Let us begin with an example and then discuss the general principles.

**An Example: Bicycling in Circles.** Suppose someone were riding a bicycle along a circular path, contemplating the physics of his or her motion. An interesting fact is that the bicycle naturally leans over, making an angle $A$ with the vertical as it goes around a circle of radius $R$ with steady speed $v$ (see figure on page 72). It occurs to the rider that a circular bicycle track banked at angle $A$ would be particularly safe to ride on: the bicycle would be at right angles to the track and could not possibly skid sideways as it can when tight turns are made on flat ground. Suppose one were going at a certain speed around a turn of a certain radius—what would the angle $A$ be?

After orbiting around on the flat ground for a while, the rider decides to stop contemplating and to start experimenting. With the aid of a friend, who operates a camera and times the circuits with a watch, the rider runs a series of trials in which $R$ is held constant but $v$ changes. Then they do a set of experiments in which $v$ is held constant but $R$ changes. When the pictures come back from the drugstore, they measure angle $A$ with a protractor (pictures showing the bicycle directly head-on or tail-on are essential for this) and plot the results as shown in figures $a$ and $b$ on page 73.

Having come this far, they are not sure what to do next. They have two plots, one showing that angle $A$ increases with $v$, the other showing that $A$ decreases with $R$. This seems less than a general and useful knowledge of the problem. They would like to be able to predict $A$ for an arbitrary $v$ and $R$. Perhaps they should repeat the experiments again many times until it is possible to plot several constant-$R$ lines in figure $a$, or several constant-$v$ lines in figure $b$. Then finding $A$ for a given $R$ and $v$ would be a matter of interpolating between the two nearest constant-$R$ lines in figure $a$ or between the two nearest constant-$v$ lines in figure $b$. They realize that getting this technique to produce accurate results will require a great many experiments. They ask themselves whether the whole thing is worth it.

A bicyclist riding at speed $v$ around a circle of radius $R$ heels over at angle $A$ with respect to the vertical.

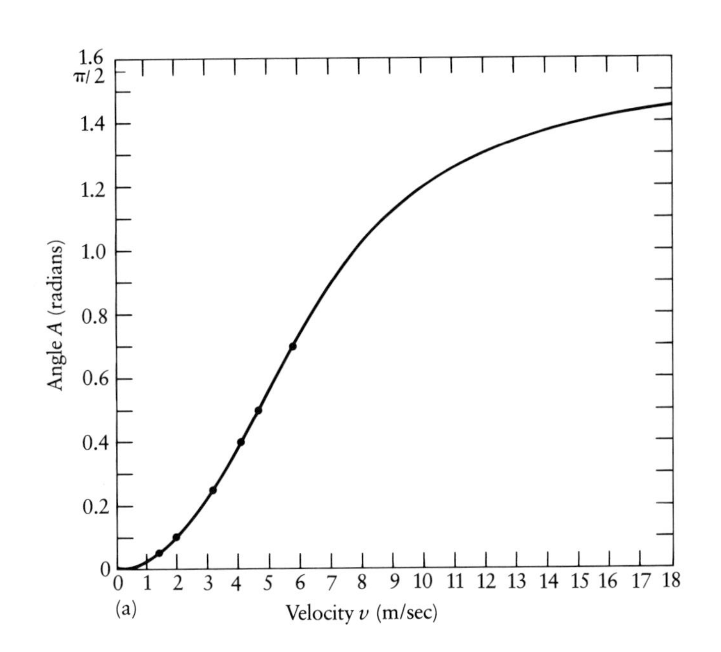


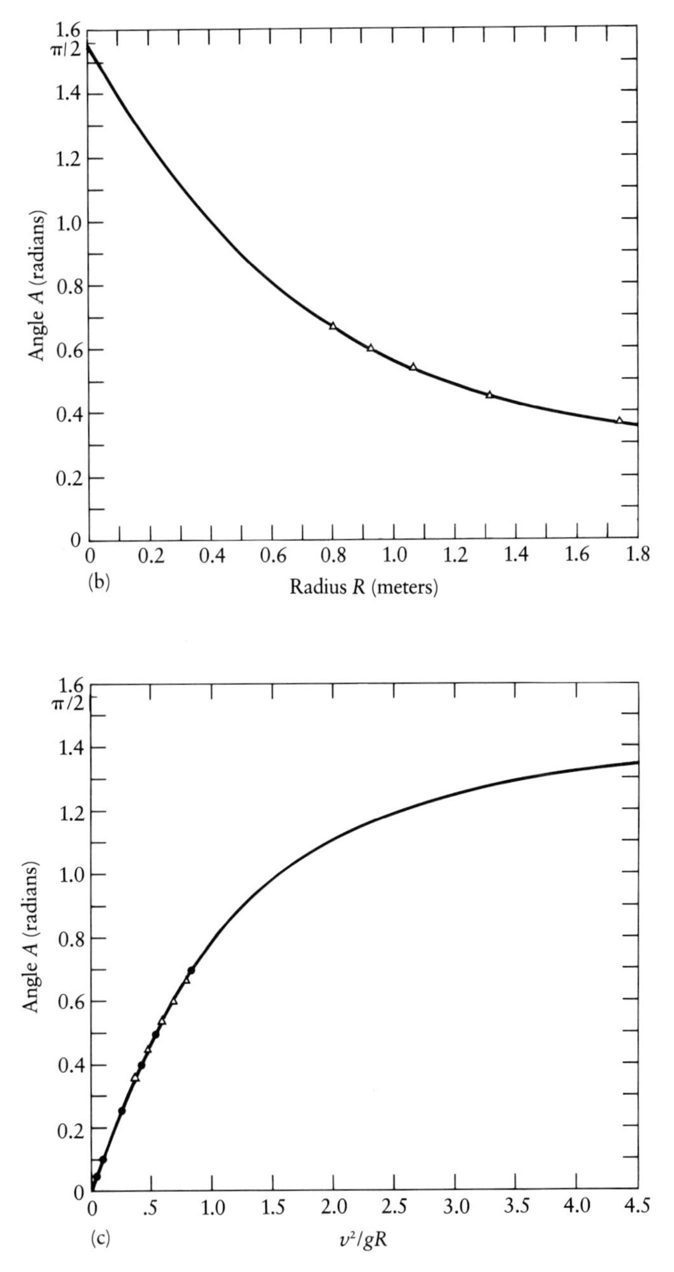


Results from the bicycling experiments, shown schematically: *a*, angle *A* plotted against *v* for constant *R*; *b*, angle *A* plotted against *R* for constant *v*; *c*, angle *A* plotted against the dimensionless group $v^2/gR$, showing how all the data fall on one line.

What they really need, at this moment, is not more experiments but an insight based on dimensional analysis.

Now an old bicycle-track designer experienced with dimensional analysis steps out of the shadows. He shows them how the data they already have will fit very nicely on one curve in a plot of *A* versus the square of the velocity divided by the product of the acceleration of gravity with the radius, or $v^2/gR$, as shown in figure *c*. He shows them how all the dimensions cancel so that $v^2/gR$ is a quantity measured not in length, or force, or time, but only as a pure number.*

This curve will furnish the general solution they are seeking. Once they have this plot, they can pick an arbitrary $v$ and $R$, calculate $v^2/gR$, and find *A* from the curve. It's that simple, and they don't need to do any more experiments.

Furthermore, the fact that the earth's gravitational acceleration, $g$, appears in the dimensionless group tells them that neither their original experiments nor the ones they planned to do later were really complete—they should have been working in a huge centrifuge, or on the moon, so that they could conduct some experiments with different values of $g$. Even if they had gone to that much trouble, dimensional analysis says that the data points that show how angle *A* changes with $g$ when $v$ and $R$ are constant would again fall precisely on the same curve shown in figure *c*.

The old bicycle-track designer has saved them a great deal of work, but how did he know to pick the group $v^2/gR$? To understand his reasoning, we need to know about fundamental quantities and dimensional formulas.

**Fundamental Quantities.** Physical variables always require instruments for their measurement. The measurement of a length requires a ruler or its equivalent. The measurement of a time requires a clock.

Using only a force scale, a ruler, and a clock, one could devise schemes for measuring most of the variables that arise in mechanics. Pressure could be measured using the force scale and the ruler; velocity could be measured using the ruler and the clock. The minimum basic instruments needed to measure a physical variable define the *fundamental quantities* for that variable. By convention, the dimensional specification of a physical variable, called its *dimensional formula*, shows the fundamental quantities (denoted by capital letters) raised to appropriate powers and enclosed in square brackets. Thus, the dimensional formula for an area is $[L^2]$, and the dimensional formula for a velocity is $[LT^{-1}]$. The dimensional formulas for several familiar physical variables are listed in the following table.

*To the reader: Do this for yourself, and remember that the Earth's gravitational acceleration, $g$, is measured in meters per second squared.

**The dimensional formulas for several physical variables.**

| | MLT (or MLTΘ) | FLT (or FLTΘ) |
|---|---|---|
| Length, $l$ | [L] | [L] |
| Time, $t$ | [T] | [T] |
| Force, $F$ | $[MLT^{-2}]$ | [F] |
| Area, $A$ | $[L^2]$ | $[L^2]$ |
| Frequency, $f$ | $[T^{-1}]$ | $[T^{-1}]$ |
| Mass, $m$ | [M] | $[FL^{-1}T^2]$ |
| Velocity, $v$ | $[LT^{-1}]$ | $[LT^{-1}]$ |
| Acceleration, $g$ | $[LT^{-2}]$ | $[LT^{-2}]$ |
| Pressure, $P$; stress, $s$ | $[ML^{-1}T^{-2}]$ | $[FL^{-2}]$ |
| Energy (work), $\mathcal{E}$ | $[ML^2T^{-2}]$ | [FL] |
| Dynamic viscosity, $\mu$ | $[ML^{-1}T^{-1}]$ | $[FL^{-2}T]$ |
| Surface tension, $\mathcal{S}$ | $[MT^{-2}]$ | $[FL^{-1}]$ |
| Mass density, $\rho$ | $[ML^{-3}]$ | $[FL^{-4}T^2]$ |
| Modulus of elasticity, $E$ | $[ML^{-1}T^{-2}]$ | $[FL^{-2}]$ |
| Specific heat at constant pressure, $C_p$ | $[L^2T^{-2}\Theta^{-1}]$ | $[L^2T^{-2}\Theta^{-1}]$ |
| Thermal conductivity, $k$ | $[MLT^{-3}\Theta^{-1}]$ | $[FT^{-1}\Theta^{-1}]$ |

As shown in the table, a physical variable may be specified in either the mass-length-time (MLT) system or in the completely equivalent force-length-time (FLT) system. These are not the only alternatives; other possibilities are a pressure-length-time system (PLT) or a density-length-time system (DLT). In the PLT system, the dimensional formula for a force would be $[PL^2]$ and that for a mass would be $[PLT^2]$. In the DLT system, a force would be given as $[DL^4T^{-2}]$ and a mass would be given as $[DL^3]$.

In problems involving heat, temperature—here denoted by the capital Greek letter theta (Θ)—must be considered a fundamental quantity. In the FLTΘ system, specific heat (the quantity of heat required to raise a unit mass one degree of temperature) has the dimensions $[L^2T^{-2}\Theta^{-1}]$. Thermal conductivity, which is defined as the rate of flow of heat through a square unit of cross-sectional area when a temperature gradient of one degree per unit of distance is established at right angles to that area, is measured by dimensions $[FT^{-1}\Theta^{-1}]$.

In the sense that they are not unique, fundamental quantities don't seem to be so fundamental after all. Whether mass, force, pressure, density, temperature, or even energy is a fundamental quantity appears to depend on what instrument or technique of measurement you prefer to call fundamental. It is necessary merely to be consistent when choosing a system of fundamental

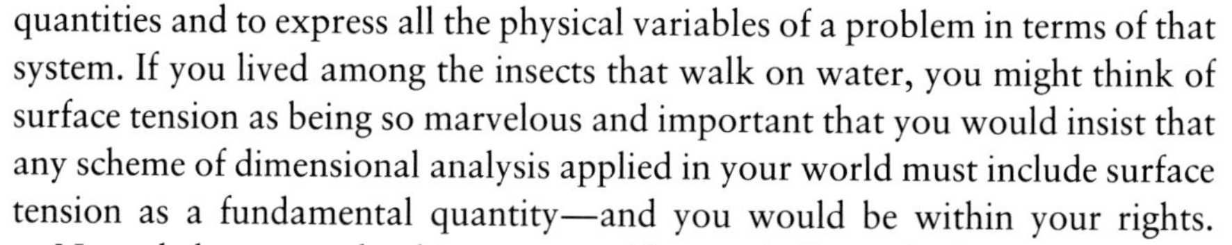

quantities and to express all the physical variables of a problem in terms of that system. If you lived among the insects that walk on water, you might think of surface tension as being so marvelous and important that you would insist that any scheme of dimensional analysis applied in your world must include surface tension as a fundamental quantity—and you would be within your rights.

Nevertheless, to make things reasonable, we shall use the force-length-time (FLT) system from now on, except in problems involving heat, in which we shall use the FLTΘ system.

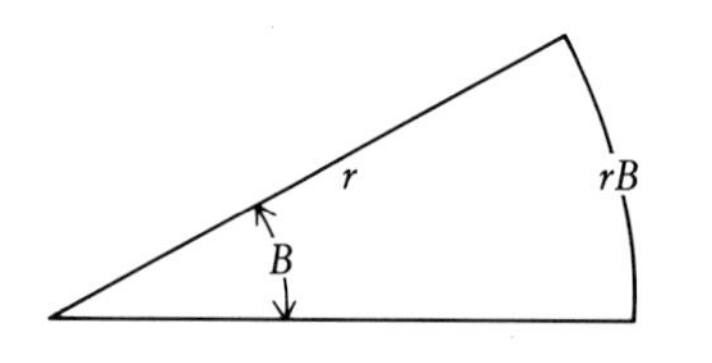


One way of understanding why an angle is a dimensionless quantity is to notice that the angle $B$ is the ratio of the arc length $rB$ to the radius $r$. The angle is therefore the ratio of two lengths.

**Dimensionless Variables and Constants.** Not all of the variables one uses in describing a physical situation are dimensional variables. For example, an angle $B$, measured in radians, is the ratio of an arc length ($rB$) to a radius ($r$) and is therefore just a number, with no dimensions at all, as is shown in the figure on the left. Because the concentration of a substance (given, for example, in parts per million) is the ratio of two volumes (or, in other circumstances, the ratio of two masses), it is also dimensionless. Another, less familiar example of a dimensionless variable is the strain of an elastic object under an applied load. The strain is defined as the change in length between the loaded and unloaded conditions divided by the initial length (see figure on page 77).

Often, when one is looking for a way to make a physical variable dimensionless, there is no such easy solution as dividing a length by a length (as in calculating strain) or dividing a velocity by a velocity (as in forming the Mach number), because the reference length or velocity one needs is not in the list of physical variables. In such cases, it is necessary to construct a length, or a velocity, or a time by artfully multiplying or dividing the physical variables that *do* appear in the list.

Consider, for example, the problem of cooking a beef roast. The roast is put in a preheated oven at 350 degrees Fahrenheit (*not* a microwave oven—an old-fashioned one). We have a meat thermometer that sticks into the interior of the roast. The directions in the cookbook say that we should take the roast out of the oven when the interior reaches a certain temperature—140 degrees for rare, 160 degrees for medium, and so on. The problem is that we don't want to be opening the oven to look at the meat thermometer all evening. We'd like to have a good estimate for the time $T_c$ required for the roast to cook. With such an estimate, we could check the roast at the appropriate time and, if the thermometer says it's ready, take it out.

Because we have a little experience with problems of this type, we know that the physical variables that matter are thermal conductivity ($k$), mass density (denoted by $\rho$, the small Greek letter rho), specific heat at constant pressure

The strain of an elastic body is defined as the change in length divided by the initial length.

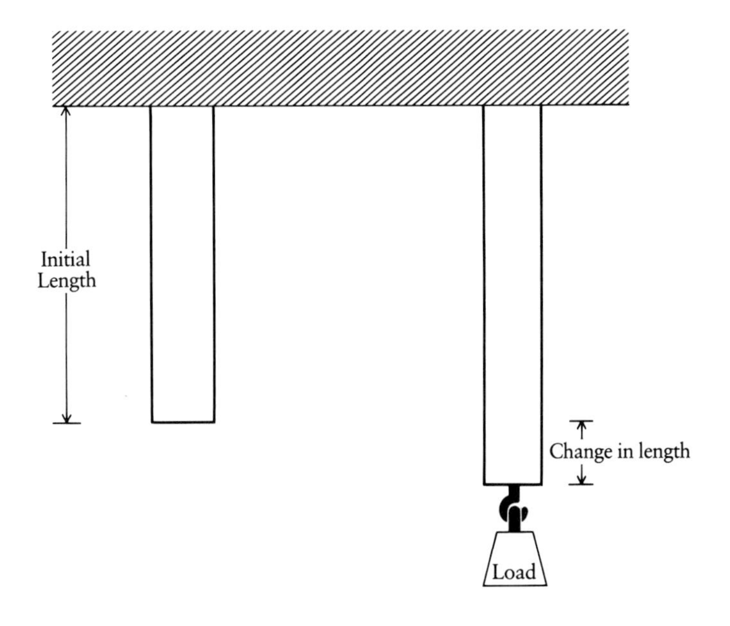


($C_p$), and a characteristic dimension ($D$) of the roast, say its greatest diameter. The thermal conductivity is a measure of how well or how poorly the layers of the meat function as insulation—a high thermal conductivity means little resistance to the flow of heat into the interior. The specific heat is a measure of how hard it is to change the temperature of a small mass of the meat—a high specific heat means a great deal of heat energy must be added to the meat to raise its temperature by one degree. Notice that nowhere in this list of variables is there a variable with the dimensions of time. With a little patience, however, we can construct one, using the table on page 75.

By trial and error, we look for a way to multiply or divide the factors $k$, $\rho$, $C_p$, and $D$ in such a way that we are left with an expression having the units of time. A clue comes from the observation that both $C_p$ and $k$ contain the factor "temperature to the minus one" $[\Theta^{-1}]$ in their dimensional formulas. If $[\Theta]$ is to be eliminated, we must divide $C_p$ by $k$ (or vice versa). Another clue comes from the fact that [F] appears in the dimensional formulas for both $\rho$ and $k$ but not in the dimensional formulas for the other variables. Clearly, dividing $\rho$ by $k$ will eliminate [F] from the dimensional formula of the product. So far this reasoning has led us to $\rho C_p/k$, which has the dimensional formula $[L^{-2}T]$. Multiplying by the square of $D$ will have the desired effect of producing a group with the dimensions of time.

So look at this! We've been able to construct a time parameter that is characteristic for the problem. It is

$$\rho C_p D^2/k.$$

If we divide the cooking time $T_c$ by this characteristic time, we get

$$kT_c/\rho C_p D^2$$

which is a dimensionless constant. The assertion that this group is a constant rather than a variable is a consequence of the fact that all the physical variables thought to be important to the problem have been used up in creating this one dimensionless group. (More about this later.) An argument using quite different reasoning but more physics is presented in the appendix at the end of this chapter; it leads to the same conclusion.

We have now come to a valuable result, particularly if you happen to be a cook. Because of the requirement of keeping $T_c/D^2$ a constant, the time for roasting increases with the square of the dimension, or (for roasts of the same shape) the time increases as the ⅔ power of the weight (recall that $D$ is proportional to the cube root of weight, so $D^2$ is proportional to the ⅔ power of weight). As any observant cook will tell you, the larger the roast, the less roasting time is required per pound. In a 350-degree oven, it is about 20 minutes per pound for roasts less than four pounds but about 18 minutes per pound for larger ones.

**Buckingham's Theorem.** We said at the beginning of this chapter that replacing the physical variables in a problem by dimensionless products or ratios always reduces the number of variables. Fine. But it turns out to be important to know how many dimensionless groups (products or ratios) there are in a problem before any of those groups can be obtained.

Let's experiment with a simple problem having a well-known solution and see if there appears to be a general rule determining the number of dimensionless groups. Consider a pendulum of length $l$ and period $T_p$. Throughout this discussion, the effects of joint friction and air resistance will be neglected. The fact that the period is determined by the length of the pendulum is both important and convenient—the rate of ticking of a grandfather clock or a pianist's metronome can be adjusted by moving the weight closer to or farther away from the axis of swinging. Furthermore, anyone who has ever pushed a child in a swing knows that both large and small children move back and forth with the same period in the same swing, demonstrating that the mass of a pendulum is not the factor that determines the period. It is true, however, that the period

Four clocks showing pendulums of varying lengths.

A 255 pound brass ball suspended on a 73 foot guide wire swings back and forth in the center hall of the National Museum of American History. A similar pendulum was built in 1851 by Léon Foucault, a Frenchman, to demonstrate the daily rotation of the earth on its axis. Although the pendulum appears to be changing its path in relation to a pointer beneath it, it is actually swinging stably in space while the floor is being carried around by the rotation of the earth.

lengthens somewhat as the amplitude of the swinging motion is made greater, but this effect is unimportant at small amplitudes.

Suppose the effect of the amplitude of the swinging motion is neglected for a first analysis. Then the list of physical variables includes the period ($T_p$), the length ($l$), and the acceleration due to gravity ($g$). Notice that, if gravity were not included in the list, we would have to stop right here because there is no way to make a dimensionless ratio from a period and a length alone. Moreover, it is reasonable to include gravity from a commonsense point of view—it is gravity, after all, that provides the force for bringing the pendulum back to the center when it has been displaced.

By trial and error, we find that $T_p$, $l$, and $g$ can be made into the dimensionless ratio $T_p^2g/l$. Experiments with small-amplitude swinging pendulums of a wide range of lengths give the same result, that $T_p^2g/l$ is always the same number, regardless of whether the length $l$ is large or small.

But this result is based on the assumption that the amplitude of swing is small. What happens at large amplitudes? Let the distance along the path of the pendulum between the two extremes of its motion be $a$, and let $a$ be added to the list of dimensional variables. Now there are two dimensionless groups, $T_p^2g/l$ and $a/l$. Experiments show that, for a given $a/l$, $T_p^2g/l$ is again a fixed number, but that number is larger as $a/l$ becomes greater. One set of experiments on a pendulum of a single length will be sufficient to establish a table of values relating $T_p^2g/l$ and $a/l$.

Let's review the results. We started with three physical variables defined by the two fundamental quantities length and time. Notice that force does not appear in the dimensional formulas for period ($T_p$), length ($l$), or gravitational acceleration ($g$). By multiplying $g$ by the square of $T_p$, a product was created whose dimension was length; by dividing by $l$, this product was made dimensionless. One multiplication or division operation was required to cancel each of the two fundamental quantities, with the result that a single dimensionless group was formed from the three physical variables.

There is a useful theorem, first stated by E. Buckingham in 1914, that generalizes this result. It says essentially that, if there are $m$ physical variables defined in terms of $n$ independent fundamental quantities, there are $m - n$ independent dimensionless groups.

The systems of fundamental quantities we have considered (FLT, MLT, DLT, PLT, FLTΘ, and so on) all obey the "independent" rule. For a fundamental quantity to be independent of the others, it must not be capable of being formed from the sum, difference, or product of any two or more of the other fundamental quantities. Thus, force, length, and time are all independent, but length, time, and acceleration are not because acceleration is dimensionally equivalent to length divided by the square of time. Similarly, a dimensionless group is independent only if it cannot be formed from the sum, product, or power of other groups.

In the problem just considered, the three physical variables $T_p$, $l$, and $g$ specified in terms of the two fundamental quantities [L] and [T] gave one dimensionless group. When the additional variable $a$ was added to the list, there were still only two fundamental quantities necessary to specify the physical variables, so the number of dimensionless groups became $4 - 2 = 2$.

In the bicycle problem, the variables were the angle $A$, the speed $v$, the radius $R$, and $g$. These variables require only the fundamental quantities of length and time for their specification; hence, Buckingham's theorem says that the number of dimensionless product groups is $4 - 2 = 2$. One of the groups is the angle $A$, which is dimensionless as it stands. The other group can be formed by trial and error, guessing various combinations of powers of $v$, $R$, and $g$ and seeing whether the dimensions cancel. This guessing process might have produced $v^2/gR$, $v/(gR)^{1/2}$, $gR/v^2$, or $g^2R^2/v^4$. All of these are valid dimensionless groups; any one of them could have been used to bring all the data points onto a single curve as was done in figure *c* on page 73.

**A Method for Obtaining Dimensionless Groups.** When there are more than just a few physical variables involved, the guessing process can be very time-consuming. In such cases, it's best to follow a simple method. Here are the steps:

1. Make a list of all the physical variables in the problem and identify their dimensional formulas.
2. Use Buckingham's theorem to determine the number of independent dimensionless groups.
3. Form a set of product groups, assuming arbitrary exponents for each of the physical variables. By requiring each product group to be dimensionless, it will be possible to solve for the arbitrary exponents.

The bicycle problem will again serve as an example. Steps 1 and 2 were carried out above. Step 3 can begin by assuming the form

$$vg^aR^b. \tag{3.1}$$

Notice that we have chosen one of the variables (in this case $v$) to appear raised to the power 1. This is good technique; it recognizes that any product group multiplied or divided by itself creates another perfectly good group. Because the group $v^cg^dR^e$ can be put into the form of expression 3.1 by taking the $c^{th}$ root, it is always permissible to let one of the variables (it doesn't matter which one) appear without an exponent in the assumed form.

Next, we write the dimensional formula for the product group

$$vg^aR^b = [LT^{-1}][LT^{-2}]^a[L]^b \tag{3.2}$$

and finally we require that the exponents of each of the fundamental quantities in the group be zero. For the [L] dimension, this means

$$1 + a + b = 0 \tag{3.3}$$

and, for the [T] dimension, it means

$$-1 - 2a = 0. \tag{3.4}$$

Solving equation 3.4 first, we can see that $a = -\frac{1}{2}$. Substituting this result into equation 3.3, we get $b = -\frac{1}{2}$. Hence, the product group is $v/(gR)^{1/2}$. If we had taken $g$ to be the variable with an exponent of 1, the group would have been $gR/v^2$. One is just as good as the other, as we have said before.

In order that you may gain some experience with the method, we now present a number of examples.

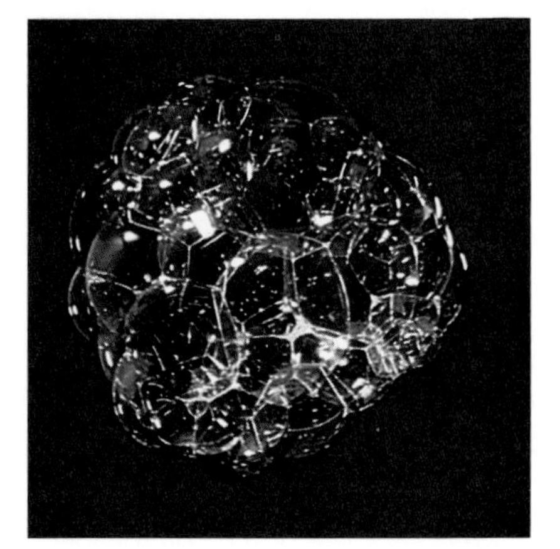

A cluster of soap bubbles. For each separate bubble within the cluster, there is a simple relation between the surface tension, the pressure inside, and the radius of curvature.

**The Pressure Within a Soap Bubble.** In each of the soap bubbles shown here, the pressure difference across the membrane is determined by the surface tension ($\mathcal{s}$) and the radius of curvature of the bubble ($R$). Thus, there are three physical variables: $P$, $\mathcal{s}$, and $R$. From the table on page 75, you can see that, in the FLT system, all three variables may be expressed in terms of force and length. Therefore, there is only one ($3 - 2 = 1$) dimensionless group. Taking that group as $P\mathcal{s}^aR^b$ and writing the dimensional formula

$$P\mathcal{s}^aR^b = [\mathrm{FL}^{-2}][\mathrm{FL}^{-1}]^a[\mathrm{L}]^b \tag{3.5}$$

we obtain

$$1 + a = 0 \tag{3.6}$$

and

$$-2 - a + b = 0 \tag{3.7}$$

giving $a = -1$ and $b = 1$. Therefore, the dimensionless group in this problem is $P\mathcal{s}^{-1}R$. As was true with the roasting problem, the fact that only one dimensionless group appears means that the group is a constant. A single measurement of the pressure, surface tension, and radius of curvature of one soap bubble (or the application of a physical principle) will establish the value of that constant forever. It turns out that the constant has the value 4 (when working this out for yourself, remember that the soap film has two surfaces), so that $P = 4\mathcal{s}/R$. Small bubbles make a louder sound when they burst than large ones do because the pressure impulse that occurs when the membrane ruptures is inversely proportional to the radius. The tiny bubbles bursting at the surface of a glass of champagne can make quite a racket.

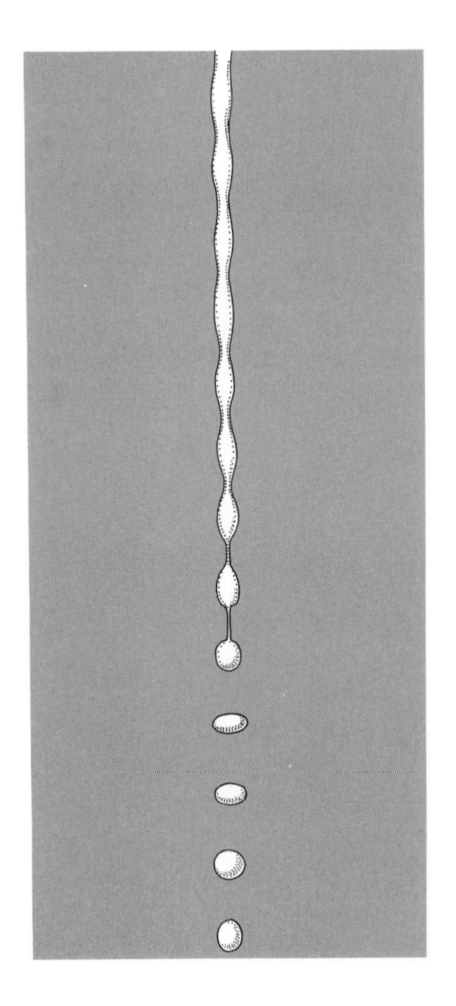

Drawing made from a flash photograph of a stream of water flowing from a tap. Instability due to surface tension causes the water column to form a series of necks and bulges, which finally become individual droplets vibrating back and forth between prolate-spheroidal and oblate-spheroidal shapes. To facilitate the formation of uniform droplets, the water tap was given a small mechanical vibration.

**The Vibration of a Water Drop.** Water drops can vibrate under the action of their own surface tension if they are disturbed from their normally spherical shape. When a drop is in the shape of a sphere, its surface area is at a minimum. To disturb it into a prolate (football-shaped) or oblate (disc-shaped) mass requires that work be done against surface-tension forces, because the surface area is increased by comparison with the sphere. A drop that has been deformed into a prolate shape and released first returns to the sherical shape but then overshoots and passes into the oblate shape. This is because the fluid motion has inertia, just as the mass of a swinging pendulum has inertia. The drop then returns through the spherical shape and becomes prolate again. The frequency of oscillation ($f$) should depend on the surface tension ($\mathfrak{s}$), the mass density ($\rho$), and the radius of the drop in its spherical shape ($R$). Because there are four physical variables measured by three fundamental quantities, there will be one dimensionless group. Using the method introduced earlier, the one dimensionless group, which must be a constant, can be identified as $f^2\rho R^3/\mathfrak{s}$, so that

$$(\text{frequency})^2 \times \text{density} \times (\text{radius})^3/\text{surface tension} = \text{constant}. \quad (3.8)$$

In his *Theory of Sound* (1878), the great English physicist Lord Rayleigh observed that a jet of liquid coming out of a tap always breaks up into a series of drops, each of which vibrates as shown in the figure on the left. Under the proper conditions, each drop is in the same phase of its motion as it passes through a point at a given distance from the tap, making it possible to see the drops apparently fixed in the air by means of a stroboscope that flashes each time a drop goes by. Using these observations, it is possible to measure the frequency of vibration for drops of a known size. Knowing the density and surface tension for water in air, one may fix the value of the constant. Rayleigh found that the experimental value for the constant agreed quite well with the theoretical value that he had determined by formulating and solving the equations of motion. A simple result from these studies is that the period of vibration is proportional to the square root of the drop volume for a set of drops having the same density and surface tension. Rearranging equation 3.8 gives the same conclusion.

**The Vibration of a Star.** In 1915, Lord Rayleigh applied dimensional analysis to the problem of a vibrating star. In that age long before thermonuclear physics, Rayleigh considered a star to be merely a liquid body held together by its own gravity. The physical variables of the problem would be the frequency of

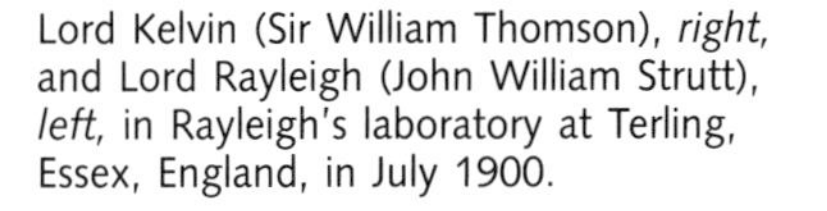

Lord Kelvin (Sir William Thomson), *right*, and Lord Rayleigh (John William Strutt), *left*, in Rayleigh's laboratory at Terling, Essex, England, in July 1900.

vibration ($f$), the radius of the star ($R$), its density ($\rho$), and the gravitational constant ($\kappa$), which has the dimensional formula $[F^{-1}L^4T^{-4}]$. Again we expect a single dimensionless group to be made from these four physical variables specified by three fundamental quantities. Writing the dimensionless group as

$$f\kappa^a\rho^bR^c$$

and, as usual, giving the dimensional formula of the group as

$$[T^{-1}][F^{-1}L^4T^{-4}]^a[FL^{-4}T^2]^b[L]^c$$

and requiring the exponents for F, L, and T to be zero gives

$$-a + b = 0 \tag{3.9}$$
$$4a - 4b + c = 0 \tag{3.10}$$
$$-1 - 4a + 2b = 0. \tag{3.11}$$

If we proceed to solve these equations as we have done before, something unexpected happens: From equation 3.9, we can see that $a = b$, but, according to equation 3.10, if $a = b$, $c = 0$. The conclusion must be that only $f$, $k$, and $\rho$ can enter the dimensionless product; $R$ cannot be accommodated. In fact, this always happens when one tries to include an inappropriate physical variable in a dimensionless group. The period $T_p$ of the pendulum discussed earlier, for example, depends on the pendulum length $l$ and on the gravitational acceleration $g$, but not on the pendulum mass $m$. Hence, the dimensionless group $T_p^2 g/l$ cannot be made to include $m$. Evidently, the frequency of vibration of a star does not depend on the size of the star any more than the period of a pendulum depends on its mass.

From equation 3.11, we can see that $a = b = -\frac{1}{2}$. It follows that the dimensionless product, a constant for all stars obeying the liquid-body model, is $f\kappa^{-1/2}\rho^{-1/2}$, so that

$$\text{frequency}/\sqrt{\text{gravitational constant} \times \text{mass density}} = \text{constant}. \tag{3.12}$$

Reasoning this way, Rayleigh predicted that the frequency of any natural mode of vibration of a star is directly proportional to the square root of the mass density of the star but independent of its size.

There is a clear difference between this result and the result for the water drop vibrating under surface-tension forces. For the water drop, the ratio of surface to volume is high, and the surface-tension skin is responsible for a size-dependent frequency of vibration. For the star, the ratio of surface to volume is very small, and gravitational forces become the important restoring forces. Because both the gravitational forces and the inertial forces acting on an element of mass in a star increase in the same way as the size of the star increases, the frequency of vibration is not a function of size.

**The Energy of an Atomic Bomb.** In 1950, the English fluid dynamicist G. I. Taylor published a paper explaining how one could calculate the energy yield of an atomic explosion merely by taking a photograph of the fireball a short time after the detonation.

From his knowledge of gas dynamics, Taylor assumed that the velocity of the gas behind the shock wave would be practically independent of the exact

A nuclear explosion.

(*Below*) Photographs of an atomic bomb explosion in 1945. The left photo shows the fireball at $t = 15 \times 10^{-3}$ seconds; the photo on the right shows it at $t = 127 \times 10^{-3}$ seconds.

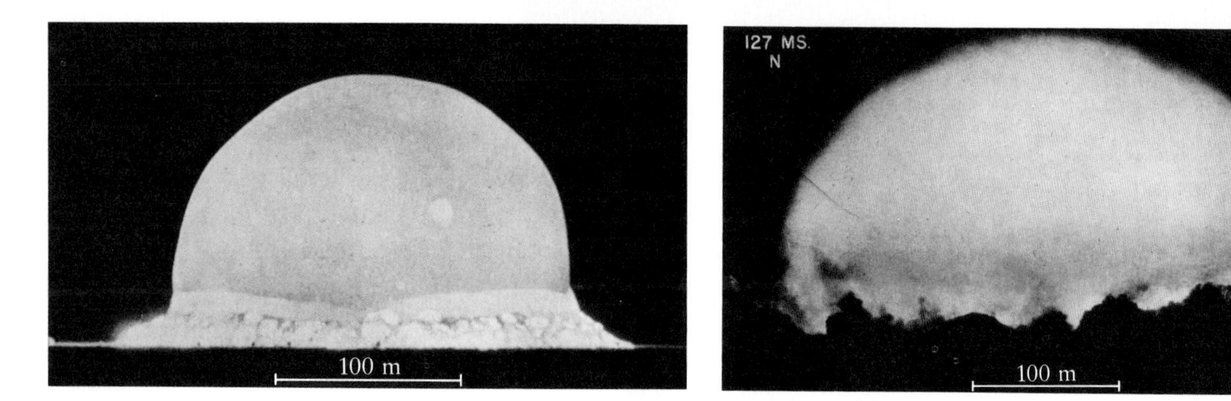

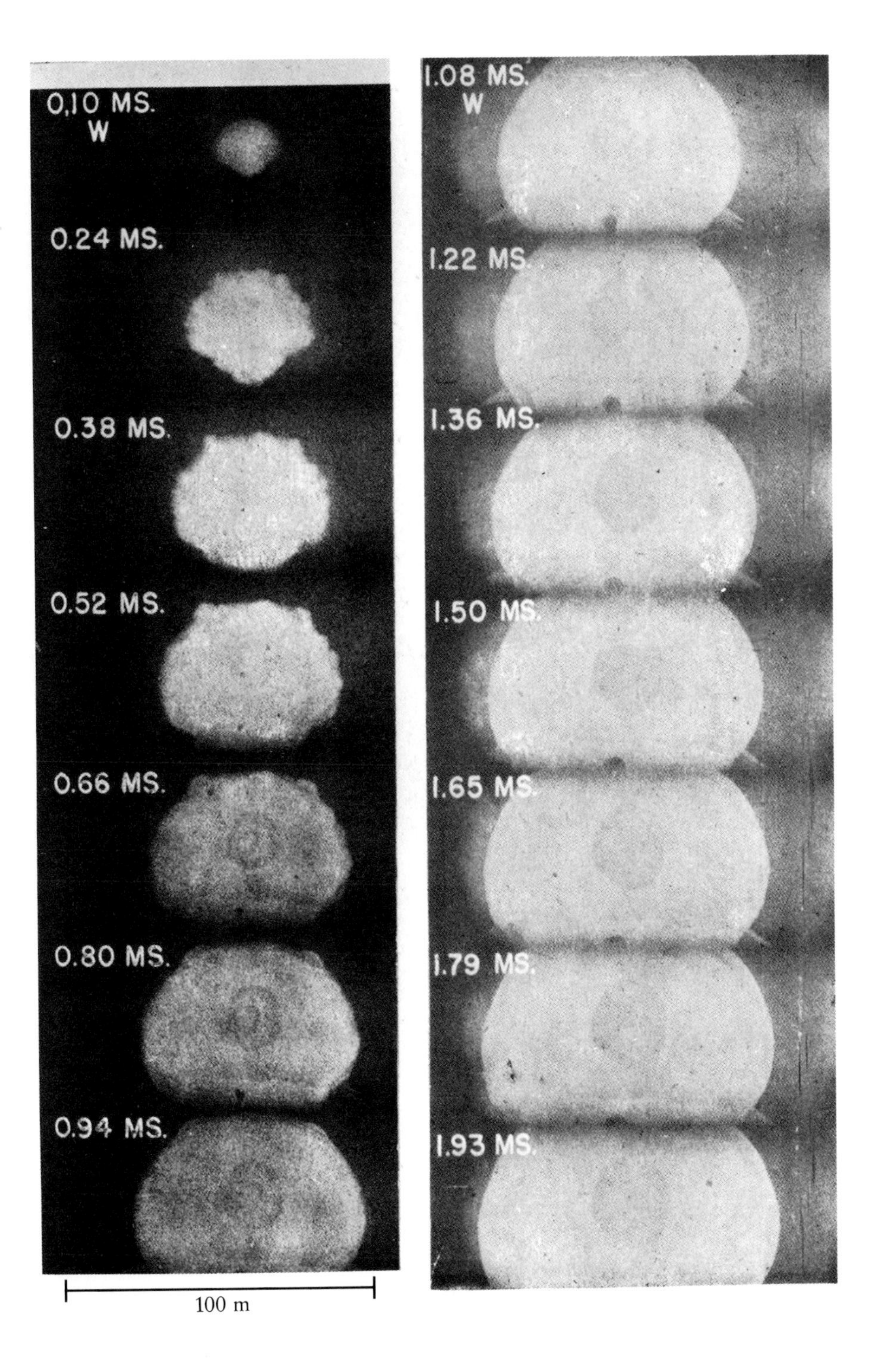


A series of photographs showing the growth over time of the fireball produced by an atomic-bomb explosion in New Mexico. The numbers at the upper left of each frame show the number of milliseconds elapsed since detonation.

Observations of the radius of the shock wave (*circles*) at various times soon after the beginning of the explosion are in good agreement with the theoretical result, $r = 1.03(\mathcal{E}/\rho)^{1/5}t^{2/5}$ (*solid line*). Dimensional analysis correctly predicts that the radius is proportional to time raised to the ⅖ power.

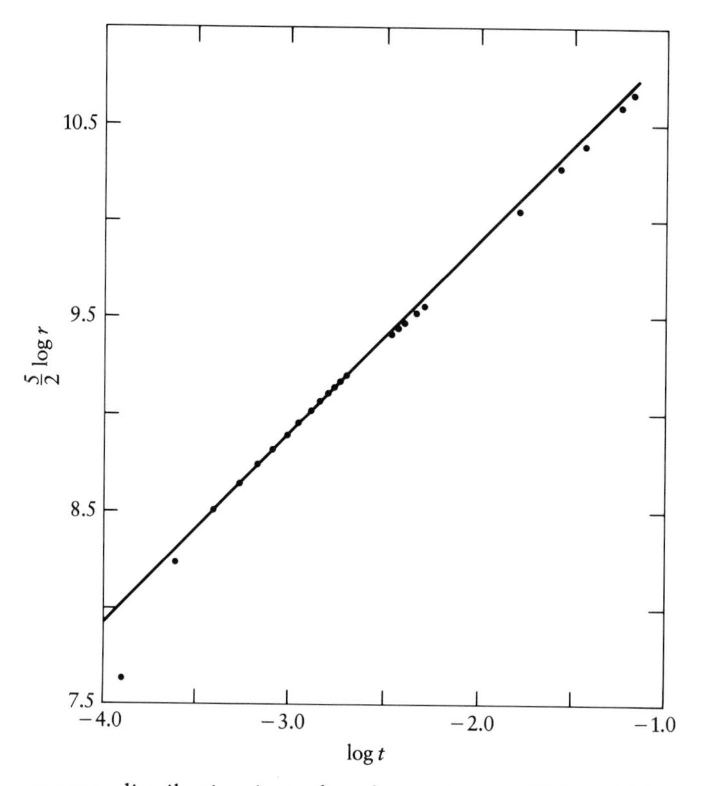


pressure distribution just after the explosion. This led him to believe that the radius ($r$) of the shock wave at a given time ($t$) would depend only on the density ($\rho$) of the undisturbed air ahead of the shock wave and the energy ($\mathcal{E}$) liberated instantaneously by the bomb at $t = 0$. Because there are four physical variables here specified by three fundamental qualities, there is one dimensionless group, $r^5\rho/\mathcal{E}t^2$. Therefore, dimensional analysis predicts that the radius of the shock wave should grow as the ⅖ power of time immediately after detonation.

Applying physical principles, including the conservation of momentum, mass, and energy, Taylor deduced that the numerical value of this dimensionless group was very close to 1. Then he examined photographs like those shown at the bottom of page 86 and plotted ⁵⁄₂ log $r$ against log $t$, as shown above. The photographic data confirmed the theoretical result quite well over a range of times shortly after the explosion. This confirmation was satisfying—and also very important. Taking the density of the undisturbed air outside of the blast wave as $\rho = 1.25 \times 10^{-3}$ gram per cubic centimeter, Taylor was able to use just one photograph (giving $r$ at a particular $t$) to calculate that the energy yield of this explosion was $\mathcal{E} = 8.45 \times 10^{20}$ ergs ($8.45 \times 10^{13}$ in Joules or watt-seconds).

**The Drag on a Submarine.** The speed of a submarine moving underwater is determined by the thrust of its engines and by the drag force ($D$) due to its motion through the water. In general, $D$ is some function of the submarine's speed ($v$). We shall also have to take into account the possibility that the drag force depends on the physical (including inertial) properties of the water, which are dependent on its density ($\rho$), and the frictional properties, which are measured by its viscosity (denoted by $\mu$, the small Greek letter mu). There is also the size of the submarine to be considered. Let's assume that the submarines we compare will always have the same shape, so we can simply give the overall length, $l$, in the confident belief that all other lengths (including the height of the conning tower, the diameter of the hull, and so on) will be proportional to $l$.

Because there are five physical variables specified in terms of three fundamental quantities, we expect there to be two ($5 - 3 = 2$) dimensionless groups. If we define one of the groups as

$$\rho l^a v^b \mu^c$$

application of the standard procedure gives $c = -1$, $b = 1$, and $a = 1$, so that the group becomes

$$\rho l v / \mu$$

or, stated in words,

density × length × velocity / viscosity.

The second dimensionless group should be chosen in such a way that it is certain to be independent of the first. One way to do this is to make sure that at least one physical parameter that was present in the first group is missing from the second, and vice versa. Therefore, we will define the second group as

$$D \rho^d l^e v^f$$

so that the drag force ($D$) is substituted for the viscosity, guaranteeing that the second group will be independent of the first. Solving for the exponents, we find that the second dimensionless group is

$$\frac{D}{\rho l^2 v^2}, \quad \text{or in words,} \quad \frac{\text{drag force}}{\text{density} \times \text{length squared} \times \text{velocity squared}}$$

Let's pause for a moment and see what we have. The first dimensionless group, $\rho l v/\mu$, could be called a relative velocity or, what is equivalent, a dimensionless velocity. This makes sense because $\rho l v/\mu$ is proportional to $v$. It dou-

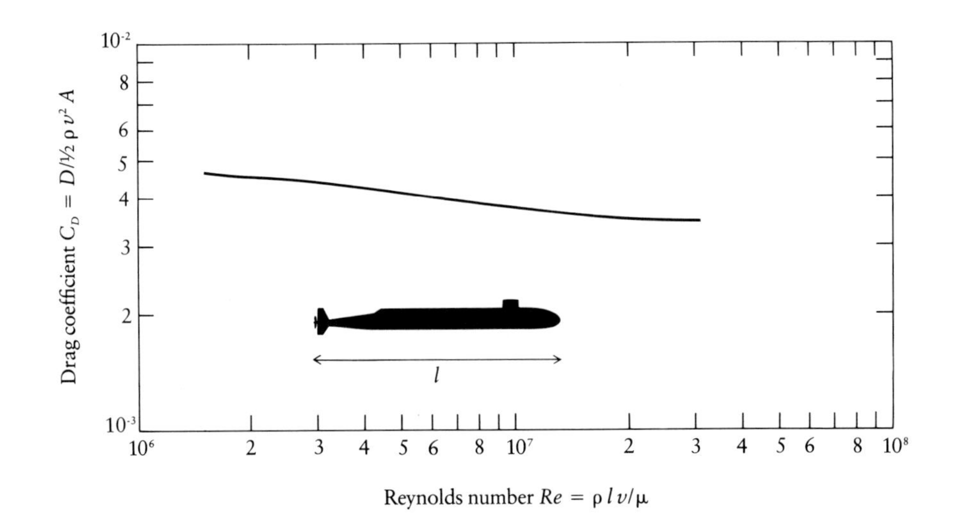


The drag coefficient of a submarine as a function of velocity given in dimensionless terms.

bles when the velocity doubles, yet it is dimensionless. It could also be thought of as a dimensionless length or even as a dimensionless density, but in this problem it makes more sense to think of it as a dimensionless velocity, because then we can imagine a simple experiment to give the data for the curve in the figure above. Suppose a model submarine is running under the water, and it is known, by measuring the force on the supporting pylon, how much thrust is required to maintain a given speed. This thrust is equal to the drag force $D$ when the submarine is cruising at a constant speed. Thus, everything is known which enters into the two groups, and for every cruising speed $v$, a separate point may be plotted. The group $\rho lv/\mu$ is called the *Reynolds number* (abbreviated *Re*), after the British mechanical engineer Osborne Reynolds, who applied an equivalent group to the dynamic similarity of pipe flows in 1883. The Reynolds number shows up in all problems in fluid mechanics in which both fluid inertia and viscosity play a role. A more complete discussion of the Reynolds number will be given later.

The second group, $D/\rho v^2 l^2$, may be considered a dimensionless drag force in this problem. Using $A$ as the submarine's wetted surface area, this dimensionless group may be rewritten as a drag coefficient

$$C_D = \frac{D}{½\rho v^2 A}.$$

This is the form used in the figure above. At a given Reynolds number, the best design for a submarine (or an aircraft, or a racing car) is the one that minimizes the drag coefficient.

The marvelous thing about the above figure, once it has been obtained through the model trials just described, is that it may be applied to any other submarine of the same shape, including a full-size prototype. Thus, it is a

(*Left*) A model of the nuclear-powered fleet ballistic missile submarine USS George Washington. (*Right*) At sea, the actual submarine underway.

simple matter to calculate the drag and therefore the power required to allow the prototype to cruise submerged at a given speed.

**Review of the Principles.** Before going on, it will be worthwhile to review the main points covered so far.

Physical variables have dimensions—that is, they are measured with basic instruments. The basic instruments necessary to measure all the physical variables in a given problem identify a set of fundamental quantities for that problem. In this book, we use the FLT and FLTΘ systems of fundamental quantities, but there are others. Buckingham's theorem says that $(m - n)$ independent dimensionless product groups can be formed from $m$ physical variables expressed in terms of $n$ independent fundamental quantities. Any set of dimensionless product groups is not unique; a given group may be replaced by its inverse, by a power of that group, or by a product of that group with another one.

Once the set of product groups has been identified, that's as far as dimensional analysis can go. The functional relation between the several product groups must now be established experimentally (as in the submarine problem) or by invoking physical principles and laws. Where a single product group appears, that product group must be a constant. The value of the constant may be determined by a single experiment in which all the variables entering into the product group are known (as in the vibration of a raindrop) or by bringing in more physics (as in the atomic bomb). Perhaps it is clearer now why we said, at the beginning of the chapter, that dimensional analysis is Physics Made Easy, but only up to a point. Using the methods of dimensional analysis, one may establish relations between the physical variables—but not quite completely. There will still be an unknown constant (in problems described by a single product group) or one or more unknown functions (when there are two, three, or more product groups) that must be determined by investigating the physical nature of the problem beyond its dimensions.

## Some Special Remarks About the Reynolds Number

Of all the dimensionless groups discussed in this chapter, by far the most important is the Reynolds number. We said earlier that the Reynolds number may be termed a dimensionless or relative velocity, and this is true; but an alternative way of thinking about the Reynolds number reveals that it is the ratio of the inertial force per unit volume to the viscous force per unit volume acting on the fluid particles. When the Reynolds number is small, viscous forces dominate and the flow is smooth and sluggish. When the Reynolds number is large, inertial forces dominate and the flow is swift and often rough.

By way of illustration, consider what happens when a colored liquid is allowed to fall into a clear one as is shown on page 93. Except for their colors, the two liquids in each experiment are identical. The Reynolds number here may be calculated from the velocity (measured, for example, in centimeters per second) with which the colored liquid issues from the supply pipe, the diameter of the supply pipe, and the viscosity and density of the liquid: Reynolds number $=$density $\times$ diameter $\times$ velocity / viscosity.

The figure shows what happens when the velocity and diameter of colored fluid is the same in each experiment but fluids of different viscosities (honey, glycerine, glycerine-and-water, and water) have been chosen so as to give a range of Reynolds numbers.

When the Reynolds number is low (in the honey experiment), the effects of viscosity are very significant. The entering liquid is brought to rest by viscous friction near the surface; it does not penetrate very far at all into the main body of the liquid in the tank.

By contrast, as the Reynolds number increases in the range from 10 to 200, the falling liquid penetrates deeper into the liquid in the tank, and a mushroom-shaped cloud forms. Not only is penetration deeper, but mixing is much more complete than it was with the honey. As the glycerine is thinned with water still further until a Reynolds number of 3,000 is reached, the

(*Right*) Allowing a liquid to flow from a pipe into a tank containing the same liquid produces a flow pattern that illustrates the concept of Reynolds number (*Re*). The velocity and diameter of the stream of liquid leaving the pipe are the same in each experiment, but the liquids in *b*, *c*, and *d* (various mixtures of glycerine and water) are less viscous than the liquid (honey) in *a*. The Reynolds numbers based on the diameter and velocity of the stream of liquid from the pipe are 0.05, 10, 200, and 3000 in *a*, *b*, *c*, and *d*, respectively.

(a)

(b)

(c)

(d)

(*Above*) The smoke plume from a taper has both laminar and turbulent portions. The laminar portion is at the bottom, where the paths followed by the smoke particles are nearly straight and parallel. When the plume becomes turbulent, the smoke particles follow erratic paths.

smooth motions characteristic of lower Reynolds numbers (called *laminar* flow) are replaced by a rough and uneven flow pattern. Flows of this type are called *turbulent*.

Fluid flows at moderate and high Reynolds numbers often contain both laminar and turbulent portions. In regions of laminar flow, layers of fluid particles move past one another as if in smooth sheets or lamina. In turbulent regions, the fluid particles follow erratic paths and become thoroughly mixed. The plume of smoke from a smoldering taper shown on the left has a laminar portion below, in which the smoke particles move on fairly straight paths, and a turbulent portion above, in which the smoke moves in swirls and eddies. This photograph was taken in absolutely still air. There was no wind blowing to

A dye filament in tube flow. In *a*, the dye filament enters the tube through a smoothly converging bellmouth protruding into the reservoir of liquid. In *b*, the liquid flows out of the tube into air. In *a* and *b*, the flow is laminar (the Reynolds number based on the tube diameter is less than 2,100) and the dye filament flows smoothly, without mixing. In *c* (page 95), the flow in the tube becomes turbulent (the Reynolds number is very much greater than 2,100) and the dyed liquid mixes with the clear liquid.

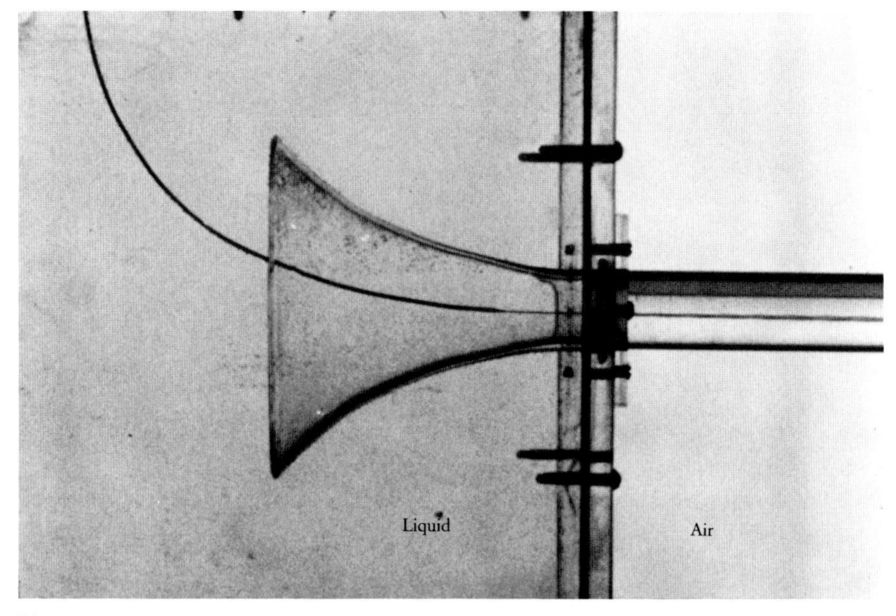


(a)

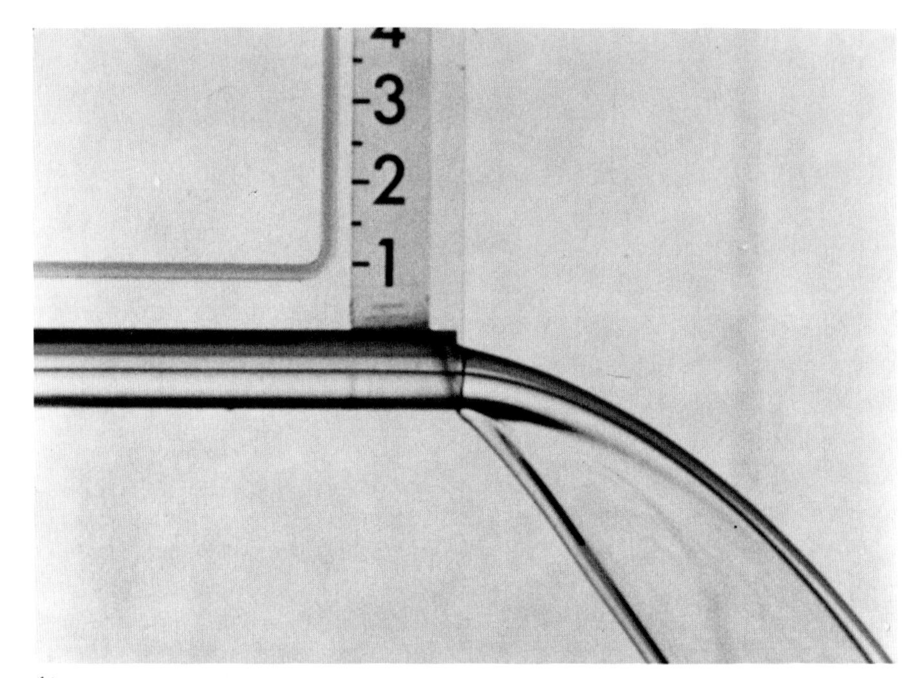


(b)

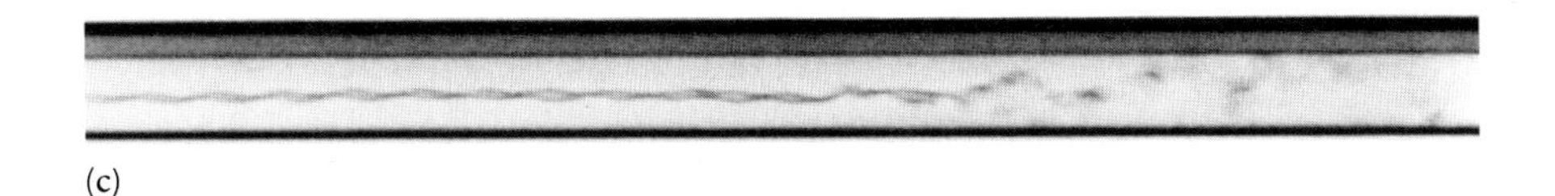

(c)

cause the swirling. Instead, a fluid-mechanical instability, due to the sliding motions between the faster (central) and slower (peripheral) layers of fluid within the smoke plume, caused the laminar flow to become turbulent.

In the early 1880s, Osborne Reynolds conducted experiments on the flow of water through glass tubes. He mounted a glass tube horizontally with one end in a tank of water and the other end open. A smooth bellmouth entrance, like that shown in the photograph on the facing page, was attached to the upstream end of the tube, and a fine stream of dye was fed into the mouth to make the fluid motions visible. He found that, when the Reynolds number based on the tube diameter and the average velocity of the liquid in the pipe was less than about 2,100, the dye streamer was confined to a single streamline and did not mix. At higher Reynolds numbers, he found that the flow may or may not be turbulent, depending on the length of the pipe and the nature and amplitude of small swirling motions present in the water reservoir. In some experiments with very smooth tubes, he found that the Reynolds number could be increased to 12,000 before turbulence occured spontaneously. Later, another investigator used Reynolds' original equipment and obtained laminar flow for a Reynolds number of 40,000 by letting the water stand in the tank for several days beforehand and by avoiding vibration of the equipment. Modern results show that, with sufficient care, the Reynolds number may be pushed above 100,000 without the spontaneous occurence of turbulence. The care is necessary because, at Reynolds numbers above 2,100, any small disturbance is amplified by the unstable flow. As turbulence begins, the dye streamer abruptly begins swirling and mixing with the clear fluid.

The flow of air in the narrow layer adjacent to the skin of an aircraft (the boundary layer) is turbulent for most aircraft at most flying speeds (see figure at the top of page 96), although, in common with the plume of smoke from a taper, there may be a region of laminar flow in the boundary layer near the nose. Some of the roaring sound one hears in an aircraft cabin, which can be particularly loud toward the tail, is not caused by the engines but is due, rather, to the turbulent boundary layer pounding on the skin of the aircraft.

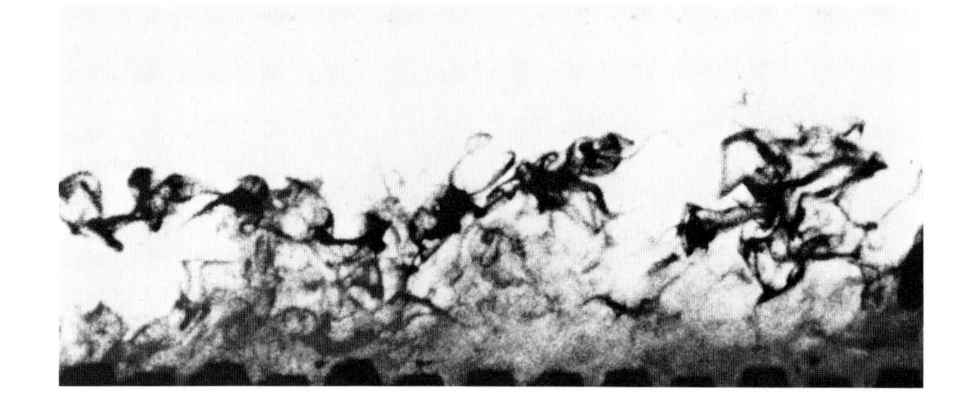

When a fluid brushes past a flat surface at a high Reynolds number, the portion of the fluid near the surface (the boundary layer) can become turbulent, even though the flow outside of the boundary layer may remain laminar. Dye injected into the turbulent boundary layer above this flat plate is thoroughly mixed as it moves downstream (from left to right).

The Reynolds number will play an important role in the remainder of this book, particularly in Chapters 5 and 6. Small organisms swim at very low Reynolds numbers—bacteria, for example, swim at a Reynolds number of about $10^{-6}$, based on their length and cruising speed.* Small organisms come by their low Reynolds numbers honestly on two accounts—because of their small sizes and because of their low speeds. On the other end of the size scale, we find large shore birds flying at Reynolds numbers near $10^4$ and large fish and porpoises swimming at Reynolds numbers in the $10^5$ range. The Reynolds number based on length and cruising speed for the blue whale is near $10^8$.

Reviewing, here are a few points about the Reynolds number that will be useful to remember until we can gain more experience with it in later chapters. First, it is defined as the fluid density times a characteristic length times a characteristic speed divided by the fluid viscosity. Just which length and which speed are chosen depends on convenience, but, once a particular definition has been given, one must stick to that definition throughout the discussion of that problem.

Second, the Reynolds number is dimensionless. Its units are neither grams per cubic centimeter nor centimeters per second; it is given merely as a number, with no units. We have seen that the Reynolds number is small when the flow speed is small, or the size is small, or both—and this was the justification for regarding the Reynolds number as either a dimensionless speed or a dimensionless length. Notice that the Reynolds number describes a whole fluid-flow situation and not merely an object or an organism; when the flow speed is

*Notice that it is conventional, when giving a Reynolds number, to explain which length and which velocity are used in the definition.

zero, the Reynolds number is zero. Nevertheless, since large organisms tend to move at large speeds as well as to be of large sizes, it is almost invariably true that a knowledge of the Reynolds number of an animal while it is in typical motion will tell a great deal about its size.

Finally, the Reynolds number is a ratio between inertial and viscous forces per unit volume. This will have important consequences when we come to discuss life at small scale. Among other things, it will mean that a miniaturized whale would get absolutely nowhere if it were reduced to the size of a sperm cell, unless it were to modify its swimming motions properly to take account of the fact that inertial forces are of little use in generating thrust when the Reynolds number is small.

## The Theory of Geometrically Similar Models

Suppose a model of a large airplane were built, preserving the mass density and all the proportions of the full-size prototype according to a 1:4 rule. The take-off speed of the model is observed to be 22 miles per hour. What is the take-off speed of the prototype? The maximum thrust of a 1:8 scale model of a ship's propeller measured in a water tunnel is 65 pounds. What is the maximum thrust of the prototype? Problems like these frequently arise in engineering practice. An exact scale model, geometrically identical to some full-size proto-

Scale model aircraft operated by radio control.

The Sperry M-1 Messinger was the first full-scale airplane tested in the propeller research tunnel in mid-1927. Note the removal of the outer wing panel which would have extended beyond the 20-foot throat diameter.

type, has been tested under controlled conditions. The task is to calculate a factor (known as a *scale factor*) for the velocity, thrust, pressure, or some other variable when the scale factor for the length is known.

From our discussions of dimensional analysis, we know that every one of the physical problems mentioned earlier reduces to a functional relation between the various dimensionless product groups that enter the problem. An example is the submarine problem, in which the drag coefficient was found to be a unique function of the Reynolds number. It can be shown that geometrically similar submarines at the same Reynolds number not only have the same drag coefficient but the same streamline pattern around their hulls and the same (scaled) pattern of pressure over their surfaces. Keeping the Reynolds number the same between the two submarines of different sizes keeps the flow patterns around them the same because the ratio of the forces acting on analogous fluid particles is the same. That is, the fluid particles one hull diameter ahead of the conning tower experience the same ratio of inertial and viscous forces, whether we consider a large or a small submarine, provided the Reynolds number is the same.

**Dynamic Similarity and Scale Factors.** We are now in a position to state the following principle, which will serve as a definition of *dynamic similarity.*

> All dimensionless groups must be kept the same in model and prototype to achieve perfect dynamic similarity.

One usually has control over only the dimensionless groups that contain the independent variables (for example, the Reynolds number in the submarine problem). Even so, this much control is sufficient, because keeping the groups containing the independent variables the same in model and prototype automatically keeps the groups containing the dependent variables (such as the drag coefficient) the same.

The constancy of the dimensionless groups allows one to compute the various scale factors. Suppose a model pendulum one-ninth the length of a full-size prototype were to be tested by timing its swing. The period of the model pendulum is $T_m = 0.5$ seconds. Since the dimensionless quantity $T_p^2 g/l$ is the same in model and prototype (that is, $T_p^2 g/l_p = T_m^2 g/l_m$), the period of the prototype must be $T_p = T_m \sqrt{l_p/l_m} = 0.5\sqrt{9} = 1.5$ seconds. A scale factor is the ratio of a certain variable for the model to the same variable for the prototype. Thus, the length scale factor is 1/9 in this problem and the time scale factor is 1/3.

A number of other examples follow.

**Musical Instruments and Shaking Bridges.** The whole reason for making musical instruments of various sizes is to give them different frequency ranges. In the figure on page 100, stringed and reed instruments of different sizes are compared. In any family of instruments, the smaller ones always produce sounds at higher frequencies.

This is not to say that the physical principles of all musical instruments are identical. In instruments of the violin family, the sound is created by bowing—the bow alternately pulls the strings and releases them under a complex stick-slip frictional mechanism. In brass instruments, the lips stretched across the mouthpiece create a vibrating stream of air as the space between them opens and closes in a fluttering mechanism exactly analogous to the action of the vocal cords in the larynx. In spite of these very great differences in the way the sound is generated, most musical instruments are alike in the way they employ the physical principle of resonance to emphasize some frequencies and attenuate others.

One way to describe this similarity is to notice that it is possible to form a dimensionless group $fl/v$ that is characteristic of each family of instruments. For the strings, $l$ is the length of the strings (or even the length of the acoustic resonating chamber) and $v$ is the speed of a lateral displacement wave traveling on one of the strings. For the brasses, $l$ can be the length of the tubing from the mouthpiece to the end of the horn and $v$ can be the speed of sound. For comparisons within a family of instruments, then, the scale factor for frequency is the reciprocal of the scale factor for length. For example, if the length is doubled, the frequency is halved.

The same principles apply to a bridge vibrating under the fluctuating forces of highway traffic or wind loads. In the bridge, as in a stringed instrument, the

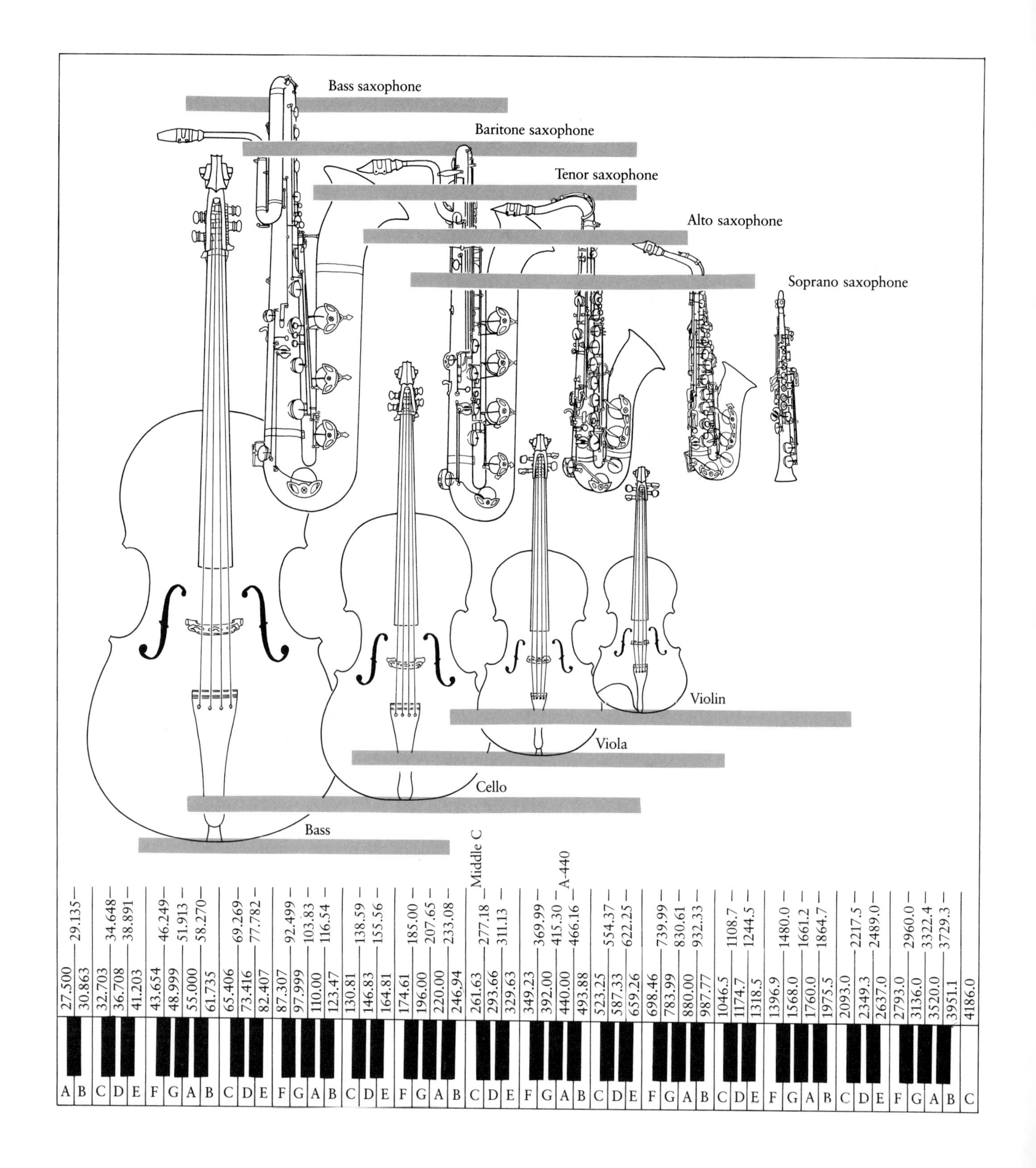

Bass saxophone
Baritone saxophone
Tenor saxophone
Alto saxophone
Soprano saxophone
Violin
Viola
Cello
Bass
Middle C
A-440
29.135
34.648
38.891
46.249
51.913
58.270
69.269
77.782
92.499
103.83
116.54
138.59
155.56
185.00
207.65
233.08
277.18
311.13
369.99
415.30
466.16
554.37
622.25
739.99
830.61
932.33
1108.7
1244.5
1480.0
1661.2
1864.7
2217.5
2489.0
2960.0
3322.4
3729.3
27.500
30.863
32.703
36.708
41.203
43.654
48.999
55.000
61.735
65.406
73.416
82.407
87.307
97.999
110.00
123.47
130.81
146.83
164.81
174.61
196.00
220.00
246.94
261.63
293.66
329.63
349.23
392.00
440.00
493.88
523.25
587.33
659.26
698.46
783.99
880.00
987.77
1046.5
1174.7
1318.5
1396.9
1568.0
1760.0
1975.5
2093.0
2349.3
2637.0
2793.0
3136.0
3520.0
3951.1
4186.0
A B C D E F G A B C D E F G A B C D E F G A B C D E F G A B C D E F G A B C D E F G A B C D E F G A B C

(*Right*) Brooklyn bridge. (*Left*) Golden Gate Bridge under construction. The Brooklyn Bridge, Brooklyn, New York, built between 1869 and 1883, is suspended 135 feet above the East River between Brooklyn and Manhattan by four steel cables hung from two towers set on caissons. At the time of its completion in 1883, the Brooklyn Bridge was the longest bridge in the world, being half again as long as the longest bridge at that time. The Golden Gate bridge, whose central span is 4,200 feet long, could boast the longest single span in the world at the time of its completion in 1937.

transient storage of elastic energy in the cables (if it is a suspension bridge) or the girders is important in determining the natural frequency. The variables that enter the bridge problem are a characteristic length, $l$ (the overall length of the bridge will do), the frequency of vibration, $f$, and the mass density, $\rho$, and modulus of elasticity, $E$, of the construction materials (often steel). The modulus of elasticity, frequently called Young's modulus after the English physicist Thomas Young (1773–1829), is defined as the ratio of the stress (force/cross-sectional area) to the strain (change in length/initial length) for a piece of material subjected to a tensile test. It therefore gives a measure of the intrinsic stiffness of a material—for example, the $E$ for steel is about $10^5$ times that for rubber. The dimensional formula for $E$ is $[FL^{-2}]$, the same as that for stress (see table on page 75), because strain is dimensionless.

Using the same methods we have employed before, we deduce that there is one dimensionless product group—namely, $fl\rho^{1/2}E^{-1/2}$. Therefore, a model bridge made of the same materials as the prototype will vibrate at a frequency 20 times higher than the prototype if the length scale factor is $1/20$.

**A Ship Collision.** A collision between ships is a nasty thing. It can often end in great damage or even in loss of life. The problem of making a theoretical prediction of the extent of damage for a given speed of collision is also nasty in its own way, because so many complicated matters have to be taken into account in the analysis. An alternative approach is to crash two model ships

(*Page 100*) Musical instruments of different sizes. The smaller instruments in a family have higher frequency ranges.

Overturned lifeboats and debris from the luxury liner *Andrea Doria,* early on the morning of July 26, 1956, as the Italian ship takes its final plunge into the Atlantic Ocean off the coast of Massachusetts. Late the night before, the ship had been in collision with the Swedish motorship *Stockholm.*

together in order to see how much damage results under various circumstances. Let's see what would be involved in studying ship collisions through the use of geometrically similar model ships made of the same materials as the prototype. We will assume that every major stuctural detail, including the thickness of the hull plates, bulkheads, keel, frames, decks, and so on, is scaled down by the same factor in going from the prototype to the model.

The physical variables in the problem are assumed to be a characteristic length, $l$ (for example, the overall length of one of the ships), the mass density, $\rho$, and the modulus of elasticity, $E$, of the construction materials, the relative speed of the collision, $v$, and a characteristic stress, $s$, at an arbitrary point in one of the structures that determines whether yielding or fracture has occurred. We expect to be able to form two dimensionless groups from these five physi-

(*Left*) High-speed container ship model evaluated on Carriage I in the David Taylor Model Basin in Bethesda, Maryland. (*Right*) Aluminum ship model under simulated sea loading. Subjected to the stresses of a lifetime at sea, this model provides information on the use of lightweight materials for construction.

cal variables specified by three fundamental quantities. One choice for the two groups is $s/E$ and $s/\rho v^2$. Notice that the length ($l$) does not appear. This is another one of those situations, like the problem of the vibrating star, in which one of the variables originally listed cannot be accommodated into a dimensionless group unless we add one or more additional variables to the list. For example, if the mass ($m$) of one of the ships were included in the original list, an additional dimensionless group, $m/\rho l^3$, could be formed, but this would provide no new information, because $m/\rho l^3$ is always strictly a constant under the assumptions we are making here of geometric similarity (isometry) and use of the same materials in the models and prototypes.

The conclusions are simple. Because the models and prototypes are made from the same materials, $\rho$ and $E$ are constants. Thus, for the stress $s$ (and therefore the damage) to be the same in model and prototype, the speed of collision $v$ must be the same. The models crashing at a speed of 40 feet per second do about the same damage to each other as the real ships would do also crashing at 40 feet per second. An important limitation of this approach is that many materials can withstand higher stresses without yielding as the rate of application of the stress is raised. This may mean that the model can escape certain types of damage that might affect the prototype, even when the peak stresses are the same in both.

## Scale Effects and Scale Distortions

Scale effects are errors in the modeling process. They arise when one or more dimensionless groups are not kept the same between model and prototype. In the ship-collision problem, it is not possible to keep constant both the rate of application of stress and a dimensionless group involving that rate. The result is that, although the maximum stress during a collision might be the same in model and prototype, the model materials appear somewhat stronger because the yield stress for sudden loading is higher than for slow loading, as mentioned in the preceding paragraph. Almost all model studies involve some scale effects. The objective is to keep them small, or at least to understand them well enough to be able to correct for them.

**The Submarine on the Surface: The Froude Number.** Suppose that the submarine considered earlier comes up to the surface. It is a well-known fact that nuclear submarines, which develop the same power above and below the surface, are capable of greater speed submerged than on the surface. Why?

The answer depends on the fact that ships on the surface, even submarines, produce waves. The waves are responsible for a substantial part of the total drag. As long as the submarine stays far below the surface, the drag is determined by skin friction and pressure forces, with the result that the drag coefficient is a unique function of the Reynolds number. For the submarine (or any ship) at the surface, the presence of surface waves means that gravitational forces become important, and another dimensionless group, the Froude number, $v^2/gl$, enters the problem.

The Froude number is named for the nineteenth-century English engineer and naval architect William Froude. Froude is generally credited with having contributed one of the great breakthroughs in ship design, a rational method for using the results obtained from small models to estimate the drag on full-scale ship hulls before they are built. He began his work after a particularly expensive mistake demonstrated the need for such a method.

The mistake was involved in the design of a huge iron ocean liner, the *Great Eastern*, in its day the largest ship in the world. Despite the fact that the designers had provided paddle wheels, a screw propeller, and even auxiliary sails, the size of the ship was too great for her power; she was so slow that she could never earn enough to pay for the cost of her fuel. Froude had been involved, in a small way, with the engineering on the *Great Eastern* and had therefore comprehended at first hand the poor state of knowledge of ship wave resistance.

He became intrigued with ship waves and began by towing models in a creek. Later, he was provided with a towing tank more than 300 feet long, one of the first of its kind. A steam-powered winch pulled a carriage down a track suspended over the tank. This equipment allowed him to tow models at a known speed through still water. He measured the drag force using a spring scale, not essentially different from the ones used now to weigh letters for postage.

Froude noticed that large and small geometrically similar models of the same hull produced different wave patterns when towed at the same speed,

A scale model of the *Great Eastern* (1858). This was the largest ship of its day. An engineering error caused it to be severely underpowered. At the time it was built, ship design was largely a matter of guesswork.

but, when the larger hulls were towed at greater speeds, he could find a particular speed for each hull at which the wave patterns were nearly identical. Under those circumstances, the dimensionless group $v^2/gl$, which has since been called the Froude number, was the same for both large and small hulls. He reasoned that, when the wave patterns were similar, geometrically similar hulls would also be dynamically similar, at least with respect to the part of the resistance contributed by the waves. He showed that it follows that the hull resistance due to wave-making is proportional to the displacement of the hull at speeds at which the wave patterns are the same.

Now consider this dilemma. A model submarine $1/50$ the length of the prototype is to be towed at the surface in a towing tank. If we use water in the towing tank and keep the Reynolds number, $\rho lv/\mu$, the same in model and prototype, the speed of the model will have to be 50 times the speed of the prototype. Alternatively, if we keep the Froude number, $v^2/gl$, the same in model and prototype, the speed of the model should be $1/\sqrt{50} = 0.14$ times the speed of the prototype. Clearly, both conditions cannot be met.

Because the drag due to wave-making is usually the most important part of the total drag for a vessel on the surface, and because this depends strongly on the Froude number, the usual procedure is to choose the speed of the model such that the Froude number is the same for model and prototype. The ratio of the depth of the water to the length of the submarine must also be the same for model and prototype, especially if the water is shallow. This will guarantee that the wave patterns generated by the model and the prototype will be the same.* A photograph taken from an airplane flying over the real submarine will show waves streaming away from the hull in a pattern practically identical to that

*This is true, provided that the model is sufficiently large—say, 15 feet long. If the model is too small, surface tension may also play a role in the wave pattern.

Nuclear submarine cruising on the surface.

seen in a photograph taken above the model in its towing tank. The fact that the Reynolds number is not correct in the model introduces a scale effect that may or may not create a significant problem.

There is a technique commonly used to provide a correction for this particular scale effect. The technique, which dates back to William Froude, entails making the assumption that the drag due to skin friction is a separate part of

Photo of the ship-wave pattern seen from an airplane. The wave pattern is confined to a wedge-shaped region bounded by straight lines.

the total drag. First, a model is towed in a towing tank, using the same Froude number as the prototype. It is assumed that the skin-friction drag on both the model and the prototype can be estimated accurately by using the skin-friction coefficients for flow past a flat plate. One then treats the submarine (as far as frictional drag is concerned) as if it were one single flat plate. The total drag is measured in the model, and the wave drag is obtained by subtracting the (estimated) skin-friction drag. Froude-number scaling (that is, the result that drag is proportional to displacement for a constant Froude number) is then used to give the wave drag in the prototype. Finally, the skin-friction drag of the prototype is calculated for an appropriately sized plate and added in to give the total drag of the prototype.

**Deliberately Introduced Roughness.** There exists another technique for compensating for the fact that Reynolds numbers and Froude numbers cannot both be kept the same in model and prototype when both are moving through the same fluid. Recall that an important difference between flows at low and high Reynolds numbers has to do with turbulence. Flow in the boundary layer around a model (at a low Reynolds number) may lack the turbulence known to be present in the flow in the boundary layer surrounding the prototype (at a high Reynolds number).

When engineers test ship models, they often deliberately introduce turbulence into the boundary layer by gluing a strip of waterproof sandpaper to the model near its bow. The rough surface of the sandpaper promotes turbulent flow in the boundary layer at a much lower Reynolds number than would be the case without the sandpaper. This turbulent boundary layer persists far downstream, even as the water flows over the smooth parts of the model, creating a flow pattern that mimics that of the full-scale prototype quite faithfully.

The same procedure may be used when a scale model of a high-speed airplane is tested in a wind tunnel. Here, perfect dynamic similarity would require that both the Mach number and the Reynolds number be kept the same for model and prototype. When air is used as the fluid in both cases, this cannot be accomplished. Instead, a turbulent boundary layer is created on the model aircraft by installing a small protuberance, a "trip wire," near the leading edges of the wings and body. Once the high Reynolds number of the prototype has thus been simulated, the Mach numbers of the model and prototype are matched by adjusting the velocity and sometimes the density of the air flowing in the wind tunnel. In this way, a small distortion in shape is used to offset the scale effects that would otherwise have resulted from the fact that both the Reynolds number and the Mach number cannot be correct simultaneously in the flow past the small model.

In the next chapter, we will employ many of the techniques and results of the present chapter to identify dimensionless groups in physical problems concerning animals and plants. We will discover that some important dimensionless groups are size-independent and others, inevitably, are not. We shall see that the scale distortions that arise in biology are not limited to the presence or absence of "trip wires." In terrestrial life forms of large size, a change in scale almost invariably produces a substantial change in the shape of the whole organism.

## Appendix: Dimensionless Groups from Physical Equations

Here is an alternative to the methods presented in this chapter for finding dimensionless groups. It demonstrates that, when physical principles are used to obtain a mathematical solution to a problem, dimensionless groups frequently show up automatically. It turns out that it is not even necessary to solve any equations to obtain the dimensionless variables and constants. It is enough to be able to write the algebraic or differential equations that govern the process.

Consider, for example, the problem of cooking the beef roast discussed in Chapter 3. Invoking two physical principles, conservation of energy (heat) and the diffusion equation, one may write down the general partial differential equation

$$\frac{\partial \Theta}{\partial t} = \frac{k}{\rho C_p} \frac{\partial^2 \Theta}{\partial x^2} \tag{1}$$

in which $t$ is time, $x$ measures the distance in from the surface, $\Theta$ is the temperature at $x$ and $t$, $k$ is thermal conductivity, $\rho$ is mass density, and $C_p$ is the specific heat at constant pressure. As they stand, the terms on the left and the right sides of this equation are not dimensionless; they have the units $[\Theta \mathrm{T}^{-1}]$. We can make equation 1 dimensionless by defining:

$$\Theta' = \frac{\Theta}{\Theta_c}, \; t' = \frac{t}{T_c} \text{ and } x' = \frac{x}{D}. \tag{2}$$

From this it follows that

$$\frac{\partial \Theta}{\partial t} = \frac{\Theta_c}{\mathrm{T_c}} \frac{\partial \Theta'}{\partial t'} \text{ and } \frac{\partial^2 \Theta}{\partial x^2} = \frac{\Theta_c}{D^2} \frac{\partial^2 \Theta'}{\partial x'^2}. \tag{3}$$

In these definitions, $\Theta_c$ is a reference temperature (the temperature the interior of the roast is supposed to reach when it's cooked), $T_c$ is the cooking time required to reach this temperature, and $D$ is a characteristic dimension (the greatest diameter of the roast). Substituting equations 2 and 3 into equation 1, we obtain

$$\frac{\partial \Theta'}{\partial t'} = \frac{kT_c}{\rho C_p D^2} \frac{\partial^2 \Theta'}{\partial x'^2}. \tag{4}$$

Let's see what we can get out of a comparison between equations 1 and 4.

The variables in equation 1 are $\Theta$, $t$, and $x$—all of these are what we have been calling physical dimensional variables. The various derivatives of $\Theta$ with respect to $t$ and $x$ are related by equation 1, containing a physical constant, $k/\rho C_p$, which has the dimensional formula $[\mathrm{L}^2 \cdot \mathrm{T}^{-1}]$. By comparison, all the variables in equation 4, $\Theta'$, $x'$, and $t'$, are dimensionless, and, as a consequence, both $\partial \Theta'/\partial t'$ and $\partial^2 \Theta'/\partial x'^2$ are dimensionless. Hence we expect the constant in equation 4, $kT_c/\rho C_p D^2$ to be dimensionless, and in fact it is. At this point, the reader may wish to return to the conclusions of the roasting problem in this chapter.

G. Doré
E. Deschamps.

# Chapter 4

# The Biology of Dimensions

In biology, the effects of size are certainly striking. Large animals and plants live longer, the activities of their individual cells are slower, and their external forms may be very different from those of smaller organisms. In Chapter 3, the methods of dimensional analysis were used to predict the physical implications of changes in size. Can the same methods be used to understand the effects of size in biology?

## Size-Invariant Dimensionless Groups

In principle, the same basic rules should apply in biological scaling as applied in the scaling of physical and engineering systems. For complete dynamical and physiological similarity, all the dimensionless products important to the functioning of a group of organisms should be kept the same in both large and small members of that group.

There do appear to be a few dimensionless groups that are relatively constant over a wide range of body sizes in mammals. In the table on page 112, a number of dimensionless groups are defined and the numerical value is given for an animal weighing 1 kilogram. The allometric mass exponent for each dimensionless group is also given. In these groups, the mass exponent is generally quite small, meaning that the product group is, for all practical purposes, independent of body mass.

**Volumes, Flow Rates, and Masses of the Organs.** The first group in the list is the ratio of two volume flow rates. On the top of this fraction is the volume of air moving in and out of the lungs per breath (tidal volume) divided by the time required for a breath cycle (breath time). On the bottom of the fraction is the volume of blood ejected by the heart per beat (stroke volume) divided by the period of a heart cycle (pulse time). Therefore, this group gives the ratio of the volume flow rate of air moving in and out of the lungs to the volume flow rate of blood pumped by the heart. This ratio takes the value 0.90, whether the animal is a mouse, a dog, or a horse. As a consequence, the ratio of air flow (ventilation) to blood flow through the lungs (perfusion) is quite strictly independent of animal body size, as is required to ensure that the blood is adequately oxygenated both in large and in small animals.

Considering only the size of organs, there are a great many ratios that could be formed by dividing the mass of one organ system by another. Many of these turn out to be independent of body size, although there are important exceptions. Group 2, the ratio of blood mass to heart mass, is an example of a

Engraving by Gustave Doré, "The Enchanted Sleep." This may well be the view we present to animals smaller than ourselves. To them, we appear to lead sleepy, almost paralyzed lives.

**Size-independent dimensionless groups in mammals.**

| Dimensionless group | Numerical value for an animal weighing 1 kilogram | Allometric mass exponent |
|---|---|---|
| 1. $\dfrac{\text{tidal volume}}{\text{breath time}} \Big/ \dfrac{\text{heart stroke volume}}{\text{pulse time}}$ | 0.90 | 0.03 |
| 2. $\dfrac{\text{mass of blood}}{\text{mass of heart}}$ | 8.3 | 0.01 |
| 3. $\dfrac{\text{velocity of pulses waves in the aorta}}{\text{velocity of blood in the aorta}}$ | 26.0 | −0.05 |
| 4. $\dfrac{\text{pulse wavelength in the aorta}}{\text{length of the aorta}}$ | 8.7 | −0.05 |
| 5. $\dfrac{\text{time for 50\% of growth}}{\text{lifespan in captivity}}$ | 0.03 | 0.05 |
| 6. $\dfrac{\text{gestation period}}{\text{lifespan in captivity}}$ | 0.015 | 0.05 |
| 7. $\dfrac{\text{respiratory cycle}}{\text{lifespan in captivity}}$ | $3.0 \times 10^{-9}$ | 0.06 |
| 8. $\dfrac{\text{cardiac cycle}}{\text{lifespan in captivity}}$ | $6.8 \times 10^{-10}$ | 0.05 |
| 9. $\dfrac{\text{half-life of drug*}}{\text{lifespan in captivity}}$ | $0.95 \times 10^{-5}$ | 0.01 |

* Methotrexate.

size-independent mass ratio. W. R. Stahl, a pioneer in these observations, noticed that the heart mass is always about 5 percent of the body mass and that the lungs always weigh about twice as much as the heart.

**Cardiovascular Parameters.** Groups 3 and 4 have a great deal to say about dynamic similarity in the circulatory system. When a physician inflates his pressure cuff around a patient's arm, he cuts off the blood flow in the arteries for a moment and then reduces the cuff pressure until it falls slightly below the peak pressure in the artery (the systolic pressure). He knows when this happens because he can hear the sound of the artery vibrating as blood is forced through during the few instants when the pressure is highest.

The sounds are a manifestation of the fact that the speed of pressure fluctuations (pulse waves) moving through the partially collapsed artery is approxi-

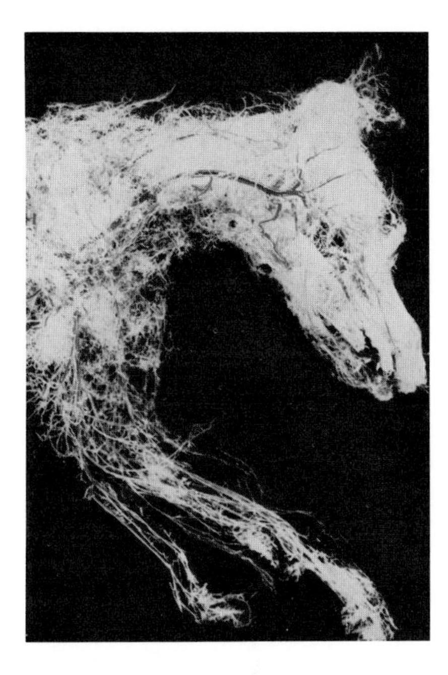

A silicone-rubber cast of the arterial system of a dog. The arterial radius decreases in the downstream direction, but the total cross-sectional area increases because of repeated branchings. The smallest arteries and capillaries appear white in this photograph.

mately equal to the speed of blood spurting through the artery. The stethoscope picks up the bumping sound of small spurts of blood in the partially collapsed artery. Under ordinary circumstances, with no pressure cuff around the artery, the speed of pulse waves is much greater than the blood velocity—in fact, about 25 times greater. This factor may be understood as giving a margin of safety with respect to dynamic collapse (closing caused by the blood spurting through), a margin that is the same in large and small animals.

The arteries function somewhat like a transmission line for pulse waves. The methods of Chapter 3 may be used to show that the speed of pulse waves, $c$, should be proportional to

$$\left(\frac{\text{modulus of elasticity} \times \text{thickness}}{\text{blood density} \times \text{arterial radius}}\right)^{1/2} = \left(\frac{Eh}{\rho R}\right)^{1/2}$$

in which $E$ is the modulus of elasticity of the arterial wall, $h$ is the wall thickness, $R$ is the arterial radius, and $\rho$ is the density of the blood. Looking at a given part of the arterial system—the ascending aorta, for example—the modulus of elasticity and the ratio of thickness of the arterial wall to the diameter of the vessel is found to be essentially size-independent. Hence, the speed of pulse waves should be size-independent. This is found to be the case: the pulse-wave speed is about 6 meters per second in the ascending aortas of mice, dogs, and humans.

As is apparent in the figure on the left, the arteries branch repeatedly and become smaller as they carry blood from the heart. When changes in all the terms determining the speed of pulse waves are taken into account, the outcome is that the speed of these waves increases by more than a factor of 10 in a given animal as the pulse travels from the ascending aorta to the small arteries at the periphery. An analysis taking these observations into account shows that the pumping effort demanded of the heart is minimized if the heart frequency corresponds to a certain "tuned" frequency. This may be described as a resonance, essentially like the resonance of the musical instruments in the previous chapter. The criterion for this type of resonance is that the ratio of the pulse wavelength to the aortic length should be a size-independent number. Group 4 in the table on page 112 is exactly this ratio, and its dependence on body mass is small enough to be neglected. Because the mean velocity of pulse waves in the aorta is not a function of body size, the resting heart frequency is inversely proportional to aortic length, just as the frequency of a musical instrument is inversely proportional to its length.

Speaking of resonance and musical instruments, we might note that the frequencies of both the first and second heart sounds scale inversely with body length. The first heart sound ("lubb"), which is caused by a resonant vibration of the ventricles as the mitral and tricuspid valves slam shut, is best heard by placing one's ear or a stethoscope to the left of the breastbone between the fifth and sixth ribs (this is just over the apex of the left ventricle). As the heart vibrates, energy oscillates rapidly between its kinetic form (mostly in the moving blood) and its elastic form (in stretched valve leaflets, heart muscle, and connective tissue). Dimensional analysis applied to a resonating fluid-filled elastic chamber produces the same result we obtained in Chapter 3 for the vibration of a bridge—namely, that the frequency of any particular mode of vibration is proportional to

$$\left(\frac{1}{\text{body length}}\right)\sqrt{\frac{\text{modulus of elasticity}}{\text{blood density}}} = l^{-1}\left(\frac{E}{\rho}\right)^{1/2}$$

where $E$ is the (size-independent) modulus of elasticity of the ventricular wall and $\rho$ is the density of blood. A similar argument applies to the frequency of the second heart sound ("dupp"), which is produced by resonance of the aorta and pulmonary artery when the aortic and pulmonary valves abruptly close at the end of the pumping period of the ventricles. Thus, the smaller the animal, the higher the pitch (frequency) of the heart sounds.

The ratio of the pressure of the blood as it leaves the ventricles to the force per unit of cross-sectional area (stress) developed in the muscular wall of the heart is another dimensionless group that is found to be independent of body size. This is not surprising because both the pressure and the stress are size-independent in their own right. The size-independence of muscle stress probably derives from the fact that there is no difference in the length or composition of the polymerized proteins (thick and thin filaments) in cardiac muscle as a function of body size. The heart muscles of smaller animals are merely faster, not stronger. The link between pressure and stress is explained in the top figure on page 115, which shows a heart sectioned at the equator. The net force acting to push the lower half downward—the pressure inside the heart multiplied by the cross-sectional area of the heart cavity ($pA_1$)—must be in equilibrium with the force acting to pull it upward—the stress in the heart muscle times the area of the muscle ($sA_2$). As mentioned, the stress, $s$, is independent of size, so if the geometric ratio of the areas, $A_1/A_2$, is independent of body size, the pressure $p$ should be independent of body size also. In the figure, the pressure within the heart has been estimated by giving the mean arterial pressure, which is approximately independent of body size.

The pressure and volume flow rate at the ascending aorta as measured in a horse, a human, and a dog are shown in the bottom figure on page 115. As explained above, we expect to see the same peak (systolic) pressure ($S$) and

Ventricular ejection pressure $p$ does not depend on body mass. The heart has been cut at the equator. $A_1$ is the cross-sectional area of the intraventricular space and $A_2$ is the area of the muscle.

Pressure and flow at the ascending aorta in horse, human, and dog. The systolic and diastolic blood pressures are indicated by $S$ and $D$, respectively. Applicable scales are shown at the right of the pressure and flow pulses.

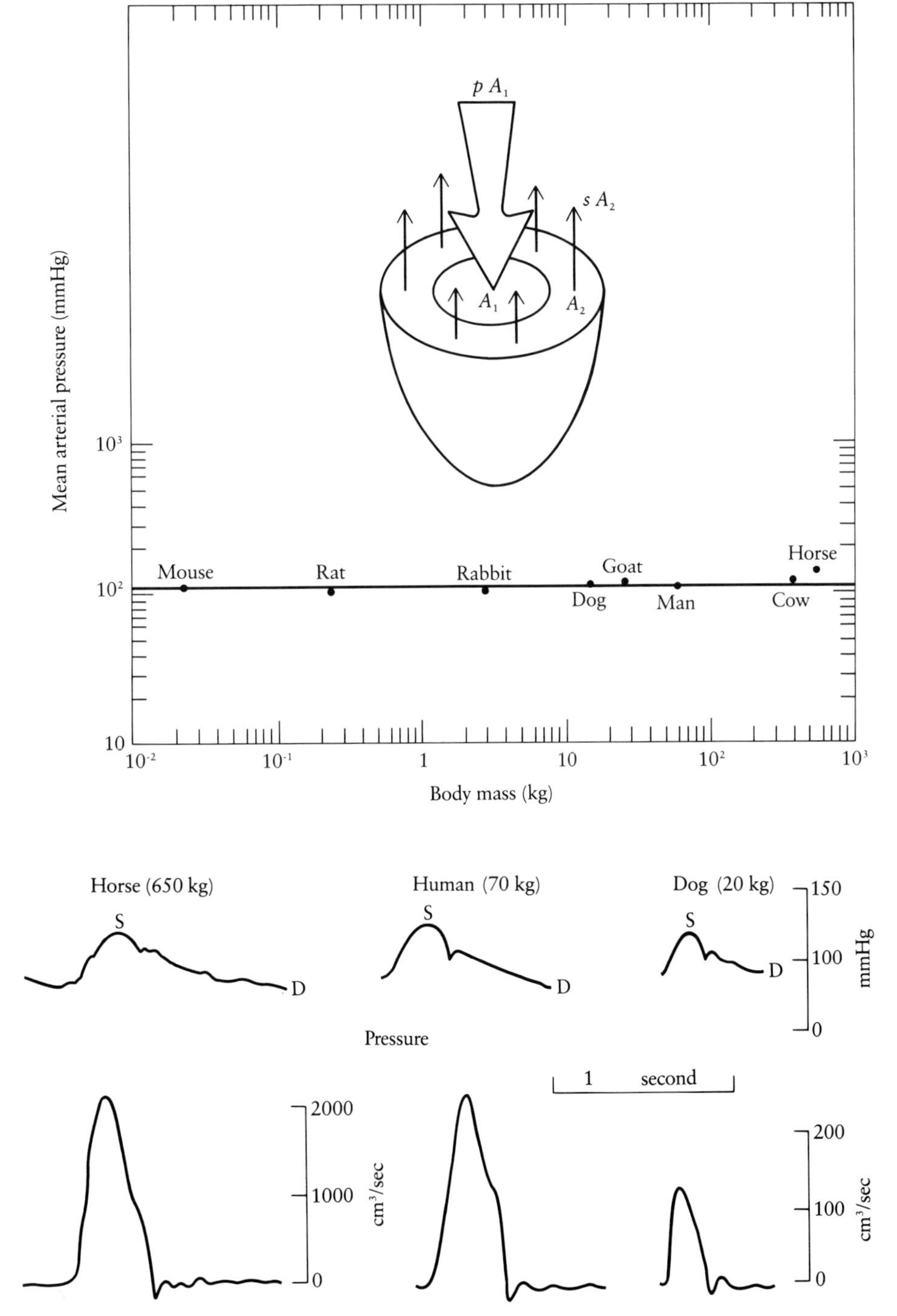

minimum (diastolic) pressure ($D$) in all three animals, and in fact we do. When the peak blood-flow velocity is calculated by dividing the peak volume flow rate of the aorta by the cross-sectional area of the ascending aorta, it is also practically the same in horses, dogs, and humans.

An equivalent argument will explain the observation that the variations in pressure that cause the respiratory flow and the air velocity (in centimeters per second) in the tracheas of mammals are both independent of body size.

**Growth Time, Cell-Cycle Time, and Lifetime.** The remaining dimensionless groups in the table on page 112 are ratios of various physiological time periods to the lifespans of mammals in captivity. Group 5 shows that time required for a mammal to complete 50 percent of its growth is about 3 percent of its life, no matter how big it is. Group 6 shows that 1.5 percent of a (placental) mammal's life is required for gestation. According to groups 7 and 8, both a respiratory cycle and a heart cycle occupy nearly size-independent fractions of a lifespan. As a consequence, every mammal can expect to live for $3.3 \times 10^8$ breath cycles and $1.5 \times 10^9$ heartbeats.

One cannot help thinking that, if we could just understand why every mammal lives for about $1.5 \times 10^9$ heartbeats, we would understand something important about aging. William A. Calder, a prolific observer of animal allometry, has noticed that the average lifetime of a red blood cell, the time it takes the body to replace half of its supply of such constituents of the plasma as albumin, the half-life of a drug (methotrexate) given intravenously, and the time required to metabolize fat stores equal to a given fraction of the body mass are all physiological periods proportional to body mass raised to an exponent near 0.25. Because lifetime scales as body mass to the power 0.20, all of these periods are approximately size-independent as a fraction of a total lifetime. Perhaps the rate of performance of biological functions in an organism, determined as it is by the activities of enzymes, is correlated with the rate of turnover of individual cells and hence the rate of accumulation of sublethal cellular defects in the form of mutations. This would suggest that some organisms, like a metal object fatiguing, can be cycled only until the concentration of defects makes some unspecified breakdown likely.

On the other hand, the approximately constant number of heartbeats and breath cycles in a mammalian lifetime may be only a coincidence resulting from other scaling considerations and not a cause, per se, of a strictly allotted lifetime. With human beings, it is not even a dependable coincidence. We fall high above the other mammals on the allometric age charts; according to what we weigh, on the average, we should only live about 33 years.

## Dimensionless Groups That Change with Body Size

The general conclusion available from the table on page 112 is that there is a great deal of similarity between mammals of different sizes, at least with respect to the timing of their lives. Seen this way, biological scaling is not essentially different from the scaling of events in the cycle of a model steam engine, in which the period of each operation (the duration of the power stroke, for example, or the opening time for a valve) is determined by cams, cranks, and gears to be strictly proportional to the overall cycle time, which, in turn, is inversely proportional to the characteristic length.

A basic feature of biological scaling, perhaps just as important as the similarity in time scaling and volume scaling noted so far, is the way in which scale effects and scale distortions appear. We shall see that some of these scale effects will mean that one or another of the dimensionless groups familiar from Chapter 3 will not be constant as size increases. Other distortions will act directly on the shape of an animal or plant.

**Reynolds Numbers.** One of the first obstacles we meet in looking for dynamic similarity among living organisms is that the Reynolds number is not constant as body size changes. The Reynolds number for a given flow problem is always larger in a larger organism.

The Reynolds number describing the flow of blood in the aorta

$$\text{Reynolds number} = \frac{\text{aortic blood velocity} \times \text{aortic diameter}}{\text{blood viscosity / blood density}}$$

is based on the time-mean aortic blood velocity and the aortic diameter at a given point—say, the ascending aorta. Observations of the type just given in the figure at the bottom of page 115 extended to mammals over the size range from guinea pigs to horses show that the velocity of blood in the ascending aorta does not change much with increasing size, nor do blood density and viscosity. (The apparent viscosity of blood does change with the size of the tube it is flowing through when the tubes are small, but not when they are as large as the aorta.) Thus the Reynolds number based on mean blood velocity is roughly proportional to the aortic diameter.

A practical consequence of the fact that the Reynolds number describing blood flow increases with body size is that so-called "innocent" systolic murmurs (those not associated with disease) may sometimes be heard through a stethoscope applied to the chests of large animals (humans and larger), particularly during and just after exercise. Recall that the Reynolds number was

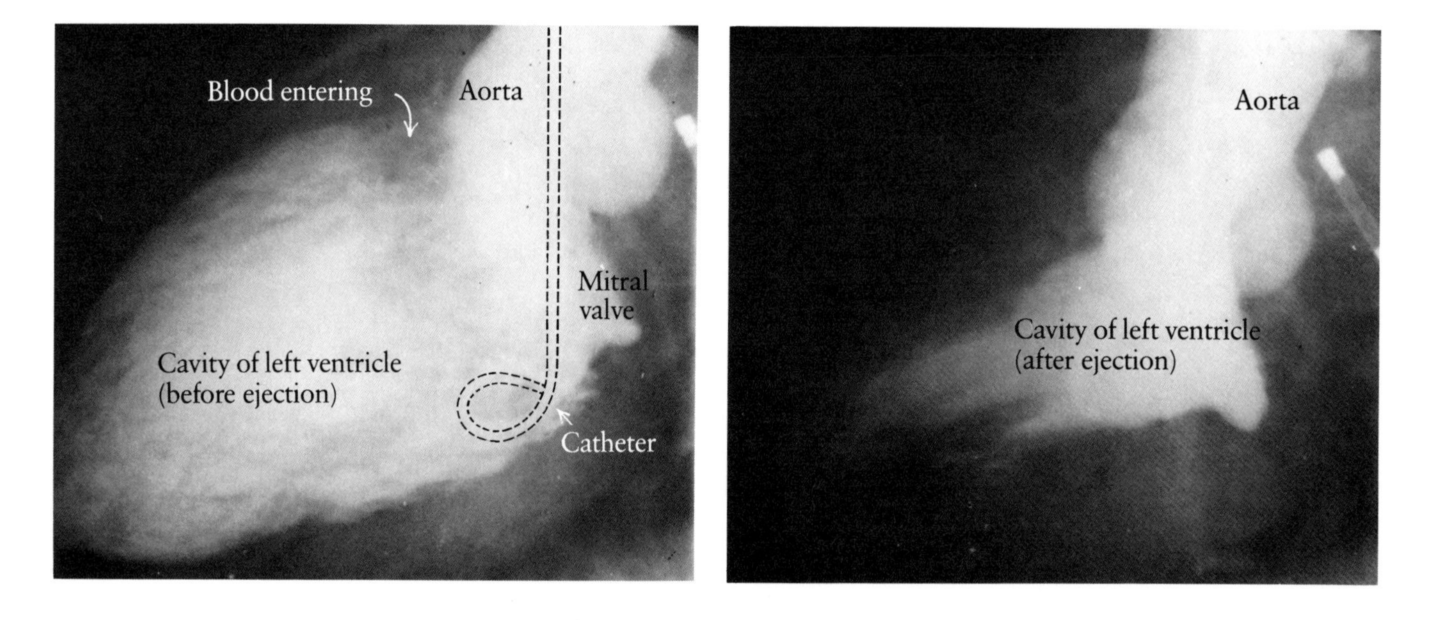


Two frames taken from an x-ray movie showing patterns of blood flow in the normal human heart. In the photograph at the left, taken before the ejection of blood from the heart, a contrast agent opaque to x-rays (appearing white) fills the left ventricular cavity. The bulge at the base of the aorta is the site of the aortic valve. The thin tube (catheter) injecting the contrast agent may be seen entering the ventricle and ending in a loop. A band of nonopacified blood (dark) is flowing into the ventricle (to the left of the aorta in the figure). In the photograph at the right, the heart has contracted, emptying the contents of the ventricle into the aorta. The turbulence of the flow moving out of the heart causes extensive mixing of the injected agent with the blood, so that the nonopacified band present in the left figure has disappeared.

The skeleton of a hummingbird hangs below the hollow thighbone (femur) of an extinct elephant bird. The proportions of the bones of larger animals are generally thicker and heavier than those of smaller animals. (The fact that the larger bird almost certainly was flightless serves to exaggerate this particular comparison.)

defined as the ratio of inertial forces to viscous forces acting on an element of fluid and that the likelihood of turbulent flow becomes greater as the Reynolds number increases. The murmurs in question are caused by the existence of boundary-layer turbulence in the ascending aorta and particularly in the pulmonary artery. Two frames from an x-ray movie showing the pattern of flow in the outflow tract of a human heart are shown in the top figure on page 118. The equivalent x-ray photo of a mouse heart would show smoother (more nearly laminar) flow.

The generalization that the Reynolds number increases with body size extends to the flow fields around swimming and flying organisms as well, because the characteristic dimension and generally also the speed increase with size. We mentioned in Chapter 3 that a sperm cell would go nowhere if it tried to swim like a whale because, given its low Reynolds number, it cannot employ the inertia of the water to propel itself. Instead, it must rely on a component of the viscous forces (which dominate at low Reynolds numbers) for propulsion. For similar reasons, a gnat cannot glide like an eagle. The relation between Reynolds number and body size will be a central theme in Chapters 5 and 6.

**Relative Proportions of the Body.** Anthropologists and paleontologists pay close attention to the relative proportions of the body. By keeping track of such quantities as the size of the grinding surfaces of teeth and the dimensions of the jaw, they can make arguments about how a fossil animal made its living.

In the drawing at the top of page 120, the skeletons of two adult primates of different species are compared. The contrasts between the small siamang and the large gorilla are reminiscent of the contrasts between the shapes of the growing human forms shown earlier on page 32. But a closer examination shows that the changes in proportions due to growth in an individual are rarely more than superficially similar to the changes based on the differences between species. There is a consistency to be found in the between-species comparison of adult forms that shows the influence of body size uncomplicated by the changes in form that accompany sexual maturity. As we illustrated with several examples in Chapter 2, once the adult form is reached, animals of different sizes but of the same species tend to obey isometry.

In the drawing on page 120, every one of the long bones of the limbs is relatively thicker and more robust in the larger animal. In each bone, the ratio of diameter to length is not a constant but increases in a regular way as related species of larger body sizes are compared.

The effects that produce a regular change in bone proportions extend to the vertebrae as well. The figure at the top of page 121 shows a log–log plot of the

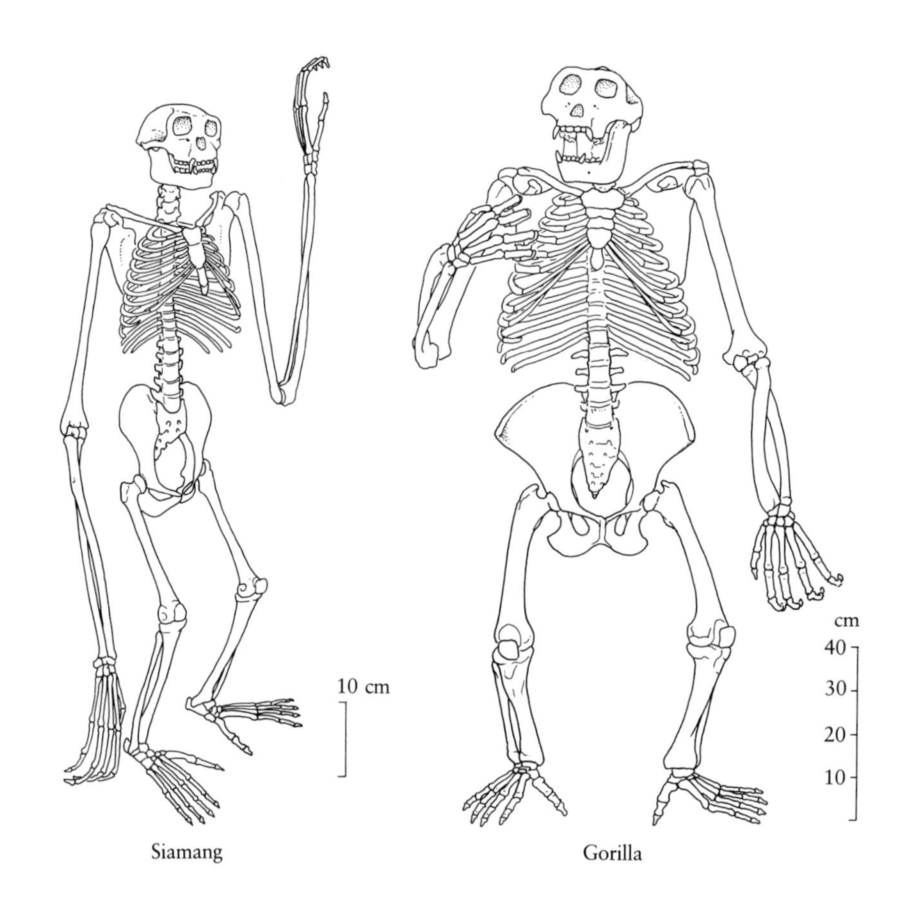


The skeletons of two primates of greatly different sizes. In the siamang (*Symphalangus syndactylus*), *left*, the long bones of the limbs are relatively more slender than they are in the gorilla (*Gorilla gorilla*), *right*.

diameter (called the "centrum height") versus the length of the indicated portion ("centrum length") of the third lumbar vertebra in a series of species of African antelopes. (The antelopes constitute a subfamily of the cattle family, Bovidae.) These measurements were made on 135 specimens representing 18 genera. Only adults are represented. The animals range in size from the tiny 2.5-kilogram Gunther's dik-dik to the 850-kilogram giant eland. In between come gazelles, impalas, and wildebeests.

Two points are worthy of special attention. The first is that the height of the centrum (the $h$ defined toward the top of the figure) increases relatively faster than the length as size increases in this series. Thus, the shape of the vertebral body is relatively thicker and wider in larger animals.

A second noteworthy feature is the way in which the vertebrae fit together. The vertebral column of the 3-kilogram Kirk's dik-dik in the lower part of the figure is curved, while the vertebral column of the 750-kilogram Cape buffalo is relatively straight. In general, small animals tend to stand and move in a crouched position, with their backs and limbs relatively more flexed than those of larger animals. In the next section, we will give a plausible organizing principle for these size-related distortions.

Comparison of the centrum height and the centrum length for the third lumbar vertebra in African bovids, shown in a log–log plot. The centrum height (the diameter) is proportional to the centrum length raised to the power 1.47. A line of slope 1 (isometry) is drawn for comparison.

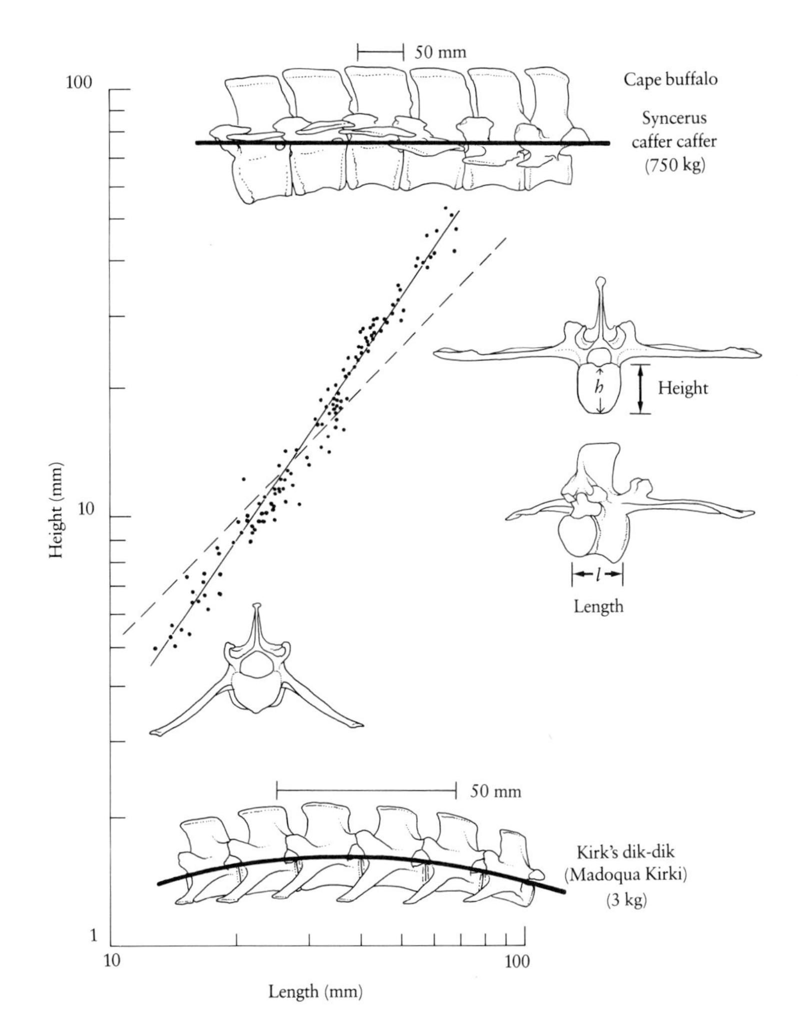


Skeletons of two members of the cat family of different sizes. The large animal is a lion, the small one is an ocelot.

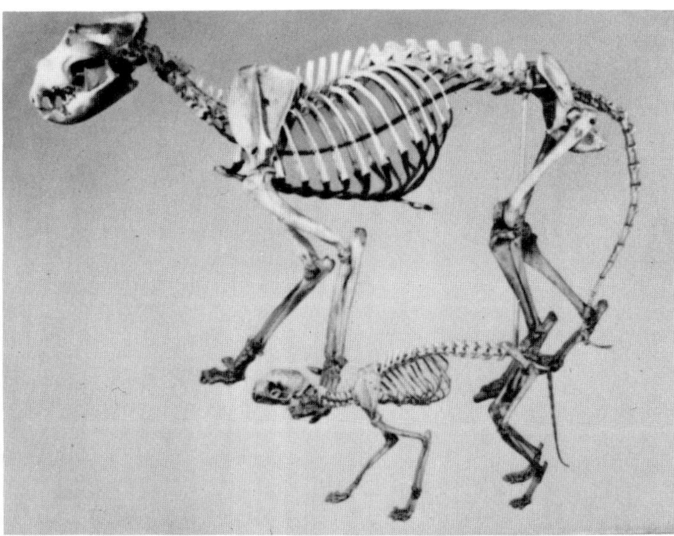

(a)

(b)

African mammals from the cattle family: *a*, springbok (*Antidorcas marsupialis*); *b*, eland (*Taurotragus* sp.). Even without a meter stick or other length scale in these photographs, it is possible to tell that the springbok is a fairly small animal and the eland is a large one, because the eland's proportions are relatively thicker than those of the smaller species.

## Shape and Strength

Large animals subject their bones and muscles to substantial forces when they run and jump. The British zoologist R. M. Alexander (1981) has estimated from experimental evidence that the ordinary loads encountered by a horse in trotting and galloping result in stresses up to one quarter as much as those needed to break the animal's bones. The tendons of a hopping kangaroo may routinely be stressed up to half their breaking strength. John Currey, a biologist at the University of York (1977), has remarked that race horses often break their legs while galloping, and x-rays of the long bones of many veteran ballet dancers reveal healed fractures.

Clearly, then, large animals risk pulling muscles, tearing tendons and ligaments, and even breaking bones when they move quickly. Although this is true, there is a more fundamental strength and stability requirement that must be satisfied before any animal can complete a single running step. As we shall see, if this requirement were not satisfied on every step, an animal would fall to its knees.

**Buckling.** When an animal runs, it actually proceeds in a series of jumps. In galloping, there are periods when all four feet are off the ground, followed by periods when the animal must bounce off the ground and into the air again.

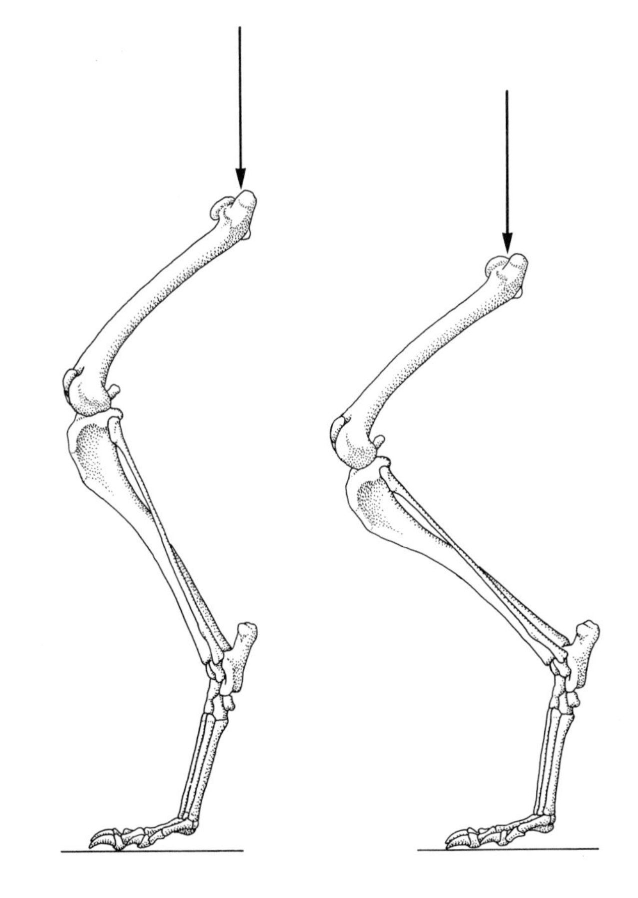

The hind leg of a dog, showing the major bones. This leg is midway through the period when the foot is on the ground during running. A vertical load that is known from experiments to be a particular multiple of the animal's weight acts to collapse the limb by flexing the knee (*arrow*). The muscles acting to extend the knee resist this load. The leg on the right is flexed more than the one on the left; for stability, the increase in extending torque due to a small flexure of the knee must exceed the increase in buckling torque caused by the same small flexure.

Observations of four-legged animals running have shown that the maximum vertical force is applied to the ground when the hind foot is almost directly under the hip joint. At this moment, the muscles acting to resist bending of the knee must counteract a large vertical force at the hip. This vertical force (shown by an arrow in the two drawings above) is a certain multiple of the animal's body weight, whether the animal is large or small.

The figure shows a dog's hind limb in two alternative configurations, one slightly more flexed than the other, but the assumption is that the vertical load indicated by the arrow is the same in both configurations. Suppose the configuration on the left represents a starting point and the one on the right shows the geometry of the limb after an outside agency forces the knee to flex a little bit. What will happen when the leg is released? Because active muscles behave like springs, generating increased forces when they are stretched to longer lengths, there will be an increase in the extending torque (force × moment arm of the muscle force) at the knee joint due to stretching of the muscles acting about the knee. At the same time, there will be an increase in the buckling torque tending to collapse the knee because the small knee flexion increases the moment arm of the applied load. If the increase in extending torque due to the small flexure is greater than the increase in buckling torque, the animal springs up when released by the outside agency. If the increase in buckling torque is greater, the leg buckles and the animal falls to the ground.

The details of this example are worked out in the appendix at the end of this chapter. An important conclusion is that a series of animals of different sizes cannot all be stable under the buckling forces of running if they are geometrically similar. If the small animals of such a series are stable, the large ones are not. It turns out that only by introducing a regular distortion in shape with increasing size can all the animals be stable. For equal stability under buckling loads in a set of animal forms, the requirement is that the ratio of the cube of the length to the square of the diameter must be kept constant in every element of the body as size increases. This result will have profound consequences for much of what follows.

**Elastic Similarity.** Let us now consider a new scaling rule as an alternative to isometry (geometric similarity), which was the main rule employed for discussing the theory of models in Chapter 3. This new scaling theory, which we shall call *elastic similarity,* uses two length scales instead of one. Longitudinal lengths, proportional to the longitudinal length scale, $\ell$, will be measured along the axes of the long bones and generally along the direction in which muscle tensions act. The transverse length scale, $d$, will be defined at right angles to $\ell$, so that bone and muscle diameters will be proportional to $d$. (The reader should note that $\ell$ and $d$ replace the single-length scale $l$ used in previous chapters.) When making the transformation of shape from a small animal to a large one, all longitudinal lengths (or simply "lengths") will be multiplied by the same factor that multiplies the basic length, $\ell$, and all diameters will be multiplied by the factor that multiplies the basic diameter, $d$. Furthermore, there will be a rule connecting $\ell$ and $d$, the same rule discovered in the previous paragraph, namely: $d \propto \ell^{3/2}$. (In later chapters where geometric similarity is discussed, $d$ will again be assumed to be proportional to $\ell$).

In the figure on page 125, the humerus bones of different species from the cattle family (Bovidae) are compared. The midshaft diameter, $d$, measured in the

The humerus bones of a series of antelopes from a 3-kilogram Kirk's dik-dik to a 750-kilogram Cape buffalo are shown. The bones follow the relation $\ell = 24.09d^{0.66}$, or, equivalently, $d = .0087\ell^{1.5}$.

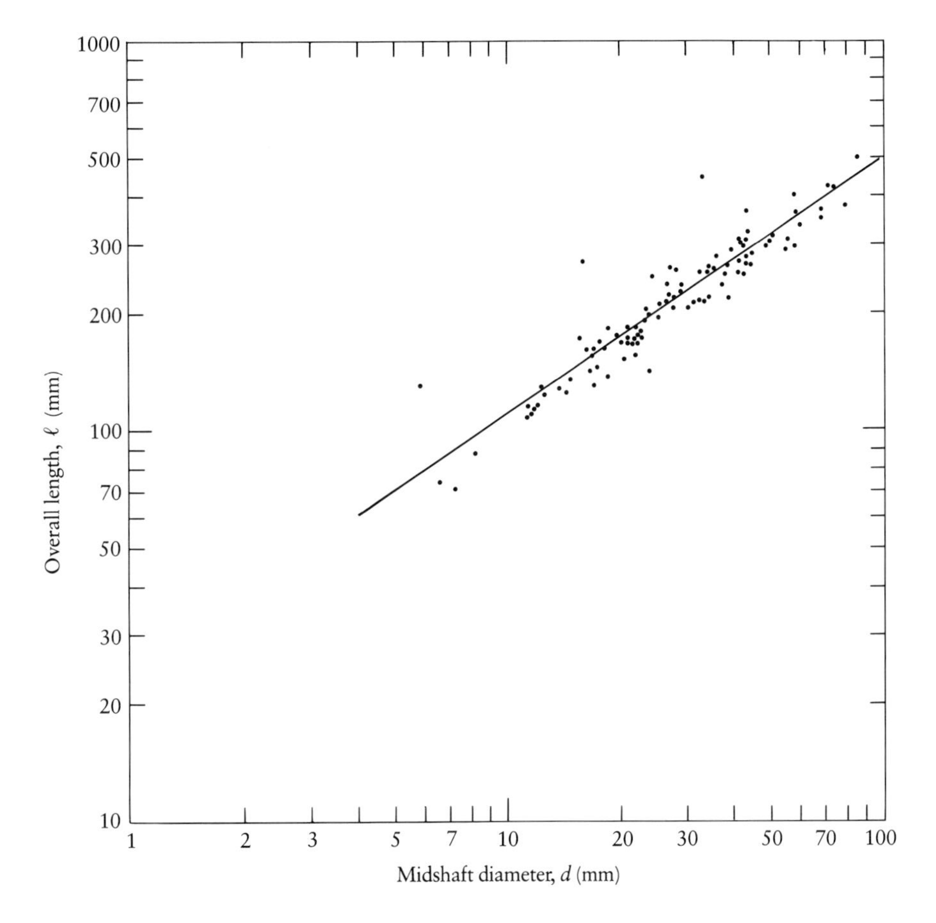


same direction the animal swings its legs (anterior–posterior direction) is compared with the overall length, $\ell$, between joint surfaces. The log–log plot shows that these bones fit rather well on a line of slope ⅔, as required by elastic similarity. (Note that $\ell$ is plotted on the vertical axis and $d$ on the horizontal. Equivalently, diameter scales as the 3⁄2 power of length.)

As it happens, one may arrive at the rule of elastic similarity, $d \propto \ell^{3/2}$, by considering bending instead of buckling. A series of rubber beams simply supported at their ends are elastically similar if the deflection at the center of the beam is proportional to the length of the beam. It is a simple matter to show that, if the ratio of the deflection to the length is a constant, then elastic similarity holds in a family of beams. It is probably the bending rather than the buckling aspect of elastic similarity that is important in the dynamics of the vertebral column in quadrupeds. Recall that centrum height (a transverse dimension in the bending direction) was proportional to centrum length to the 1.47 power for the vertebrae analyzed in the figure at the top of page 121. This is as predicted for elastic similarity, in which diameter is proportional to the $3/2 = 1.5$ power of length in any element of the skeleton.

Theropod dinosaurs of a range of body sizes. The bones of the larger animals are relatively thicker. From the smallest to the largest, these animals (all representatives of different species) are a 165-kg ornithomimid, a 735-kg tyrannosaurid, a 2,500-kg tyrannosaurid, and a 6,000-kg tyrannosaurid (*Tyrannosaurus rex*).

In many respects, the limb bones of predatory dinosaurs scale like those of kangaroos. Here, the skeleton of *Tyrannosaurus rex* is compared with that of a (much smaller) modern kangaroo. The bones of the dinosaur are thicker because it is larger; a smaller theropod dinosaur—say, the ornithomimid on page 126—bears an even closer resemblance to the kangaroo.

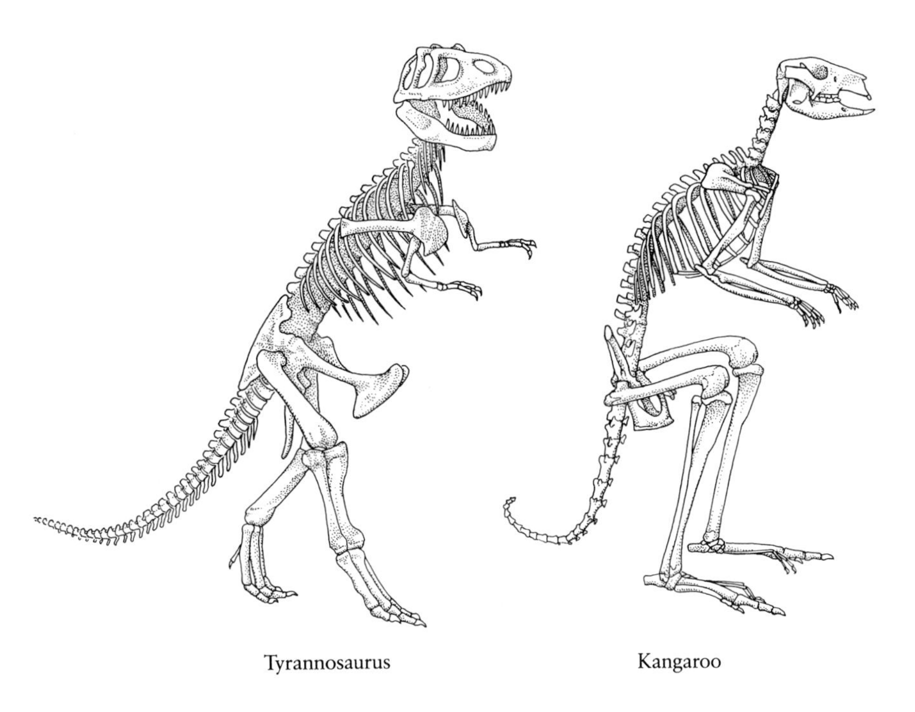


**Dinosaurs.** A marvelous potential now emerges for using the ideas and techniques of allometry to understand more about dinosaurs than one could hope to know otherwise. As an example, take the predatory theropod ("beast-foot") dinosaurs shown on page 126, which roamed around eating big animals in the latter part of the Cretaceous period (roughly between 100 million and 65 million years ago). These reconstructions drawn by the vertebrate paleontologist Gregory Paul from bones on display at the American Museum of Natural History in New York show a range of animals from a 165-kilogram ornithomimid to the 6,000-kilogram *Tyrannosaurus rex*. A striking feature of these drawings is the way in which, as size increases, the thicknesses of the limb bones increase relatively faster than their lengths. In fact, Paul has plotted diameter against length for many individual bones and for the whole hind limb and has found good agreement with most aspects of the elastic similarity model in a series of theropods from the smallest, weighing 50 kilograms, to the largest, weighing 6,000 kilograms. In many respects, the limb bones in predatory dinosaurs scale like the equivalent bones in ostriches and kangaroos. Paleontologists have used this and other evidence to suggest that some of the large dinosaurs were capable of running very fast—quite contrary to the common conception of dinosaurs as lethargic and slow.

If and when we gain confidence in the recently proposed hypothesis that dinosaurs were warm-blooded (homoiothermous), the way seems open to use allometric comparisons with modern-day warm-blooded animals, perhaps the allometric rules cited in the table at the beginning of this chapter. Only then will it be possible to arrive at reasonable estimates for the animals' lifespans, growth periods, resting heart rates, and many other details of their physiology.

**Deriving Kleiber's Law.** Before leaving the subject of scale distortions among animals, it is worthwhile pointing out that Kleiber's law may be derived from the principle of elastic similarity. Only a sketch of the theory relating body size to metabolic rate is given here; a more complete treatment may be found in McMahon (1983).

Suppose the form of an animal is represented by a set of connected cylindrical elements like those shown in the figure at the left. (There are toys for sale that are built this way, with strings running through holes drilled along the axes of the cylinders. The toy stands up when the strings are taut, but it falls over if the strings are allowed to go slack.) Typical segments representing part of the hind limb and part of the trunk are also shown. Each segment will be assumed to account for a given fraction of the total body mass. Suppose we focus on a particular segment of diameter $d$ and length $\ell$. The mass of each segment is a size-independent fraction of the total body mass $m$, so that diameter squared times length is proportional to total body mass:

$$d^2\ell \propto m. \tag{4.1}$$

Under the rules of isometry, in which $d \propto \ell$, equation 4.1 says that $\ell^3 \propto m$, so that the length of a segment is proportional to the cube root of the body mass, just as was true for the series of cubes compared in Chapter 2. But under the rules of elastic similarity, $d \propto \ell^{3/2}$, and equation 4.1 takes one of two forms, depending on whether it is solved for $d$ or for $\ell$:

$$d \propto m^{3/8} \tag{4.2}$$

or

$$\ell \propto m^{1/4}. \tag{4.3}$$

Notice that $3/8$ (= 0.375) is larger than $1/3$ (= 0.33), while $1/4$ (= 0.25) is smaller than $1/3$. This shows that the diameter increases with increasing body size faster in elastic similarity than it does in isometry, while the length dimension increases more slowly. In the figure on page 129, the chest circumference in a series of primates, from 0.25-kilogram tamarins to 25-kilogram baboons, is found to fit an allometric formula with a mass exponent of about 0.37,

Representing an animal form as a series of quasicylindrical elements. The surface area is proportional to $\ell d$, whereas the cross-sectional area is proportional to $d^2$.

Allometric plot of chest circumference against body mass for five species of adult primates. The drawings at the bottom show two primate forms preserving elastic similarity. These two forms differ in length by a factor of 2, in diameter by a factor of 2.83, and in body mass by a factor of 16.

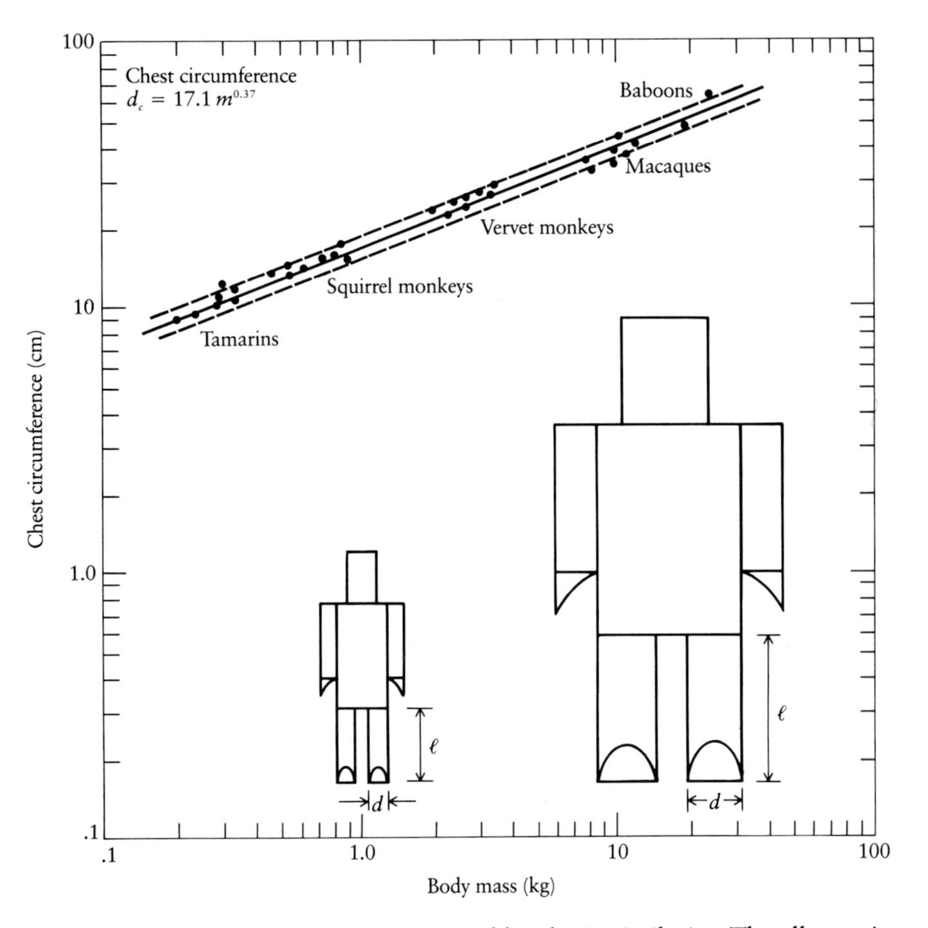


which is close to 0.375 (= ⅜), as required by elastic similarity. The allometric mass exponents in Paul's series of dinosaurs were 0.38 for the femur circumference and 0.26 for the hind-limb length, again in good agreement with the theory.

Long before Kleiber published his important paper setting out the observation that basal metabolic rate in mammals is proportional to $m^{0.75}$, other investigators had noticed that the ratio of metabolic rate to body weight was not constant, even within a series of dogs of different sizes. The German physiologist Max Rubner (1883) proposed that metabolic rate was somehow limited by body surface area, citing the fact that oxygen comes in and heat goes out through surfaces that are proportional to the total body-surface area.

The problem with this argument is that the total body-surface area measured in mammals from shrews to whales scales with an allometric mass exponent that is different from that of the basal metabolic rate. As shown in the

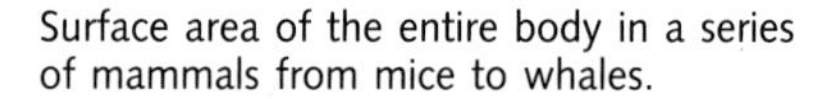

Surface area of the entire body in a series of mammals from mice to whales.

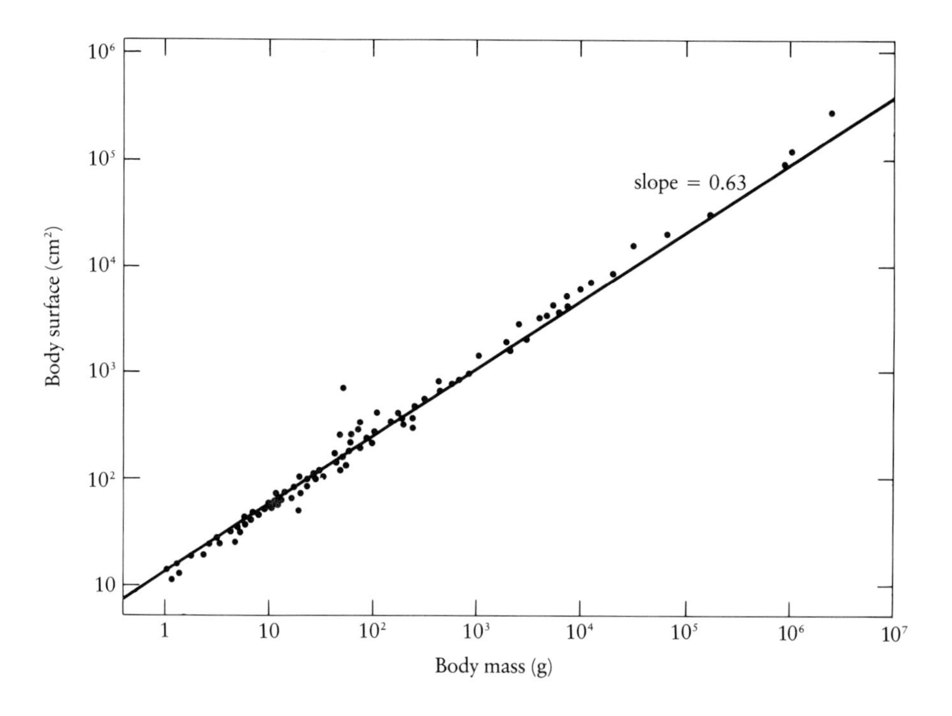


figure above, the surface area scales as $m^{0.63}$, which is quite different from the metabolic rate, which scales as $m^{0.75}$. How should we think about the surface area? Returning to the figure on page 128, it is clear that the total body-surface area for the segmented animal is proportional to the cylindrical area of any one of the segments, which is in turn proportional to the product of length and diameter. In isometric animals, length times diameter varies as $m^{2/3}$, whereas, in elastically similar animals, length times diameter is proportional to $m^{1/4}m^{3/8} = m^{5/8} = m^{0.625}$. Thus, the results in the figure above are in agreement with the predictions of elastic similarity, but they are not in harmony with the idea that the metabolic rate is limited by body-surface area.

An alternative suggestion might be that the *cross-sectional* area of the body, rather than body-surface area, provides the limitation. Elastic similarity predicts that the cross-sectional area of any anatomical feature of the body is proportional to $d^2 \propto m^{3/4}$. Note that ¾ is the exponent in Kleiber's law. Moreover, there are theoretical reasons why cross-sectional area should relate to metabolic rates. If the maximum stress developed by any muscle is independent of body size and the cross-sectional area of that muscle increases as $m^{3/4}$, then both the force developed by the muscle and the oxygen demand that is a consequence of that force also increase as $m^{3/4}$. Shown in the figure at the top of page 131 are results obtained by the comparative zoologist C. R. Taylor

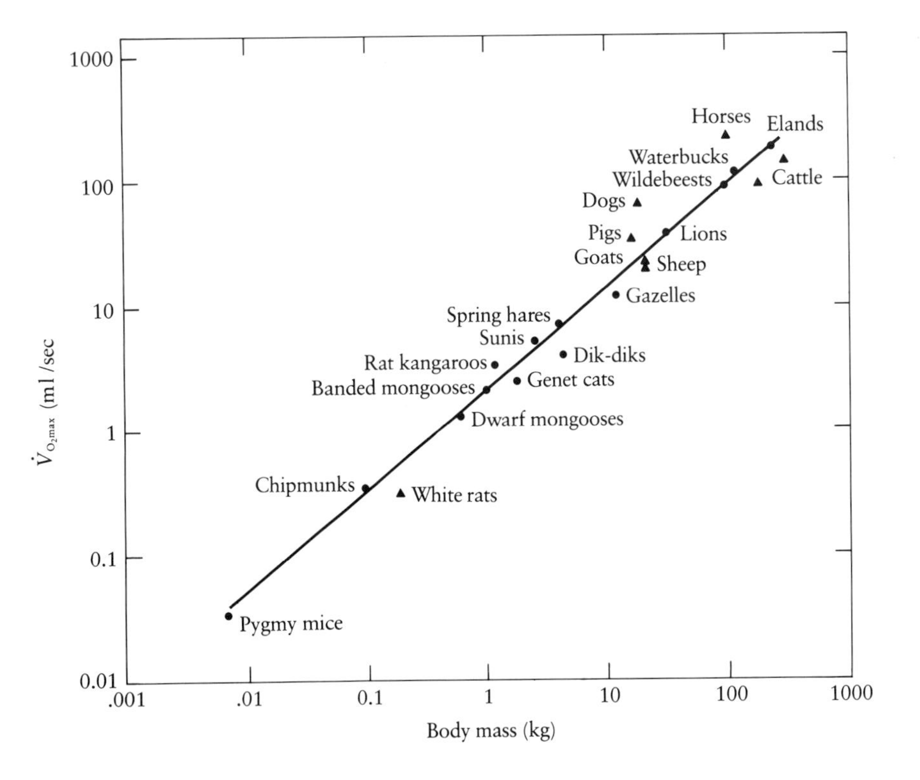


Maximal rate of oxygen consumption from pygmy mice to elands. The circles show the results for 14 species of wild animals; the triangles show results for seven species of laboratory and domestic animals. Two points have been given for cattle showing large and small sizes. The maximal rate is about 10 times the basal rate in all the animals studied.

from measurements of the peak rate of oxygen consumption as a function of body mass in laboratory, domestic, and wild animals. These animals ranged in size from pygmy mice with a body mass of 8 grams to elands with a mass of more than 250 kilograms. The line showing maximal oxygen consumption has a slightly greater slope than the basal line (0.809 rather than 0.75), but Taylor asserts that the slopes are not significantly different.

There is no logical reason why the maximal rate of oxygen consumption should be about 10 times the basal rate, and not, say two times or 100 times the basal rate. Understanding what limits this ratio of peak to basal metabolic rate (the ratio is called the *metabolic scope*) is a worthy subject for future research, inasmuch as this question is broadly involved in the principles that control the intrinsic metabolic rates of cells. It would be valuable to understand how a cell knows whether it is in a small dog or a large dog and how it adjusts its basal metabolic rate accordingly.

**Physiological Time.** Finally, we are in a position to understand why many of the physiological periods listed in the table at the beginning of this chapter have allometric mass exponents near 0.25. For both elastic similarity and isometric scaling, the tidal volume of the lung, the stroke volume of the heart, the volume of the blood, and, in fact, the volumes of every one of the internal organs are proportional to body mass. The rate at which air flows into and out of the lungs and the rate at which blood flows around the circulation are proportional to $m^{3/4}$. Thus, the time for a breath is proportional to the tidal volume divided by the air-flow rate, or $m/m^{3/4} = m^{1/4}$, and the time for a heartbeat is proportional to the ventricular stroke volume divided by the cardiac output, or $m/m^{3/4} = m^{1/4}$ again. Similar reasoning explains why the time for the circulation of the entire blood volume and the time for the passage of all the blood once through the kidneys or liver is proportional to $m^{1/4}$. The time required for a hormone released in the blood to reach its target in the body is also proportional to $m^{1/4}$, because the blood velocity is size-independent and the arterial lengths scale as $m^{1/4}$. The time of vibration of the vocal cords, because it is proportional to the length of the cords, is another period proportional to $m^{1/4}$.

By now it must be clear that looking at the world of mammals and other large animals as machines subject to gravity, as we have just done, provides a nice organizing scheme for a great many observations. Beginning with a simple rule concerning the structural stability of muscles and joints, we have derived a result that appears to control the many and various rates of living, even the intrinsic metabolic rates of the cells.

**A Few Cautions.** Even so, there are some cautions that should be kept in mind when using elastic similarity. First, elastic similarity was derived by considering comparisons between large terrestrial animals of vastly different body masses. It should not be applied to comparisons between members of the same species, because animals within a species (which are generally not different in body mass by more than a factor of 10) are observed to scale isometrically, as noted in Chapter 2. It should not be applied to comparisons between juveniles and adults, because the changes due to sexual maturity cause major changes in body shape and muscle strength. It should not be applied to swimming and flying organisms, because entirely different loadings of the skeleton and musculature determine the relation between shape and size, particularly with respect to wings (more about this in Chapter 5). It should not be applied to animals much smaller than the smallest mammal, because forces other than those due to gravity and inertia become important at smaller sizes.

Elastic similarity is a helpful model of animal scaling only in those circumstances in which it is a plausible idea to draw a picture of a limb bone, a whole limb, an organ system, or a whole animal on stretchy graph paper and to scale it up by distorting the diameter dimension by one factor and the length dimension by another. Thus, it works fairly well if the animals to be compared are in the same family, or sometimes even in different families within the same order. Even within these comparisons, it works better for the bones and muscles closer to the heart than it does for those further away (as if Nature made her occupation-related adaptations in hooves, feet, and hands but preserved a certain consistency of design in the more central portions of the body). Perhaps it is especially significant that the vertebral column, the respiratory system, and the cardiovascular system seem to follow the rules of elastic similarity so closely in mammals. The paws, claws, hands, and hooves of the mouse, cat, human, and horse do not obey any particularly strict allometric rules, and their ears and tails are almost completely incomparable, but their aortas, tracheas, and hearts are quite satisfactory elastically similar models of one another.

In the next two chapters, we shall look at the advantages and disadvantages of being large and small, with a particular emphasis on the role of locomotion in an animal's life. In those discussions, many of the results of this chapter will come in handy.

## Appendix: Deriving Elastic Similarity

The bones of the hind leg of a quadrupedal animal are shown diagrammatically in the figure at the top of page 134. Because we are focusing our attention on the knees, the main extensor muscles of the knee, the quadriceps group, is shown and labeled $M$. For the sake of argument, this muscle group is assumed to originate high up on the femur and to attach below the knee on the tibia (it actually originates on the pelvis, but taking account of this fact does not change the result). This leg is shown near the middle of its step cycle during running. It is assumed that the vertical force, $P$, a size-independent multiple of the animal's body weight, is acting to collapse the limb. Experimental observations on animals running at low galloping speeds support this assumption. We will examine the conditions under which the springlike action of the muscle is sufficiently strong to keep the limb from collapsing under the dynamic forces of running. In this argument, we will ignore the dependence of muscle force on shortening velocity, treating the active muscle as if it were a purely elastic body.

Suppose that the limb starts from the configuration shown on the left. In this argument, the hip-flexion angle $\phi$ is assumed to be small, even though it is drawn large in the figure for clarity. Imagine a small increase in $\phi$ that moves

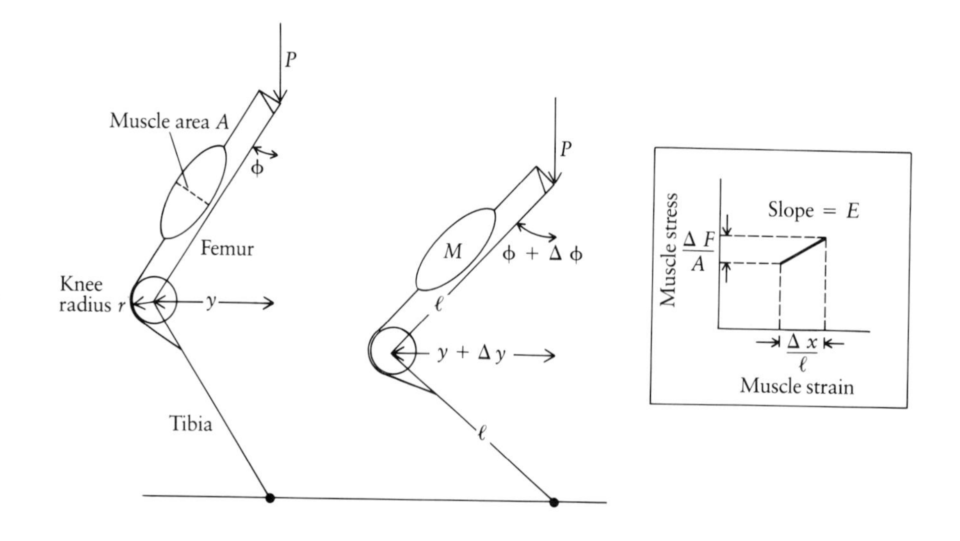


Schematic diagram showing the knee of a dog's hind leg (from the figure on page 123). The force $P$ is a constant force that is assumed to be proportional to body weight. This diagram is used to deduce the relation between $r$ and $\ell$ in animals of different sizes that places the limbs of such animals in similar danger of buckling. (An animal whose limbs have buckled has fallen to its knees.)

the limb into the position shown on the right. This small change in $\phi$ (denoted by $\Delta\phi$), causes the muscle to change its length by an amount

$$\Delta x = 2r\Delta\phi \simeq 2r\Delta y/\ell.$$

Assume that the rest length of the muscle (that is, the length it has when it develops zero force under passive conditions) is equal to the length $\ell$ of the femur. Therefore, the change in muscle strain (change in length divided by rest length) due to $\Delta\phi$ is

$$\Delta(\text{muscle strain}) = \frac{\Delta x}{\ell} \simeq \frac{2r\Delta y}{\ell^2}.$$

As shown in the inset of the figure above, active muscle is assumed to be characterized by a stress–strain curve whose local slope is $E$. In Chapter 3, we discussed the fact that all elastic materials have stress–strain curves. The curves for such materials as steel and aluminum have extensive linear regions, in which $E$ is independent of the strain level, while others, such as rubber and active muscle, have a local slope that changes with the strain. In this problem, in which we confine the change in muscle length $\Delta x$ to small values, we are justified in taking $E$ as fixed. Because $E$ is the factor that relates a change in strain to a change in stress,

$$\Delta(\text{muscle stress}) \simeq \frac{2Er\Delta y}{\ell^2}$$

and

$$\Delta(\text{muscle force}) = \Delta F \simeq \frac{2Er\Delta y}{\ell^2}A$$

in which $A$ is the cross-sectional area of the muscle.

Because the muscle is attached to the femur on one end and to the tibia on the other, it generates a torque on the femur that acts to restore the leg to its upright position:

$$\Delta(\text{extending torque}) \simeq \frac{2Er\Delta yA}{\ell^2} r. \tag{1}$$

The change in the applied torque acting on the femur as a consequence of the change in knee flexion is

$$\Delta(\text{buckling torque}) = P\Delta y. \tag{2}$$

The boundary between stable and unstable conditions is reached when the increase in extending torque generated by the small movement $\Delta\phi$ is equal to the increase in buckling torque contributed by the same movement. If the extending torque is greater, the animal springs up when released from the small disturbance; if the buckling torque is greater, it falls to the ground when released. Therefore, the condition of neutral stability is found by setting equations 1 and 2 equal to each other:

$$P = \frac{2Er^2A}{\ell^2}. \tag{3}$$

Suppose we assume that

$$P = k_1 m$$
$$A = k_2 r^2$$
$$m = k_3 r^2 \ell$$

in which $m$ is body mass as usual and $k_1$, $k_2$, and $k_3$ are all size-independent constants. Then equation 3 becomes

$$\ell^3 = \frac{2Ek_2}{k_1 k_3} r^2. \tag{4}$$

You may have been surprised when we retained the two lengths $r$ and $\ell$ in our equations rather than assuming that each was proportional to some characteristic length. The reason for doing this should now be clear in the result, equation 4. This result states that, if the limbs of a series of animals of different sizes are to resist buckling under a dynamic force $P$ that is proportional to body weight, then the knee radius, $r$, cannot be proportional to the length of the femur, $\ell$, as would be the case in a series of geometrically similar animals. Instead, the ratio $\ell^3/r^2$ must be kept constant in such a series of animals in order that they may be all under a similar threat of having their legs collapse under them every time they attempt to run. This condition gives us the allometry of elastic similarity $\ell^3 = kr^2$ or, equivalently, $\ell \propto r^{2/3}$ or $r \propto \ell^{3/2}$. The $r$ in this derivation scales exactly as the diameter, $d$, of the limb, so this gives us the $d \propto \ell^{3/2}$ of elastic similarity.

# Chapter 5

# On Being Large

In Chapter 1, we made the point that there has been a general trend toward size increase during the course of evolution. This can be seen in the fact that the largest animals and plants are the most recent, so that the upper size limits have been slowly increasing over the last 3 thousand million years. Further, many groups of vertebrates and invertebrates show striking trends of increasing size over periods spanning millions of years. We discussed why this might be, considering what forces of natural selection might be instrumental in producing the changes in size. In this chapter, we will assume all this as background and ask what problems are produced by increases in size, what the limits of size are, and what factors affect those limits.

One set of constraints on large size involves natural selection and changing ecological conditions. There are many reasons for considering very large animals or plants to be highly specialized. When we say that an organism is specialized, we mean that it occupies a highly specific ecological niche. Consider, for example, the African elephant. Because of its great size, it must consume large quantities of vegetation. It also grows slowly, usually has only one offspring at a time, and the time span between one generation and the next is on the order of 10 years. This means that, provided there is a sufficient amount of food available over long periods of time, a population of African elephants will prosper and remain stable.

But suppose there were a great drought sufficiently prolonged to cause extensive destruction of food plants. Under such circumstances, the population would be greatly reduced, and, because of the slow rate of reproduction, it would take many years for the population to recover. All of these problems are directly related to the elephant's size. To illustrate the point, let us see by comparison what would happen to a small rodent under the same environmental stress. An African field mouse would also be threatened by such a catastrophic, prolonged drought. But the needs of any individual mouse are minuscule compared to those of an elephant. If only a small amount of grass persists near springs or streams, a few mice will survive. When favorable weather returns, there will be a reservoir of individuals that can multiply with great rapidity, owing to the fact that they have a short generation time and large litters. As a result, they can repopulate quickly when their food plants reappear. In other words, there is a resilience to the small mammals in fluctuating environments that the large ones lack, especially if the fluctuation reaches a certain critical magnitude or duration.

An elephant grows slowly and usually has only one offspring at a time. As a result, elephants cannot repopulate quickly after a catastrophe.

It is for this very reason that we assume that evolution proceeds mainly from small to large. As long as environmental stability persists, large size may be an adaptive advantage, but the advantage goes to the small organisms in times of major ecological stress. This explains why it is thought that all major evolutionary steps forward have been made in small animals and plants, and that the larger ones have shown success in a particular environment for as long as that environment has lasted but have failed whenever major changes have occurred. The smaller ones not only survive successfully but also adapt and change in the new environment. Their success is a long-term success over huge periods of geological time, while large size makes for comparatively short-term successes, if we are talking in terms of many thousands or millions of years. It is presumed that within these speculations about size and environment lies the basic reason why dinosaurs disappeared at the end of the Cretaceous period (about 65 million years ago) and woolly mammoths disappeared following the ice ages. If some day our shrinking groves of redwoods should meet the same fate and the ubiquitous dandelion should survive, the argument would be the same. Large size has many immediate adaptive advantages; but, if one thinks in terms of geological time and the greater course of evolution, it is clear that small size is less risky and is therefore ultimately more successful.

If we turn now to the physical constraints of large size, clearly gravity is the key. Small terrestrial animals can run up trees—in fact, many of them can run as fast vertically as they can horizontally. Horses might enjoy climbing trees as fast as squirrels do, but in fact they can't climb trees at all, or at least they never do. Furthermore, large animals run a greater risk of injury when they fall. As J. B. S. Haldane (1928, p. 21) wrote,

> You can drop a mouse down a thousand-yard mine shaft; and, on arriving at the bottom, it gets a slight shock and walks away, provided that the ground is fairly soft. A rat is killed, a man is broken, a horse splashes.

It is not even necessary to imagine falling down a mine shaft to see that a larger animal runs a relatively greater risk from falls. Suppose a series of geometrically similar animals simply topple over from a standing position and hit their heads on the ground. During the fall, the gravitational potential energy of the head is first transformed into kinetic energy and then, as the head hits the ground, into strain energy. Because the specific gravitational energy ($mgh/m = gh$, in which $m$ is the mass of the head) is directly proportional to the height $h$ of the head above the ground, the specific strain energy of the head when it is maximally deformed is also proportional to $h$. As we discussed in Chapter 3 in speaking of ship collisions, the specific strain energy in an impact controls the extent of damage. Hence, we can expect the extent of head injuries suffered by an animal in a simple fall to be proportional to body height. According to these arguments, the accidental fall of a dinosaur as large as the 10-ton *Tyrannosaurus rex* might have been one of its most dangerous pieces of bad luck.

## Trees

The largest, most massive, and most long-lived organisms that ever existed were not dinosaurs or whales or even animals at all, but trees. Giant sequoias (*Sequoiadendron giganteum*), like the one shown below, have grown to heights of more than 110 meters (360 feet) and reached girths of as much as 30 meters

During the course of evolution, there has been something resembling an arms race, in which each tree struggles with its neighbors to get to the light. A stand of giant sequoias (*left*). A young tree among the giants (*right*).

A tree weighed down by snow. Some trees spring up again after losing their load of snow. Others are not so fortunate; the large strains caused by buckling can lead to broken limbs and even broken trunks.

(100 feet). The oldest sequoias are more than 4,000 years old, which means that the ones living in California at the present time were there before the discovery of the American forests by even the oldest of the lumber and paper companies.

A tree is a fixed, standing structure, but it must be mechanically clever enough to spring back when the wind disturbs it, and it must not buckle under its own weight. Presumably, there is a competition among trees in certain forest environments to become as tall as possible so as to catch as much of the sun as possible for photosynthesis. One imagines that, during the course of evolution, there has been something resembling an arms race in which each tree struggles with its neighbors to see which can pierce through the top of the canopy and win the sure prize. That prize means the tallest tree reproduces most successfully and its genes prosper in its tall offspring. Of course, not all trees will participate in this particular competition; others will compete by prospering in the shade or by waiting for opportunities in open clearings where a big tree has blown over. There is a succession of types of trees that occupies newly invaded land, and each wave of trees succeeds the previous one until the tallest trees of a mature forest appear.

**The Proportions of Trees.** There is a pattern in the proportions of the large trees (shown on page 141) that is worth investigating. The ratio of the diameter (measured a given distance up the trunk) to the overall height of the tree is greater in the Douglas fir than it is in the ponderosa pine and greater again in the giant sequoia than it is in the Douglas fir. We can understand this effect by applying dimensional analysis, following the same rules we used so often in Chapter 3. We shall assume that the physical variables of importance to the problem are the diameter at the base, the height, the modulus of elasticity, and the weight density. This choice of variables is based on our observation that the elastic properties of trees seem more important than their strength properties, at least with respect to supporting their weight. In the winter, one can often see birches and pines bent over by the weight of snow, so that their branches and sometimes even their tops drag on the ground. After they lose their snow loads, many of those trees spring up again. Some others do not—being less flexible, they have broken under their burdens of snow. Apparently, the danger of breaking under added weight is not as great as the danger of buckling, although buckling can be a prelude to breaking in tall trees.

Using the methods of Chapter 3, we find one dimensionless group, which may be written in the form

$$\frac{\text{modulus of elasticity} \times \text{diameter}^2}{\text{gravity} \times \text{density} \times (\text{height})^3}.$$

A giant sequoia (*left*), a Douglas fir (*center*), and a ponderosa pine (*right*) show how the diameter of a tree trunk at its base increases as the height of the tree increases. The sequoia has lost its top; its base diameter suggests that undamaged it would have reached a height of 300 feet.

Inspecting a table of the physical properties of timber in a handbook tells us that the ratio of the modulus of elasticity to the mass density is approximately constant across the range of green woods, an expression of the fact that increasing the fraction of the wood volume occupied by fiber increases both its modulus of elasticity and its mass density. Thus, we are left with the conclu-

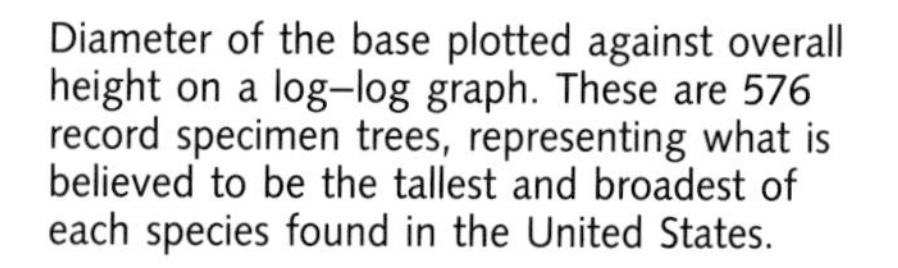
Diameter of the base plotted against overall height on a log–log graph. These are 576 record specimen trees, representing what is believed to be the tallest and broadest of each species found in the United States.

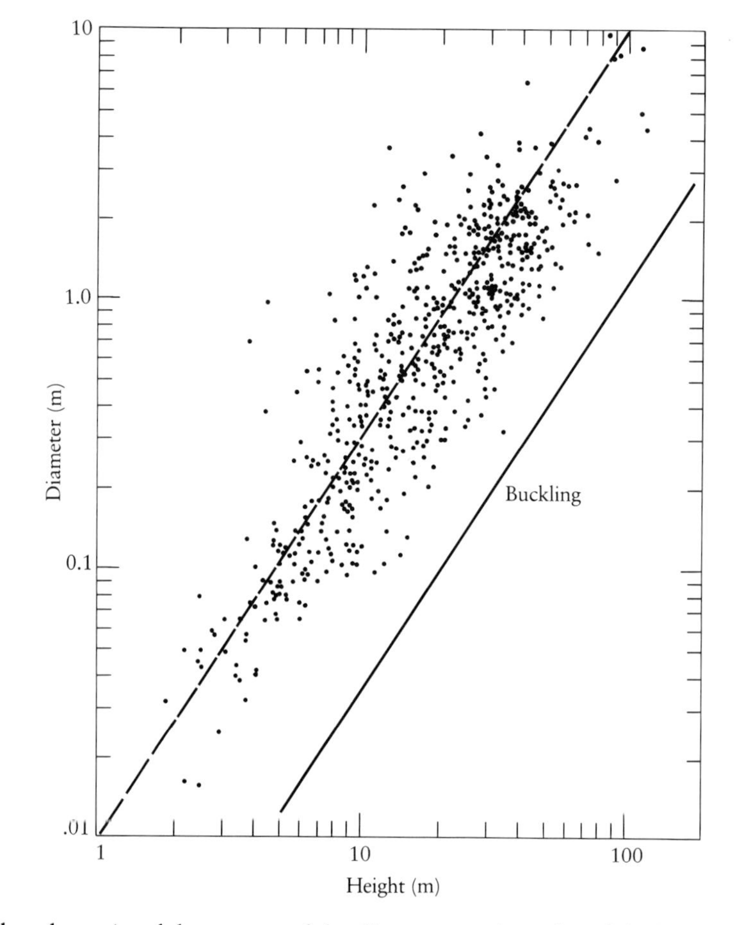


sion that the ratio of the square of the diameter to the cube of the height should be approximately constant in tall trees.

The mathematician G. Greenhill came to this same conclusion in 1881, following a somewhat different analysis. He looked at the flagstaff in Kew Gardens, which was 221 feet (about 67 meters) tall and 21 inches (about 53 centimeters) in diameter at the base, and asked himself how tall could a cylindrical flagstaff of the same diameter be before buckling under its own weight. Applying principles from solid mechanics, he obtained a result that told him that the flagstaff at Kew could not exceed a height of 300 feet (a little more than 91 meters). His result, like the one we obtained from dimensional analysis, specified that the square of the diameter should be proportional to the cube of the height, so that the diameter should increase as the $3/2$ power of the height in a set of very tall flagstaffs or trees, provided that the ratio of the modulus of elasticity to the mass density was a constant.

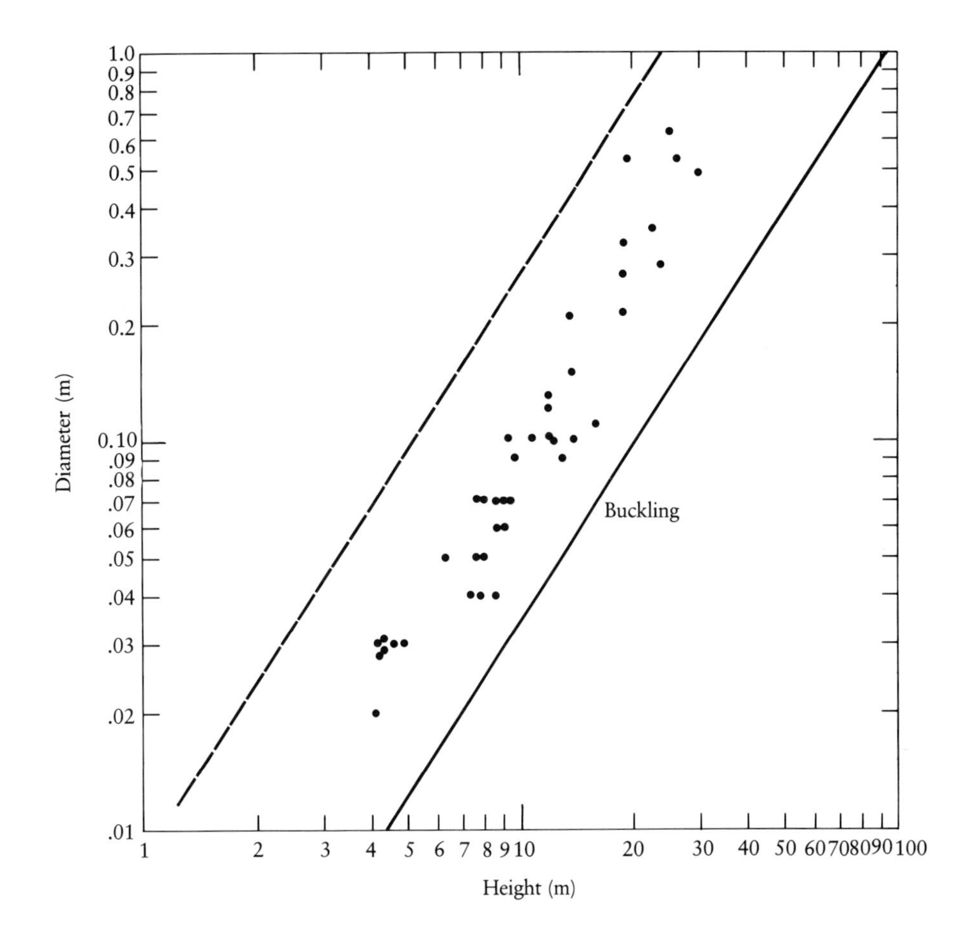


Diameter vs. height in a series of American beech trees (*Fagus grandifolia*) of different ages, showing that diameter increases faster than height with increasing age. Both the broken line and the solid line are the same ones plotted in the figure on the opposite page.

Greenhill's result for buckling is shown in the figure on page 142 along with points giving the height and diameter 5 feet up from the ground in 576 record tree specimens representing most of the common species found in the United States. These data points come from the American Forestry Association's Social Register of Big Trees. A specimen gets into this register by being very tall, or very broad, or both, so that a given tree species may be represented in the figure by more than one point. As a consequence of the way in which the register is kept, we can say with some confidence that the cloud of points in the figure brackets the proportions of most of the very big trees that didn't get in the register as well as those that did.

As may be seen in the figure, a broken line drawn by eye parallel to the buckling line fits near the center of the cloud of points. This is evidence that the proportions of trees, like the proportions of large animals, are determined by the rules of elastic similarity. A tree on or near this line would have to be made about four times taller to buckle under its own weight, presuming that it remains a cylinder of constant diameter all the way to the top. Equivalently, a tree on this line of any particular height could still stand up even if its diameter were reduced to only one-tenth of its original value. There are some trees, however (almost certainly ones grown in dense forests), that lie much closer to the buckling line than to the broken line. These are the risk takers that are asking for trouble in an ice storm.

The proportions of a given tree during growth also follow the rules of elastic similarity reasonably well, as shown in the figure on the previous page. Here a series of trees of the same species (American beech) but different sizes growing in the Institute Woods on the grounds of the Institute for Advanced Study in Princeton, New Jersey, have been measured by Richard Kiltie and Henry S. Horn. The data points lie roughly parallel to the buckling line transferred from the figure on page 142. They indicate that a beech tree of any particular diameter is about twice as tall as the average American tree of that same diameter (broken line). Albert Einstein and J. Robert Oppenheimer walked through these woods many times, but apparently they never commented on the allometry all around them.

A rubber rod treated as a cantilever beam sags under its own weight. Here rods of the same diameter but different lengths are compared. A rod of a particular intermediate length (a length between that of the third and the fourth rods from the top) reaches the farthest lateral distance away from the point of support. The white line indicates the vertical, for reference.

**Branches.** There is more to a tree than its trunk. The branches are also a part of its mechanical design. In the figure on the left, a much simplified model of a tree branch is shown. It is a solid rubber rod held fixed at one end and allowed to droop under its own weight. The rod is tried out at several different lengths in this multiple-exposure photograph so that, in effect, several rods of the same diameter but different lengths are compared.

Starting from the shortest length, it is apparent that, the longer the rod is made, the farther it reaches out to the left, away from the supporting structure. This trend continues only up to a point. It appears that there is one intermediate length that maximizes the lateral distance that a branch reaches out from the tree trunk and, therefore, out of the shade of higher branches. This critical length for an elastic branch plays the same role as the critical height for an elastic column, in the sense that it represents a practical limit. Just as the trunk height stays safely below the buckling height, we expect the branch length to stay safely below the length of greatest lateral extent. Using exactly the same dimensional-analysis argument employed earlier for buckling, we can argue that the ratio of the square of the diameter to the cube of the length should be the same in a group of springy beams whose lengths are a given fraction of the critical length.

Except for some weeping forms (various willows, for example, or weeping beech), the branches of real trees curve upwards toward the tip, not downwards. Both light-sensing and gravity-sensing mechanisms cause a differential growth across the branch, and the fact that the bottom part grows more than the top gives the upward curvature. It can be shown that the arguments presented so far do not change when we assume an upward-curving shape for the unloaded (weightless) branch, nor do they change when the angle the branch makes with the trunk is assumed to be other than 90 degrees.

Testing this part of the elastic-similarity theory applied to trees required a set of very tedious measurements. Several whole trees (including a white oak with more than 3,000 segments linking the various branch points) were measured in a methodical way, and the data were entered into a computer. The computer then traced out every possible pathway from the end of each one of the more than 1,100 twigs to the main trunk, keeping track of the relation between the local diameter and the length to the tip. The results confirmed that local diameter was proportional to the $3/2$ power of length to the tip, as predicted by elastic-similarity theory, provided that a small correction was added to the measured length to the tip to account for the fact that the twigs of real trees end at some minimum diameter rather than tapering off to the vanishing point.

**Shaking.** There is an alternative test of the theory, one that is much easier to do. Applying the theory of vibrations to the problem of a tapering cantilever beam, one may derive a simple relation between the frequency of vibration for any normal mode and the length to the tip of the tapering beam. If the tapering rule is

$$\text{local diameter} = \text{constant} \times (\text{length to the tip})^{b}, \qquad (5.1)$$

then the natural frequency is given by:

$$\text{natural frequency} = C \times (\text{length to the tip})^{b-2} \qquad (5.2)$$

in which $C$ is another constant dependent on the cross-sectional shape of the beam and the ratio of the beam material's modulus of elasticity to its mass density.

There are several cases we should distinguish. First, when $b = 0$, we have the case of no taper at all, and the natural frequency is inversely proportional to the square of the length. You can test this prediction by holding a ruler made out of any material—steel, wood, or plastic—firmly against a table top, leaving a short portion of it hanging over the edge. Plucking the unsupported portion will generate a buzzing sound as the ruler vibrates, and, if you have a musical ear, you can verify that halving the length increases the frequency of the buzzing by a factor of 4.

The case of $b = 1$ corresponds to a linear taper with length, and the natural frequency is inversely proportional to length. This result should be familiar by

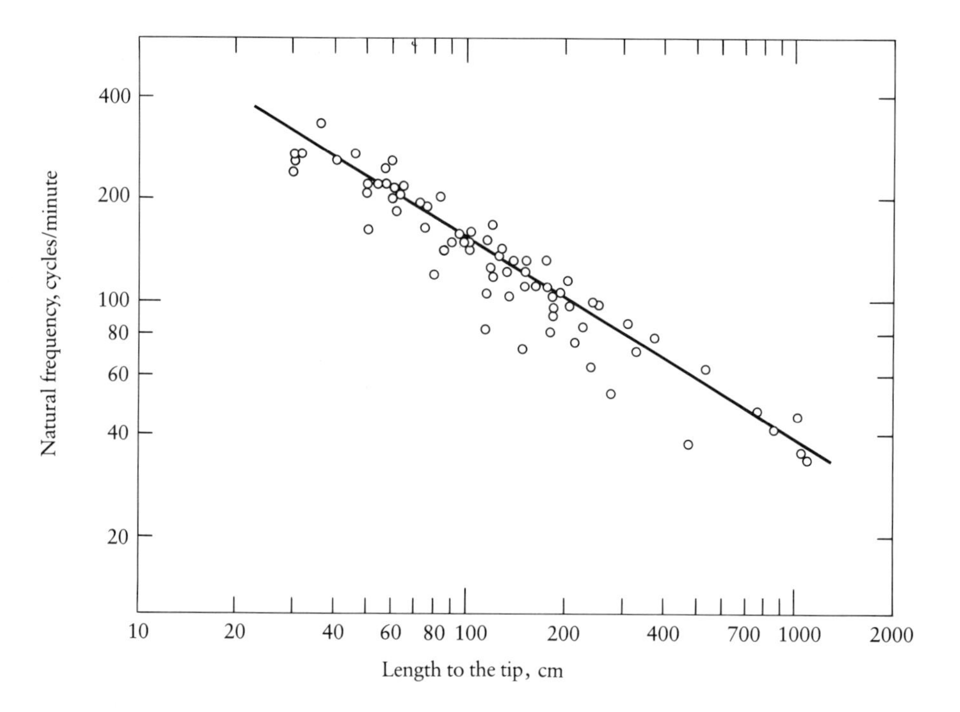


Relation between natural frequency of vibration and length for whole trees and individual branches in poplar trees (*Populus tremuloides*). The least-squares fit gives natural frequency equal to 2,248 times the length to the tip raised to the power −0.59. The exponent (−0.59) is in fairly close agreement with the exponent predicted by the theory of elastic similarity (−½).

now, since it has arisen twice before: once when we discussed the vibration of isometrically proportioned bridges in Chapter 2 and again in connection with the frequency of heart sounds in Chapter 4.

Finally, when the exponent $b$ is $3/2$, we have elastic similarity. Here, the natural frequency is predicted to be inversely proportional to the square root of the length.

A tree without its leaves makes a fairly satisfactory tuning fork, as shown on page 147. Because of the low natural frequency and the substantial damping due to air resistance and internal friction, the best way to get a tree into resonant vibration is to push on it rhythmically, the way you would push a child in a swing, applying a little shove each time it gets to the same point in its oscillatory motion. One may then use an ordinary watch to measure the time it takes to go through a certain number of vibrations in order to determine its natural frequency.

When measuring the natural frequency of a whole tree, one simply pushes the trunk. When measuring the natural frequency of a branch or a segment of a branch, it is necessary to clamp the branch in a strong vise attached to the ground by means of a rigid jig. The portion of the branch extending beyond the vise is now a cantilever beam, and one can push on it at a point approximately

A small willow tree is shown at rest (*left*) and in a double-exposure photograph during shaking (*right*). The natural frequency of a whole tree or a single branch is found to be a regular function of its length.

one-third of the way from the clamp to the tip in order to excite lowest-mode resonant vibrations.

Typical results from such tree-shaking experiments are depicted in the figure on page 146. The data points there represent the natural frequencies of whole trees (points toward the right in the figure) and branch segments (toward the left) in poplar trees (*Populus tremuloides*). The experiments were done in Vermont in the early spring, before the leaves appeared. Other experiments on larches, red maples, and red and white oaks, both with and without leaves, gave similar results. The natural frequency was always inversely proportional to the square root of length, as predicted by the elastic similarity model. Trees and branches with their leaves on had a (damped) natural frequency about half that appropriate for a tree with its leaves off; the difference was mostly due to the increase in air drag caused by the leaves.

**Minute "Plants."** To underscore the fact that the elastic similarity of trees comes about by the interaction of gravity with their large size, we consider briefly an example with a different result taken from the world of tiny plantlike organisms.

Using drawings made with a microscope, it is possible to measure the height and diameter of the minute fruiting bodies of cellular slime molds. These organisms are especially suited for such measurements not only because they are small (measured in millimeters) but also because they normally occur in a wide range of sizes. They look like small pins; they consist of a delicate, tapering

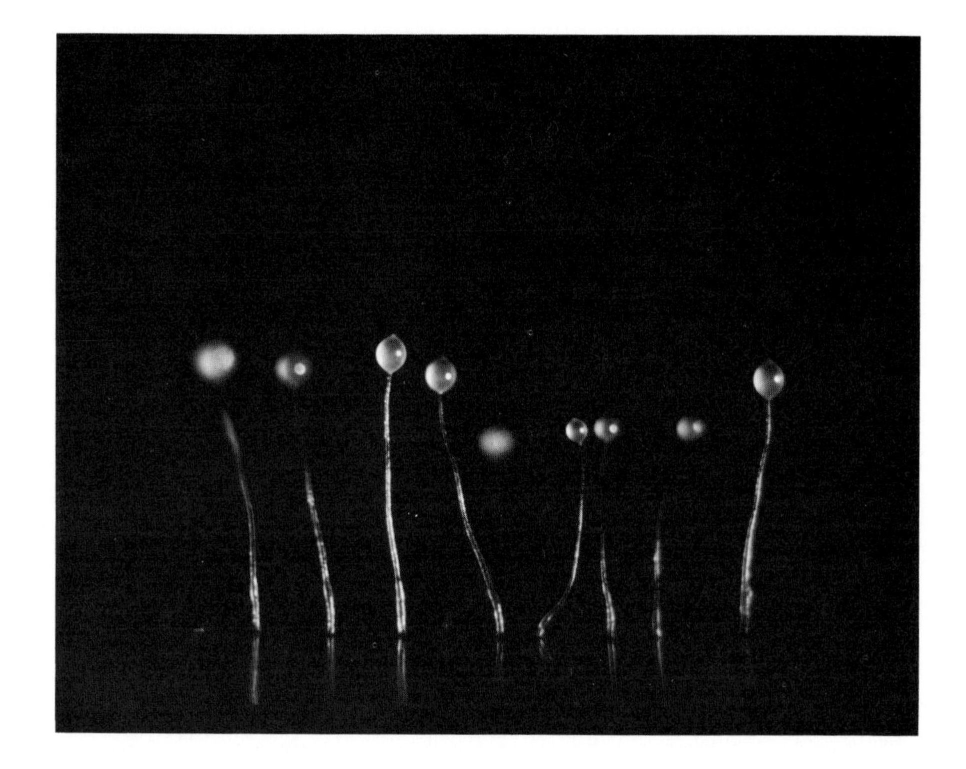

Fruiting bodies of the cellular slime mold *Dictyostelium discoideum*. The principal cell unit is an amoeba. In the early part of their life history, the amoebae feed and grow as separate cells, but later they group into a slug-shaped mass. Finally, two cell types emerge: the stalk cells and the spores. The spores are lifted on the stalk cells to form a structure that looks like a tiny pin.

stalk that supports a spherical spore mass at its tip as can be seen in the photo above. By measuring the height of the stalk and the diameter at its midpoint, information can be obtained for the log–log plot shown in the figure on page 149. It is apparent that, in these minute organisms, the slope of the line is approximately 1. Unlike trees, they are isometric.

**Where Are the Size Sensors?** Some trees have the remarkable ability to respond to such external forces as gravity and wind so that they become correspondingly thicker. Their proportions are not directly inherited. Instead, their genes produce a system that can respond to external stresses by thickening the strained regions of the branches and trunk. In plants, the mechanisms for modifying shape during growth depend on hormones, most notably the plant hormone auxin (indoleacetic acid), which triggers growth in the cambium, the layer of living cells present in many plants beneath the bark (see figures on page 150). The cambium contains components known as ray initials, which continue radially into the sapwood as wood rays. These rays are in the correct

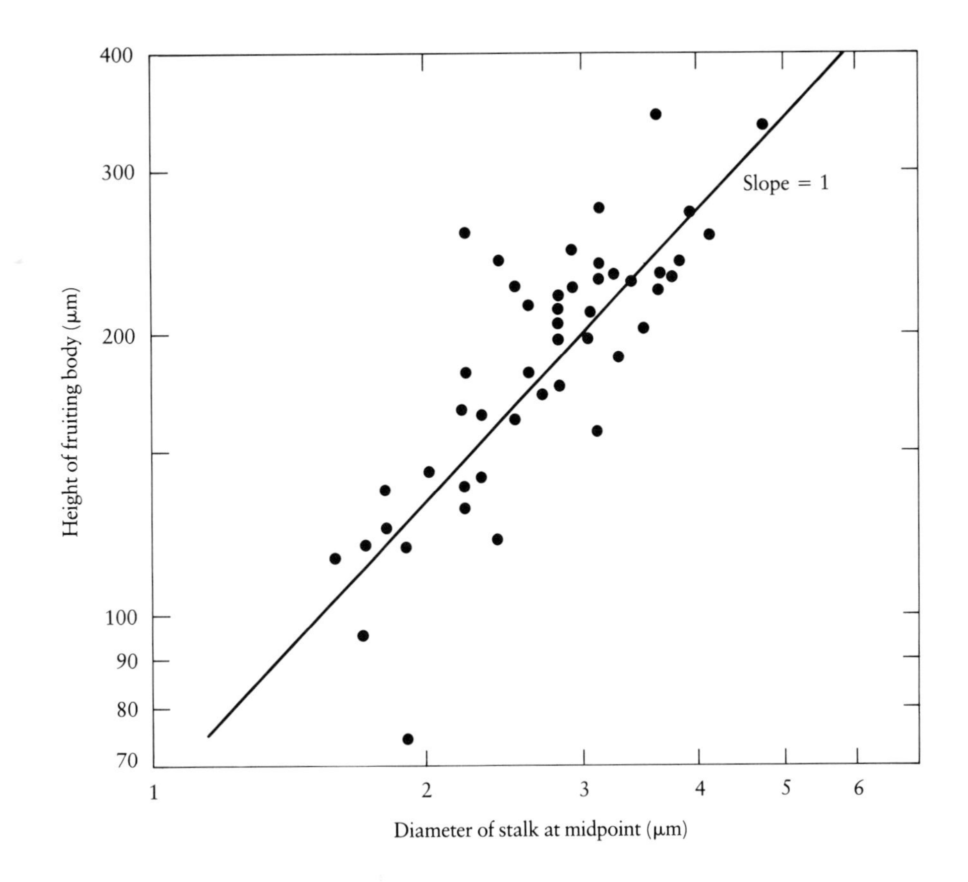


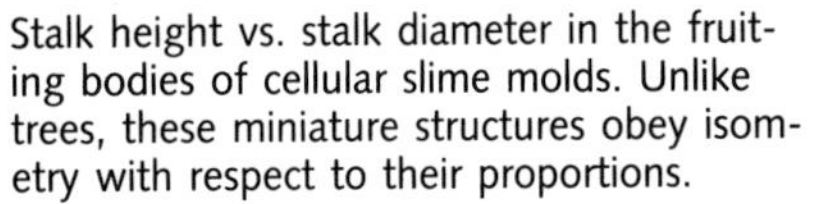
Stalk height vs. stalk diameter in the fruiting bodies of cellular slime molds. Unlike trees, these miniature structures obey isometry with respect to their proportions.

position to sense the local bending curvature caused by gravitational forces or the forces of the wind, because that bending curvature produces a squeezing of the wood on one side of the branch or trunk and a stretching on the other side. Perhaps the wood rays are sensors that tell the cambium where to speed up secondary growth and therefore thicken the stem. We are aware of no evidence for or against this particular hypothesis, but silviculturists have long known that trees raised in a greenhouse will grow more rapidly in girth if given as little as five or ten modest mechanical flexures per day. Furthermore, trees grown outdoors must not be supported by guy wires for too long or they will grow so tall and slender that they will be unable to stand up by themselves when the guy wires are removed. Whatever the sensors, whatever the hormonal intermediates, trees appear to set their rate of diameter growth partly on the basis of how much they bend and flex. If some signal says they are bending too much, they thicken themselves up to stop it. Perhaps the lack of such a mechanism is behind the fact that millimeter-size organisms, such as cellular slime molds, grow without changing shape. Perhaps the physical mechanisms of osmosis, surface tension, and even molecular adhesion are comparable in importance to gravity for the physiology of such small structures. The subject needs much more investigation before we can hope to know.

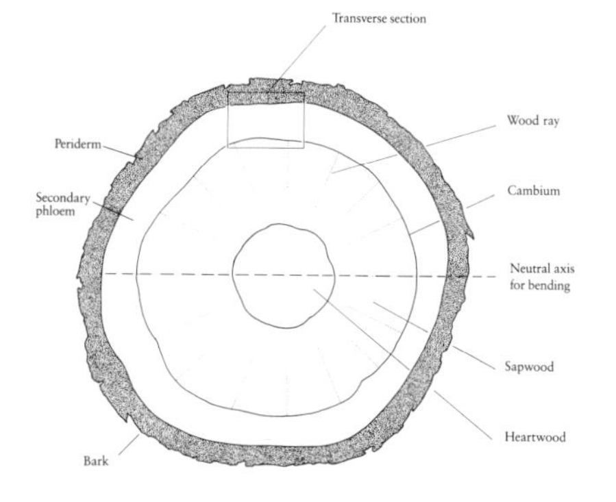


Suggested mechanism by which trees may detect bending deformations in their trunks and branches. If bending occurs, the wood below the neutral axis for bending is compressed and the wood above is subjected to tension (or vice versa). These gradients in stress may promote secondary growth in the cambium. Stress gradients along the wood rays could be involved in triggering the production of auxin or other plant hormones in the cambium. The micrograph shows a transverse section of a linden branch.

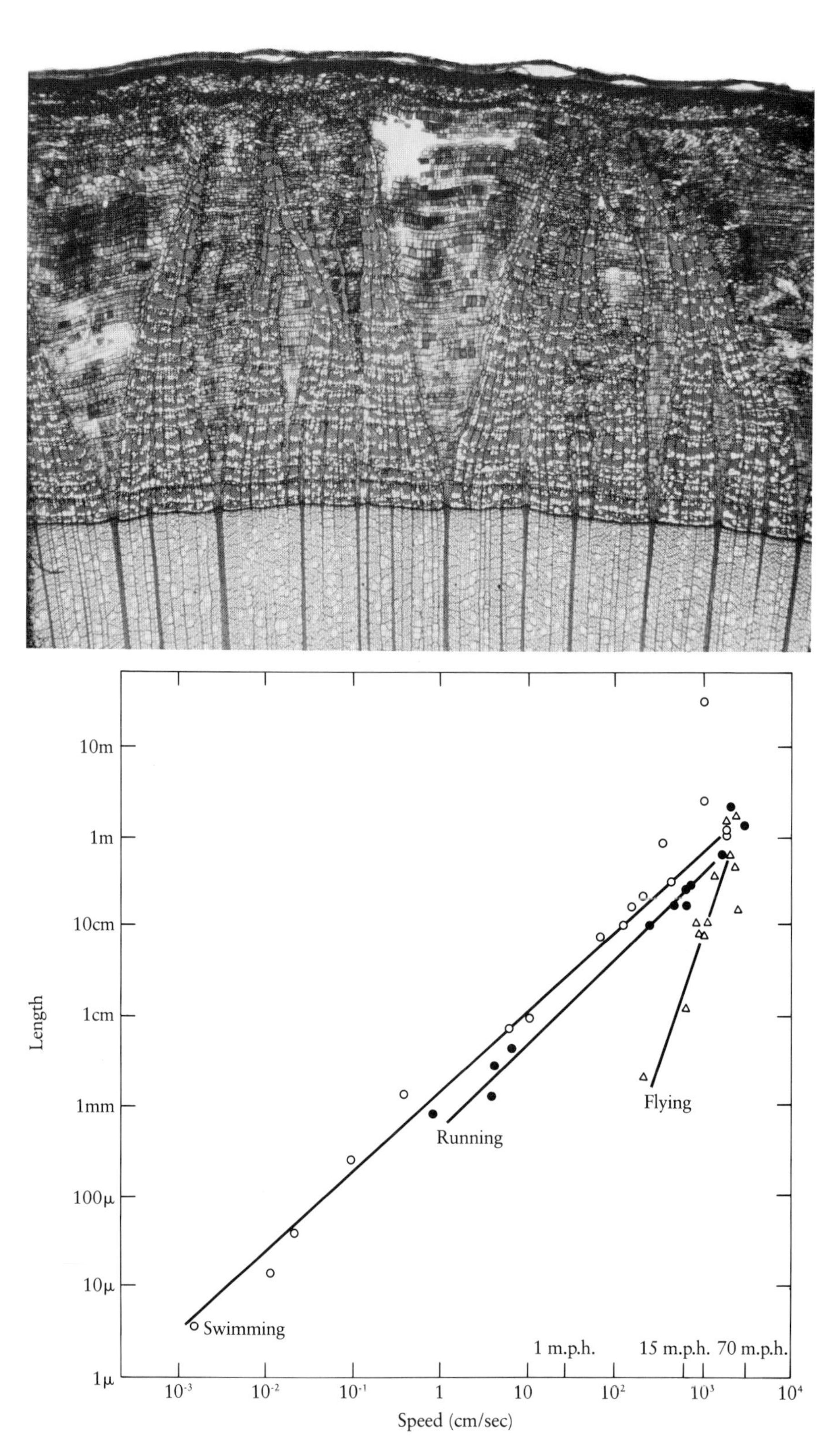


The maximum swimming, running, and flying speeds of organisms of different sizes. The length of each animal is plotted against its maximum velocity on log–log scales. The data have been selected to provide a diversity of types of organisms and to show the fastest examples of each type. It is possible to identify the animals by consulting the table on pages 152–153, which gives the data and the sources for this graph.

## Speed and Size

Just as the selective advantage in large trees is presumed to be the ability to catch more sun, so size in animals has obvious adaptive features. It is not just that large ones fight more effectively for mates and for food, it is also that larger ones can move more rapidly. This is true whether the movement is running, flying, or swimming, and it applies to all kinds of animals. This can be seen in the figure on the opposite page (and also in the tables on pages 152 and 153), in which the maximum observed speed for different kinds of organisms has been plotted against their length on a log–log grid. At this point, we are not making any distinctions about whether the maximum speed is sustainable for long periods or is only briefly attainable in a sprint; these distinctions will come later. Here, we merely observe that, if speed is a selective advantage (as we presume it is in capturing prey or avoiding predators), then this could be a reason for the general evolutionary trend of increasing size in animals.

Most of the remainder of this chapter will be concerned with understanding the physical and biological reasons behind this very general fact of life, that speed increases with size. But, before we leave this figure, it is interesting to pause for a moment and notice that the fastest runners (mammals the size of a cheetah), the fastest fliers (birds the size of a swan or a pelican), and the fastest swimmers (mammals the size of a porpoise) are all on the order of one meter long. There are several distinctive modes for each of the types of locomotion. Terrestrial locomotion includes walking, trotting, and galloping (for quadrupeds); flying includes hovering, flapping flight, and soaring; and swimming includes several modes of progression used by microscopic animals (flagellar beating, ciliary rowing) and both sustained swimming and burst swimming of the sort employed by fishes and marine mammals. The point is that the very largest animals in each group tend not to be the very fastest. It is almost as if these largest animals had the opportunity to be the fastest as well as the biggest but didn't care to try. We shall come back to this point in the discussions to follow.

Before going on to examine running, flying, and swimming in detail, it is useful to have a broad look at the relative economy of each. The costs of freight transportation by truck, railroad, airplane, and ship are often given in dollars per ton per mile. Measurements of the rate of oxygen consumption (and, therefore, of the rate of energy utilization) may be expressed in a similar way by formulating a cost of transport in terms of energy (given in watt-hours per kilogram per kilometer of distance traveled). This cost of transport is a decreasing allometric function of body size in terrestrial animals, birds, and

**Swimming speed and length in animals.**

| Species | Length | Swimming Speed (cm/sec) | Reference |
|---|---|---|---|
| 1. *Bacillus subtilus* | 2.5 μm | $1.5\times10^{-3}$ | *Tabulae Biologicae* |
| 2. *Spirillum volutans* | 13.0 μm | $1.1\times10^{-2}$ | idem |
| 3. *Euglena* sp. | 38.0 μm | $2.3\times10^{-2}$ | idem |
| 4. *Paramecium* sp. | 220.0 μm | $1.0\times10^{-1}$ | idem |
| 5. *Unionicola ypsilophorus* (water mite) | 1.3 mm | $4.0\times10^{-1}$ | Welsh (1932, *J. Gen. Physiol.* 16:349) |
| 6. *Pleuronectes platessa* (plaice, larval) | 7.6 mm | 6.4 | Boyar (1961, *Trans. Amer. Fish. Soc.* 90:21) |
| 7. *P. platessa* | 9.5 mm | 11.5 | idem |
| 8. *Carassius auratus* (goldfish) | 7.0 mm | 75 | Bainbridge (1961, *Symp. Zool. Soc. London* 5:13) |
| 9. *Leuciscus leuciscus* (European dace) | 10.0 cm | 130 | idem |
| 10. *L. leuciscus* | 15.0 cm | 175 | idem |
| 11. *L. leuciscus* | 20.0 cm | 220 | idem |
| 12. *Pomolobus pseudo harengus (river herring)* | 30.0 cm | 440 | Dow (1962, *J. Conseil Internat. Explor. Mer* 27:77) |
| 13. *Pygoscelis adeliae* (Adélie penguin) | 75.0 cm | 380 | Meinertzhagen (1955, *Ibis* 97:81) |
| 14. *Thunnus albacares* (yellowfin tuna) | 98.0 cm | 2,080 | Walters and Firestone (1964, *Nature* 202:208) |
| 15. *Acanthocybium solanderi* (wahoo) | 1.1 m | 2,150 | idem |
| 16. *Delphinus delphis* (common dolphin) | 2.2 m | 1,030 | Hill (1950, *Sci. Prog.* 38:209) |
| 17. *Sibbaldus musculus* (blue whale) | 26.0 m | 1,030 | idem |

**Running speed and length in animals.**

| Species | Length* | Running speed (cm/sec) | Reference |
|---|---|---|---|
| 1. *Bryobia* sp. (clover mite) | 0.8 mm | $8.5\times10^{-1}$ | Pillai, Nelson, and Winston (personal communication) |
| 2. Unidentified anyestid mite | 1.3 mm | 4.3 | idem |
| 3. *Iridomyrmex humilis* (Argentine ant) | 2.4 mm | 4.4 | Shapley (1920, *Proc. Nat. Acad. Sci.* 6:204; 1924; 10:436) *Proc. Nat. Acad. Sci.* 10:436 |
| 4. *Liometopum apiculatum* (ant) | 4.2 mm | 6.5 | idem |
| 5. *Peromyscus maniculatus* (deer mouse) | 9.0 cm | 250 | Layne and Benton (1954, *J. Mammal.* 35:103) |
| 6. *Callisaurus draconoides* (zebra-tailed lizard) | 15.0 cm | 720 | Belkin (1961, *Copeia* 1961:223) |
| 7. *Tamias striatus* (Eastern chipmunk) | 16.0 cm | 480 | Layne and Benton (1954, *J. Mammal.* 35:103) |
| 8. *Dipsosaurus dorsalis* (desert iguana) | 24.0 cm | 730 | Belkin (1961, *Copeia* 1961:223) |
| 9. *Sciurus carolinensis* (Eastern gray squirrel) | 25.0 cm | 760 | Layne and Benton (1954, *J. Mammal.* 35:103) |
| 10. *Vulpes fulva* (red fox) | 60.0 cm | 2,000 | Hill (1950, *Sci. Prog.* 38:209) |
| 11. *Acinonyx jubatus* (cheetah) | 1.2 m | 2,900 | idem |
| 12. *Struthio camelus* (ostrich) | 2.1 m | 2,300 | idem |

* In this table, the lengths of the vertebrates have been estimated in different ways. For mammals, lengths are from the base of the tail to the tip of the snout. For lizards, one-half of the tail is also included. For the ostrich, length is the erect height.)

**Flying speed and length in animals.**

| Species | Length | Flying speed (cm/sec) | Reference |
|---|---|---|---|
| 1. *Drosophila melanogaster* (fruit fly) | 2.0 mm | 190 | Hocking (1953, *Trans. Roy. Entomol. Soc.* 104:223) |
| 2. *Tabanus affinis* (horse fly) | 1.3 cm | 660 | idem |
| 3. *Archilochus colubris* (ruby-throated hummingbird) | 8.1 cm | 1,120 | Pearson (1961, *Condor* 63:506) |
| 4. *Anax* sp. (dragonfly) | 8.5 cm | 1,000 | Wigglesworth (1939, *Principles of Insect Physiology*) |
| 5. *Eptesicus fuscus* (big brown bat) | 11.0 cm | 690 | Hazard and Davis (1964, *J. Mammal.* 45:236) |
| 6. *Phylloscopus trochilus* (willow warbler) | 11.0 cm | 1,200 | Meinertzhagen (1955, *Ibis* 97:81) |
| 7. *Apus apus* (common swift) | 17.0 cm | 2,550 | idem |
| 8. *Cypselurus cyanopterus* (flying fish) | 34.0 cm | 1,560 | idem; Schultz and Stern (1948, *The Ways of Fishes*) |
| 9. *Numenius phaeopus* (whimbrel) | 41.0 cm | 2,320 | Meinertzhagen (1955, *Ibis* 97:81) |
| 10. *Anas acuta* (common pintail) | 56.0 cm | 2,280 | idem |
| 11. *Olor columbianus bewicki* (Bewick's swan) | 1.2 m | 1,880 | idem |
| 12. *Pelecanus onocrotalus* (Old World white pelican) | 1.6 m | 2,280 | idem |

Cost of transport vs. body mass. The cost of transport is a number based on the rate of oxygen utilization, the speed, and the size of the animal. It is the energy required to move one kilogram a distance of one kilometer.

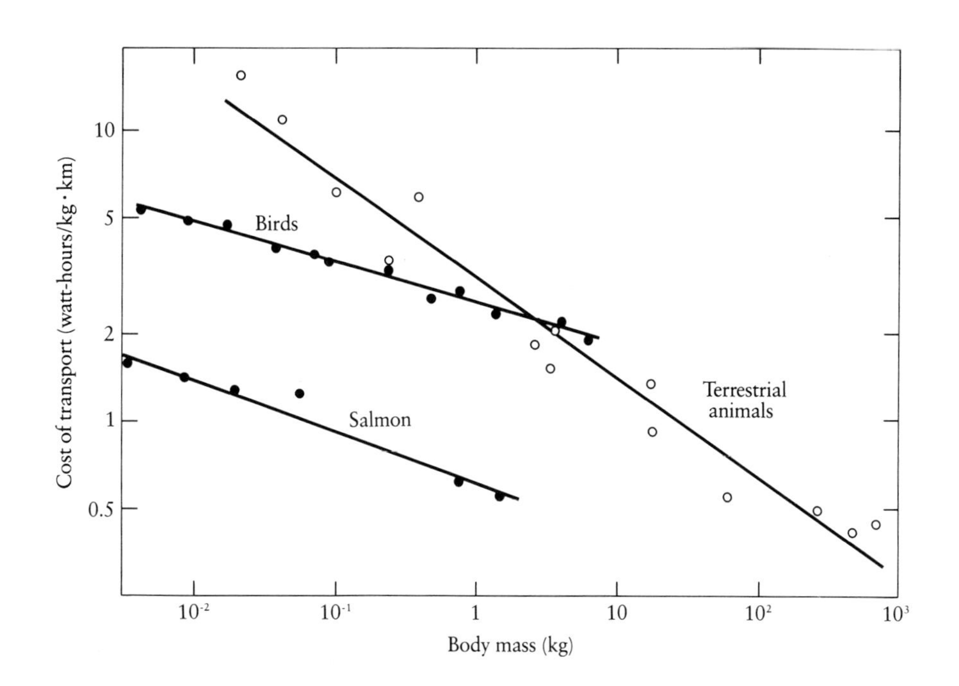

fish (see figure on page 153). Each of the points in the figure shows the minimum cost of transport at the optimum speed for a particular animal. For body masses less than about 2 kilograms, swimming is the most economical way to move, followed by flying, followed by walking or running. Perhaps it is no coincidence that birds cease to fly above a body weight in the neighborhood of 10 kilograms. The lines for birds and terrestrial animals cross at about 2 kilograms, so that birds larger than this size have an economic incentive to hang up their aviator's goggles and become flightless.

## Locomotion on Land

A main theme of this chapter is that gravity plays a central role in the lives of large animals. Even when merely standing still, a large animal is at a disadvantage with respect to a small one.

The relative disadvantages that a large animal suffers while standing still have mainly to do with circulation. If a soldier stands totally immobile at rigid attention, he may faint, because the blood may fail to reach his brain in adequate quantities. (But nature takes care of us: The circulatory system of someone who has fainted and slumped to the ground suddenly has no difficulty getting enough blood to the brain, and soon consciousness will return.) The effect of gravity on the circulation can be demonstrated dramatically by standing still in a barrel of water that comes up above one's legs. The water level will slowly rise in the barrel, because one's legs, without the benefit of muscular exercise, slowly become engorged with blood. The reason that people do not often faint in everyday life because of an inadequate supply of blood to the brain is that they are constantly moving about. Muscle contractions within the legs raise the pressure around the veins, helping to push the blood back to the heart. As William Harvey discovered in the seventeenth century, the veins are lined at intervals with thin leaflets that act as one-way valves, so that the squeezing action of the leg muscles always acts to move the blood toward the heart. These venous valves serve to make the muscles of the arms and legs function as extra blood pumps, helping the heart to do its work and lessening the effects of gravity.

**Walking and Trotting: Froude Number.** In 1887 Eadweard Muybridge published *Animal Locomotion*, a remarkable book of photographs of animals in motion that is still useful today. The sequences of photographs reproduced on pages 156 and 157 are taken from Muybridge's book. R. McNeill Alexander analyzed Muybridge's photographs to obtain information about the speed and stride length (the distance between footprints of the same foot) in horses, large cats, other mammals, and an ostrich. He also measured the hip height when the animals were standing at rest. Using dimensional analysis, he plotted the stride length divided by hip height against a dimensionless speed: speed / (gravity × hip height)$^{1/2}$ (figure below). These two dimensionless groups are reasonably successful in bringing the stride-length data from a wide range of animals onto a single line.

The dimensionless speed Alexander used is one of the several forms taken by the Froude number. When we encountered the Froude number in Chapter 3, it described the ratio of inertial forces to gravitational forces per unit volume; at the same Froude number, large and small ships of the same shape make the same pattern of waves. Here, it still describes a ratio between inertial and gravitational forces, but these are now forces applied to an animal's limbs and trunk. One need only accept the assumption that both gravity and hip height have something to do with the relation between stride length and speed for these dimensionless variables to be plausible.

In addition to presenting this dimensionless plot, Alexander noted that transitions in gait tend to occur at particular values of the Froude number. Humans, he said, break into a run at about 2.5 meters per second, a velocity that corresponds to a Froude number of 0.8. Horses go from walking to trotting at a Froude number between 0.7 and 0.9, and cats change from walking to trotting at about 0.9 meters per second, which corresponds to a Froude number of 0.6.

Although Alexander's identification of the Froude number as a critical parameter describing gait transitions does not depend on any physical model, the inverted pendulum shown in the inset of the figure below helps to explain why it seems to work. The inverted pendulum swings in an arc whose radius is the

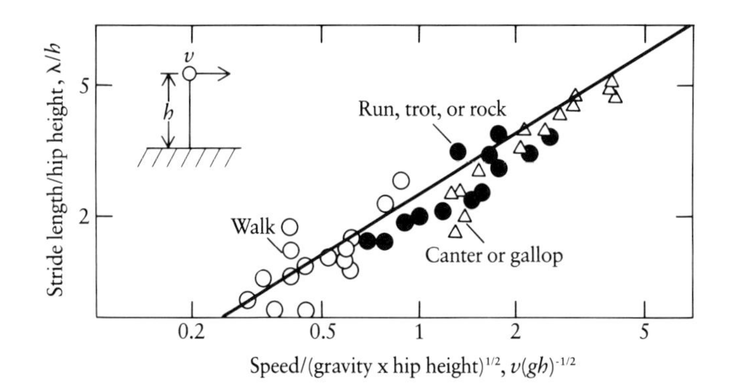


Dimensionless stride length, $\lambda/h$, vs. dimensionless speed, $v/(gh)^{1/2}$. Here, $h$ is the height of the animal's hip from the ground in normal standing and $\lambda$ is the stride length. Data are from Muybridge's photographs, some of which are shown on the next two pages. *Inset*: An inverted pendulum of length $h$ and speed $v$ tends to fly off the ground when $v^2/gh > 1$.

Photographs of animals in motion obtained by Eadweard Muybridge using a series of 24 cameras whose shutters were triggered in sequence by an electric motor closing electrical contacts.

hip height. When the centripetal acceleration (given by the square of the speed divided by the hip height) is greater than the acceleration due to gravity, the inverted pendulum will fly off the ground unless held down by its pivot. This happens when the Froude number is greater than 1.0. Because running is a gait that has an airborne phase, whereas animals walking always have at least one foot on the ground, a certain critical Froude number appears to be a realistic parameter for separating walking and running.

**A. V. Hill's Isometric Model.** The distinguished British muscle physiologist A. V. Hill gave a popular lecture in the Friday Evening Discourse Series at the Royal Institution in the fall of 1949. That lecture later became an important paper (Hill, 1950).

His subject was the performance of animals running, jumping, and swimming, and how this performance is partly determined by the animal's size. He assumed that the animals to be compared were isometric models of (geometrically similar to) each other, and he further asserted, with supporting evidence, that the peak stress a muscle can develop is independent of the size of the animal it comes from. (We have seen, in Chapter 4, how this fact is important in explaining why blood pressure does not vary with size and why the peak metabolic rate of muscles is proportional to their cross-sectional area.)

Hill reasoned that, when an animal runs on the level at constant speed, most of its energy goes into moving its limbs back and forth with respect to the body. In physics, work is measured in units of energy. For example, it takes 2,400 foot-pounds of energy to lift a 160-pound pole vaulter over a bar set at 15 feet. When the pole vaulter comes to earth again, the gravitational energy he had at the top of his arc has been converted to the energy of motion—kinetic energy—and that energy is proportional to his mass times his velocity squared. In the same way, the kinetic energy of a limb moving with respect to

Pole vaulter going over a bar. The vaulter stores part of the kinetic energy of his run in the bent pole, then recovers this energy before releasing the pole at the top of his arc.

the body (when the body is moving at constant velocity) is proportional to the mass of the limb times the square of the speed with respect to the body:

$$\text{kinetic energy of limb} \propto \text{mass of limb} \times (\text{speed of limb with respect to body})^2. \qquad (5.3)$$

The work required to give the limb this kinetic energy comes from the muscles. Consider the work done by a muscle in one stroke. The work is given by the muscle force times the change in muscle length. Suppose, as Hill did, that muscle stress is independent of body size, so that muscle force is proportional to the cross-sectional area of the muscle. Suppose also that the change in muscle length during a stroke is proportional to the limb length. We are left with the conclusion that

$$\text{Work per stroke} \propto \text{force} \times \text{change in length} \propto \text{muscle area} \times \text{limb length}$$

and, finally, because area times length is proportional to the mass of the limb,

$$\text{work per stroke} \propto \text{mass of limb}. \qquad (5.4)$$

Comparing statements 5.3 and 5.4, we conclude that the work per stroke can be proportional to the kinetic energy of the limb (as required by the laws of mechanics) only if the speed of the limb with respect to the body is independent of body size. When a limb is on the ground, the speed of the limb with respect to the body is equal and opposite to the speed of the body over the ground. Hence, Hill's arguments predict that all geometrically similar animals run at the same speed. Furthermore, because that speed is proportional to the muscle length divided by the stroke time, we have an additional conclusion: the time for completion of a muscle stroke should be proportional to the length of an animal's leg (or any other characteristic length).

Hill compared the racing performances of a series of animals of the same general design:

| | |
|---|---|
| Whippet (20–21 pounds): | 34.2 mph over 200 yards |
| Greyhound (55–60 pounds): | 37.5 mph over 525 yards |
| Horse with rider (1,600 pounds): | 42.4 mph over 660 yards |

He concluded that the speeds of the larger animals were only moderately greater than those of the smaller, in general agreement with his predictions. Once again we note that Hill assumed isometry in the scaling of animal proportions. It is also worth pointing out that Hill's predictions do not agree with the general trend that running speed increases with size, as already noted in the figure on page 150. In what follows, we will consider the dynamics of running in somewhat greater detail.

**Scaling of the Trot–Gallop Transition.** It isn't easy to compare the top speeds of running animals. In the first place, all animals are capable of higher top speeds over short distances than they are over long ones. Training and motivation are also important factors. Presumably, it would be important to compare only top animal athletes, not just any assortment of four-footed individuals who happen to wander by. Also, the highest speed attainable under aerobic metabolism is not the same as the highest sprinting speed. Which should be called the animal's "top speed"?

Much of this uncertainty is removed if one compares animal performance near the speeds at which they make a transition in gait. In the figure on page 160, stride frequency is plotted against speed for a mouse, a rat, a dog, and a horse running on a treadmill. As each animal trots faster, its stride frequency increases. After it makes a transition to galloping, its stride frequency stays relatively constant, increasing by less than 10 percent throughout the full range of galloping speeds. The speeds at which animals make the transition from trotting to galloping seem to be physiologically equivalent, because these speeds are relatively independent of training, experience, and motivation. The lower part of the figure shows that the speed at the trot–gallop transition

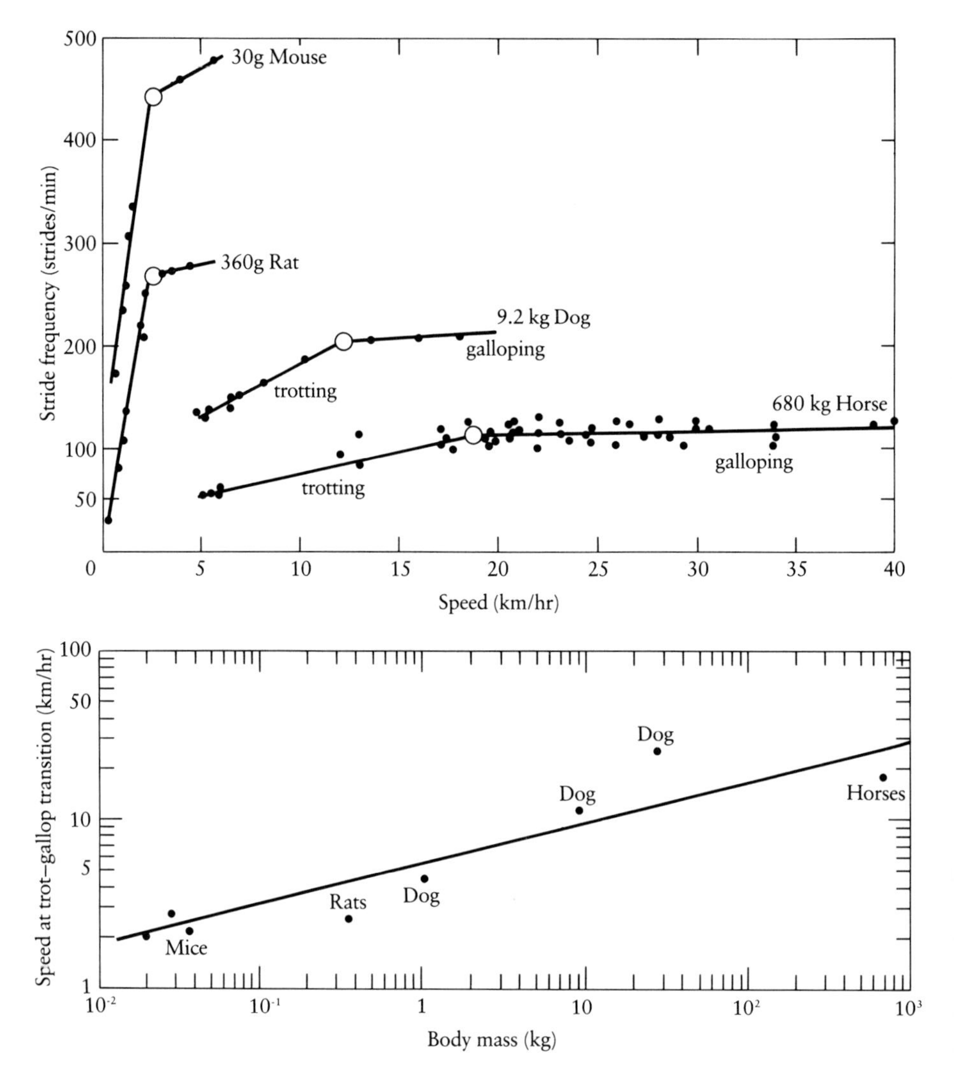


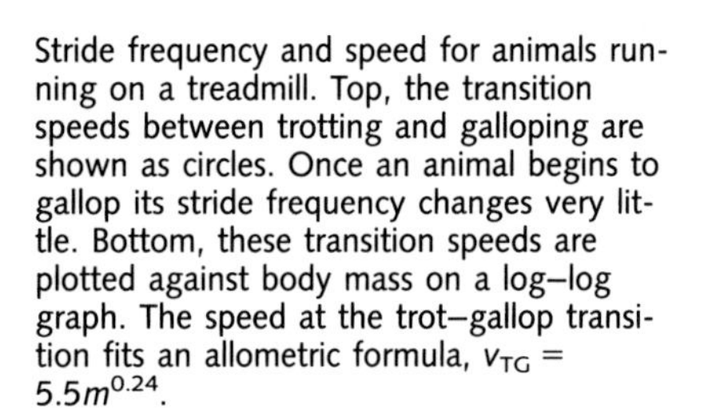
Stride frequency and speed for animals running on a treadmill. Top, the transition speeds between trotting and galloping are shown as circles. Once an animal begins to gallop its stride frequency changes very little. Bottom, these transition speeds are plotted against body mass on a log–log graph. The speed at the trot–gallop transition fits an allometric formula, $v_{TG} = 5.5m^{0.24}$.

increases with the 0.24 power of body mass—that is, roughly in proportion with the length scale of elastic similarity ($m^{1/4}$).

According to Hill's arguments, the stride frequency at top speed should be proportional to the reciprocal of the length scale for isometric scaling, $1/\ell = m^{-1/3}$. The stride frequency at the trot–gallop transition, which we have just said is not very different from the stride frequency at top speed, is shown plotted against body mass on an allometric plot in the figure at the top of page 161. The data fit very tightly on an allometric line, and the slope is not $-1/3$ ($=-0.33$) but $-0.14$.

**Resonant Rebound.** Before we attempt to understand these departures from Hill's predictions, we should mention some fascinating aspects of the mechanics of kangaroos. T. L. Dawson and C. R. Taylor have reported that kangaroos hop on a treadmill at a nearly constant frequency, just as mammals that run

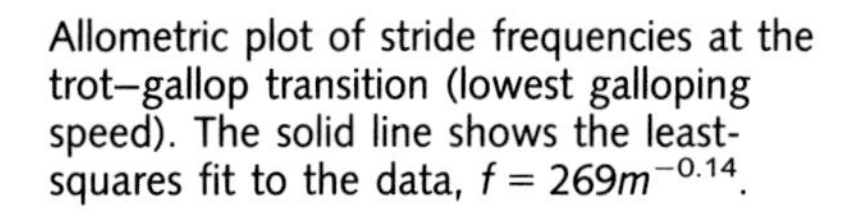
Allometric plot of stride frequencies at the trot–gallop transition (lowest galloping speed). The solid line shows the least-squares fit to the data, $f = 269m^{-0.14}$.

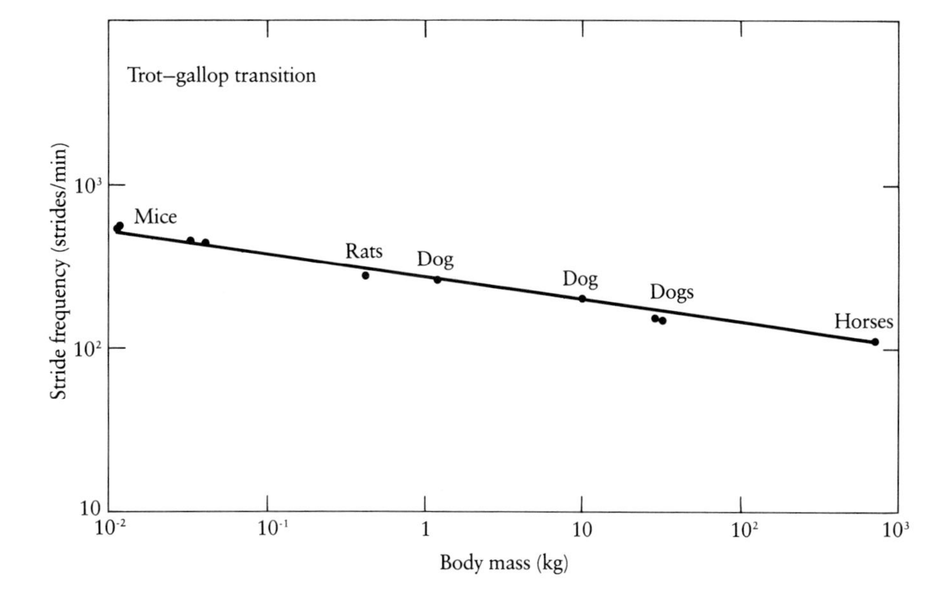


using all four legs gallop at a nearly constant frequency. In hopping kangaroos, the stride frequency was observed to increase less than 5 percent as the speed doubled. Furthermore, the rate of oxygen consumption actually showed a shallow minimum midway through the range of hopping speeds before increasing at the highest speeds.

The entire situation is reminiscent of a mass–spring system operating at its natural frequency. In this analogy, speed is increased by increasing the amplitude of vibration of the mass (this amplitude corresponds to the stride length). Mechanical energy is transiently stored in the spring, and, consequently, metabolic energy is saved because the muscles are not required to undertake the whole task of accelerating and braking the limbs and the body on each hop. The kangaroo has a large, thick Achilles tendon, which could serve as the spring in the mass–spring analogy.

A kangaroo has a large, thick Achilles tendon, which can store energy like a spring.

To explore these ideas further, Taylor and his colleagues invited an expert pogo-stick rider to hop on their treadmill (see figure on page 162). They reasoned that, by changing the stiffness of the pogo-stick spring, they should be able to change the natural frequency of the mass–spring system and therefore change the hopping frequency.

It didn't work out that way. Pogo-stick hopping, like kangaroo hopping, did show a fairly constant frequency that was independent of speed, but changing the stiffness of the spring didn't change the frequency of hopping noticeably. Finally, the investigators asked the subject to hop on the treadmill without the

Pogo-stick rider on a treadmill.

pogo stick. The frequency was still approximately the same. Taylor (1977) writes:

> Then it dawned on us that it was the bending of the whole body, both in the kangaroo and in the man on the pogo stick, which was determining the frequency. The resonant spring, therefore, was the whole body.

In later experiments, Taylor and his students showed that one of the main muscles of the trunk in the dog, the iliocostalis, which is inactive at a walk and a trot, is recruited into activity during galloping. This is evidence in favor of the idea that, as quadrupedal animals make a transition from a trot to a gallop, they use not only the muscles of the limbs but also the muscles of the abdomen and back as part of the spring. One of the advantages of switching to galloping at high speeds is that the labor of running can be spread over more of the muscles of the body.

Running may therefore be seen as a series of elastic rebounds from the ground, much as a rubber ball bounces. It is a straightforward matter to show that the period of the rebound is proportional to the square of the length divided by the diameter in the model used in the figure on page 123 to describe the contact phase of an animal's step cycle (McMahon, 1983). In isometric scaling, both length and diameter are proportional to $m^{1/3}$, so the ratio of the square of the length to the diameter is proportional to $m^{2/3}/m^{1/3} = m^{1/3}$. Thus, the period for the resonant rebound is proportional to $m^{1/3}$ in a series of isometric animals, but it is proportional to $m^{2/4}/m^{3/8} = m^{1/8}$ in animals built to maintain elastic similarity. Associating the stride frequency at the trot–gallop transition with the inverse of this period, we obtain a predicted value for stride frequency that is proportional to $m^{-1/8} = m^{-0.125}$, which is quite close to the observed $m^{-0.14}$ shown earlier in the figure on page 161.

**Flexion Angles of Joints.** Toward the end of Chapter 4, we noted that large animals tend to stand with their limbs relatively straighter than those of small animals. This trend toward greater straightness holds in dynamics as well, as shown in the running motions of animals in a size series (top of page 163). The vertebral column of the 15-kilogram Thomson's gazelle flexes a great deal more than that of the 250-kilogram eland when both are galloping at a slow speed. Furthermore, the excursion angle of the hind limb, as defined in the figure, is greater in the smaller animal.

These effects may be understood by referring to the schematic drawing of a typical joint—in this case, a knee—shown at the top of the lower figure on

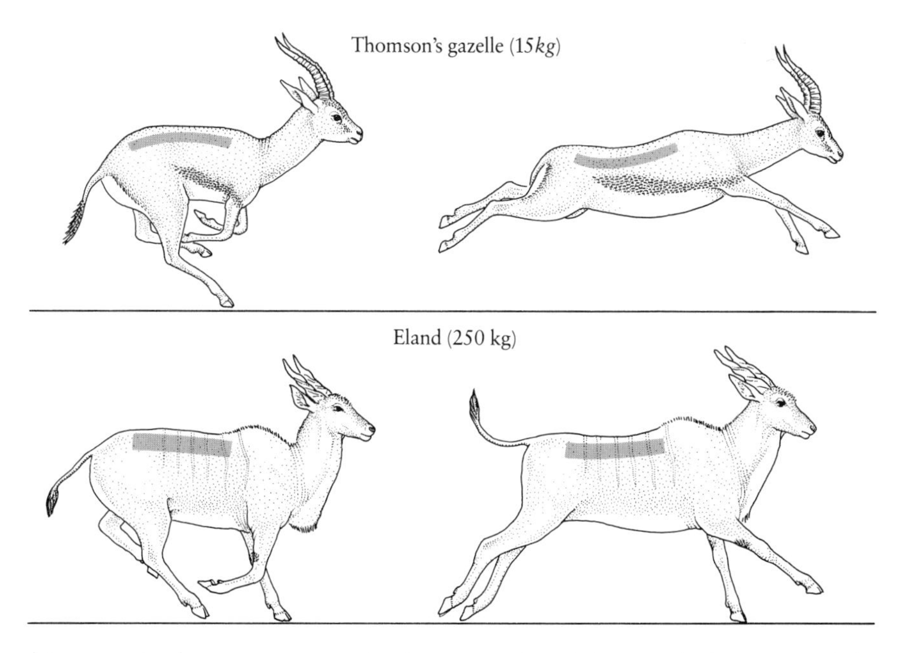


Comparison of Thomson's gazelle, *Gazella thomsoni* (15 kilograms), and eland, *Taurotragus oryx*, (250 kilograms), at intermediate galloping speeds. These pictures were traced from films showing the animals running naturally in the wild. Both the bending of the vertebral column and the angular excursion of the hind limb are shown to be smaller in the larger animal.

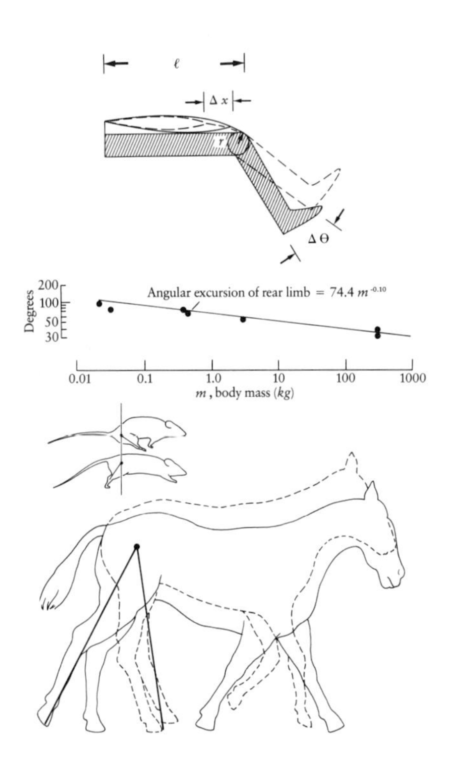


When a muscle (*top*) changes length by an amount $\Delta x$, the joint angle changes by $\Delta\theta = \Delta x/r$. The excursion angle of the hind limb is measured as shown in the mouse and horse drawings (traced from movie film) at the bottom of the figure. At the center of the figure is an allometric plot of the excursion angle of the hind limb for mouse, rat, dog, and horse at their lowest galloping speeds. The allometric equation of the line is: angle of excursion = $74.4m^{-0.10}$.

this page. As the extensor muscle changes its length by $\Delta x$, the knee extends by an angle $\Delta\Theta = \Delta x/r$, in which $r$ is the joint radius. In order to keep the individual filaments of muscle protein in the same relative positions with respect to one another in the muscles of both small and large animals, we require the muscle stroke, $\Delta x$, to be proportional to the muscle rest length, $\ell$. Therefore, $\Delta\Theta$ is proportional to $\ell/r$, which is size-independent under isometric scaling but proportional to $m^{1/4}/m^{3/8} = m^{-1/8}$ under the rules of elastic similarity.

Observations of the hind limb at the lowest galloping speed in a range of mammals from mouse to horse shows that the excursion angle is proportional to $m^{-0.10}$, which is in reasonable agreement with the $m^{-1/8} = m^{-0.125}$ predicted from elastic similarity. There is a regular change in the mechanical advantage the muscles have about the joints of the skeleton—a change that comes about as an inevitable consequence of the stability of the joints under loads determined by inertia and gravity.

**Running on Hills.** Earlier in the chapter, we speculated that horses might like to run up trees. The fact is, they have a tough enough time climbing a modest hill, and they usually slow their speed to a walk on a grade. There are consequences here for migratory behavior when animals move over snowy mountains (where their tracks can be easily followed). The tracks reveal that the larger the animal, the less likely it is to take a steep path up the grade.

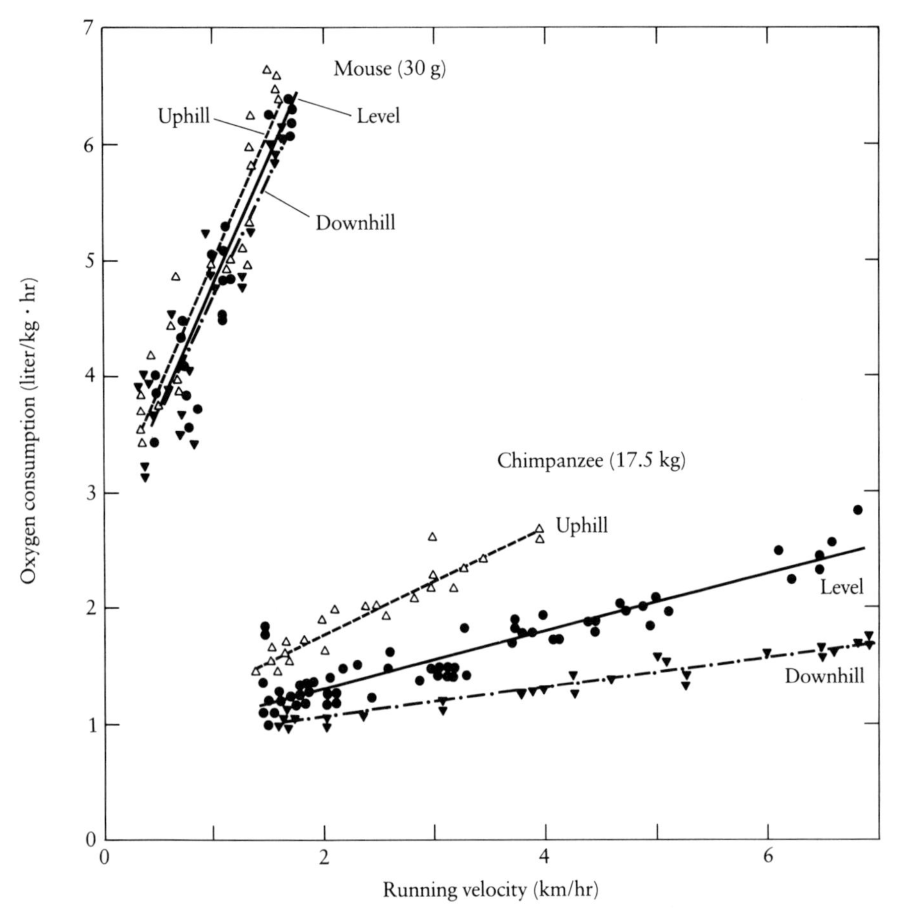


Rate of oxygen consumption plotted against running speed for a mouse and a chimpanzee running uphill, on the level, and downhill.

In a charming and important piece of research, C. R. Taylor and his collaborators trained mice and chimpanzees to run on a treadmill that could be tilted up or down to obtain the same mechanical conditions that the animals would have experienced running up or down a 15-degree incline. A fan matched the wind velocity to the tread speed. Taylor and his researchers found that the rate of oxygen consumption for the mice was only slightly higher or lower as the treadmill was pointed up or down, but the rate for chimps was much elevated for climbing and much reduced for descending, compared with that for level running, as is shown in the figure above.

They explained their results in the following way. According to Kleiber's equation (discussed in Chapter 2), each gram of tissue from a 30-gram mouse consumes oxygen at a rate about 13 times greater than the rate per gram in a 1,000-kilogram horse. Because the mechanical work required to lift 1 kilogram a distance of 1 meter vertically is the same in the mouse and the horse, then (assuming that the muscles convert metabolic energy into work with the same efficiency in the two animals) both consume the same amount of oxygen lifting 1 kilogram 1 meter vertically. It follows that the relative increase in the rate of oxygen consumption (divided by body mass) for running up the same slope at the same speed is 13 times greater in the horse.

There is an interesting sidelight in this study. As is apparent in the preceding figure, the chimps are using energy at a substantially reduced rate (compared with that for level running) when they run downhill. Thus, the total energy consumed while running uphill for a length of time and then running downhill for the same length of time is not very much greater than it would have been if the animals had run on the level throughout the whole trial. One way of putting the result quantitatively is to say that the animals, both mice and chimpanzees, were able to recover about 90 percent of the work they had done against gravity on the uphill run when they were allowed to run down for the same distance. This observation illustrates a basic energetic principle of muscle physiology, that a muscle forced to lengthen while maintaining tension consumes energy at a reduced rate (compared with that of the same muscle when it is allowed to shorten and do work).

## Flying

Fossil of a large soaring reptile, *Pterodactylus elegans*. These animals were active late in the Jurassic period (roughly 150 million years ago). They had no feathers, although true birds, which did have feathers, were evolving during the same period.

Flying is an activity generally confined to animals in the middle of the size spectrum. The largest flying animals alive today are the soaring birds. One of the champions among these, the 10-kilogram wandering albatross, has a wingspan of 3.5 meters. The extinct pterosaurs were larger. Estimates based on skeletal fragments suggest that members of the genus *Pteranodon* weighed 18 kilograms and had wingspans of 8 meters. The skeletal remains of a pterosaur found in Texas in 1971 suggest that some of these animals had wingspans greater than 11 meters.

The smallest airborne creatures are the tiny arthropods that float about in the wind like grains of pollen; their masses are usually less than 1 milligram. These aerial arthropods are not really proper fliers because they can't control where they're going. Some tiny members of the insect orders Diptera and Hymenoptera (species of flies and wasps) are almost as small; they have body masses from 1 to 600 milligrams, and all are strong fliers (although they must stay out of winds or they will be blown away).

We shall see that the effects of size give rise to very distinct limitations on performance. Large flying animals, including the albatrosses and vultures, take advantage of soaring flight wherever they can, because flapping is such an effort for them. Birds of intermediate size, such as mute swans (they are just as

Soarers: *a*, the magnificent frigatebird (*Fregata magnificens*) has a wingspan of more than 2 meters, and the weight of its skeleton can be less than the weight of its feathers; *b*, the wandering albatross (*Diomedea exulans*), can have a wingspan of as much as 3.5 meters, the greatest of any modern bird; *c* The largest flying animal of any age was the pterosaur *Quetzalcoatlus northropi,* with a wingspan estimated at 11 to 12 meters. A human figure is included for comparison.

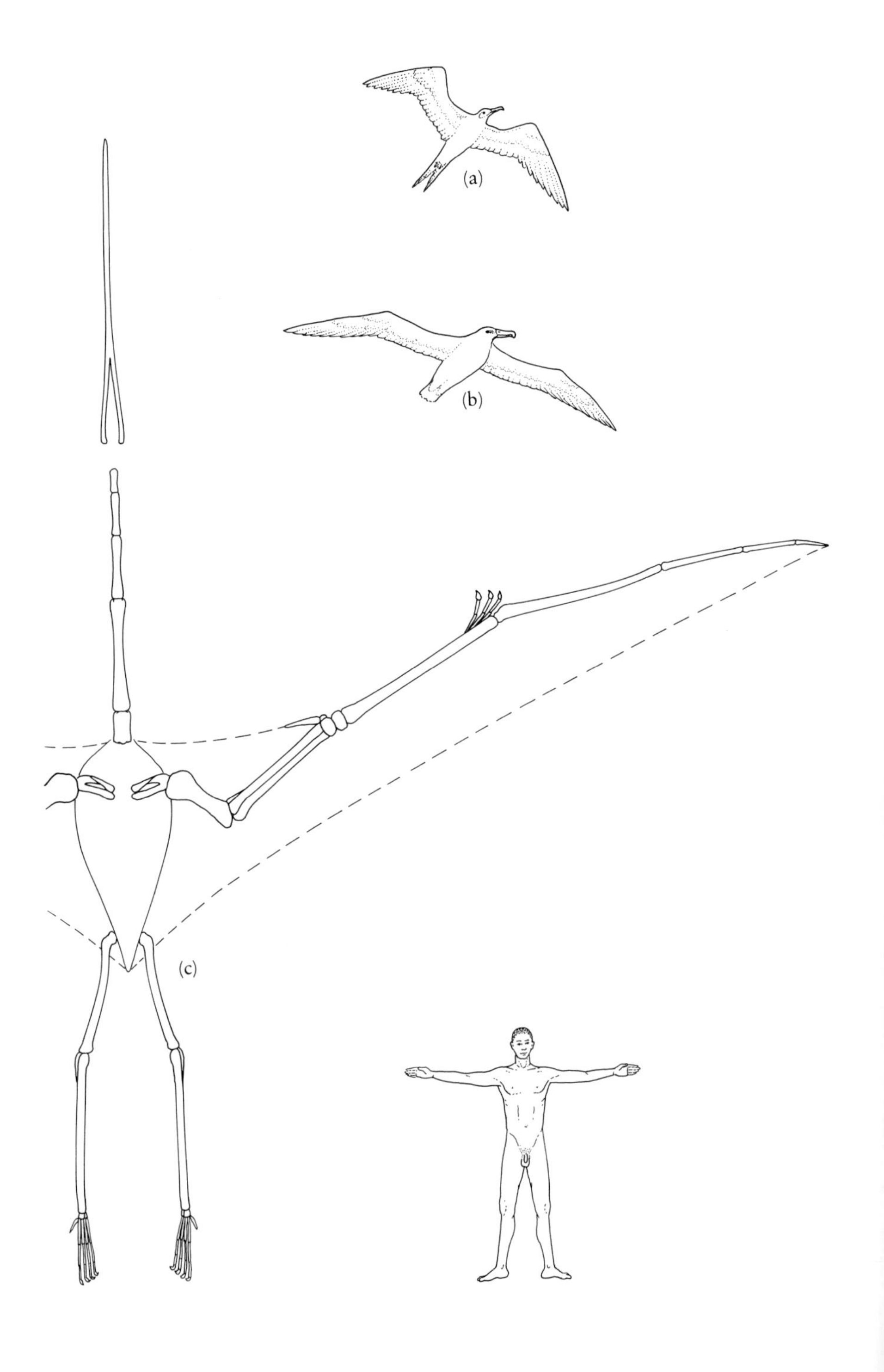

(*Above*) A hummingbird, *Topaza pella*, showing its wings.

(*Right*) Illustration by Gustave Doré for *Rime of the Ancient Mariner* by Samuel Taylor Coleridge. A ship, off course in the region of the South Pole, sights an albatross, a sign of good luck. A sailor shoots the bird with his crossbow, beginning a series of terrible ordeals.

lovely in the air as they are on the water), use continuous flapping to fly fast. Birds the size of a swallow use a mode called bounding flight. First they fly up, then they fold their wings and sail in a parabolic arc before extending their wings to fly again. Small birds may hover intermittently on take-off and landing, but hummingbirds, which are capable of continuous hovering, use this mode of flight constantly as they feed. It works, and they get a wonderfully concentrated food from flowers in this way, but it costs them dearly in energy, as we shall see.

All fliers—even the large ones, who prefer to soar—flap their wings once in a while, and it turns out that the frequency of wing beats is roughly inversely

Trumpeter swans (*left*) and great egret (*right*).

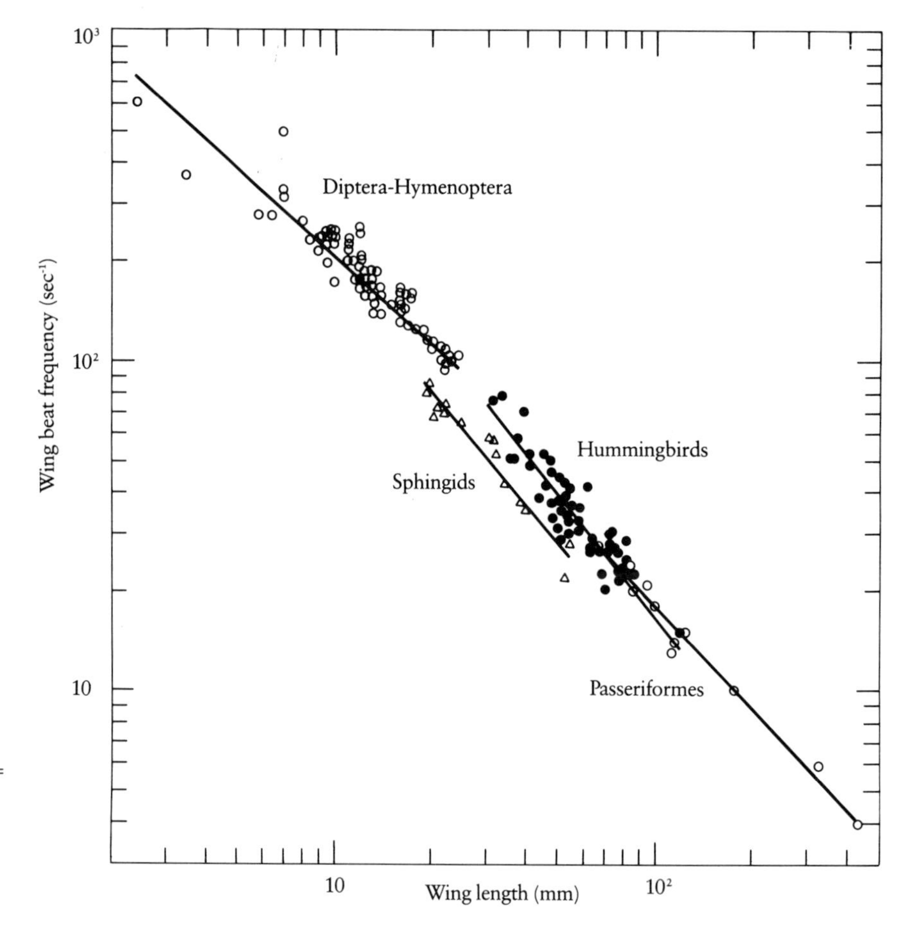


Wing-beat frequency plotted against wing length for insects and birds. (Diptera = flies, Hymenoptera = wasps and their relatives, Sphingids = sphinx moths, and Passeriformes = perching birds.) The data points in this log–log plot cluster about lines whose slopes are close to −1, showing that wing beat frequency varies practically inversely with wing length within a group.

proportional to the wing length (bottom figure on page 168). The rotational frequencies of aircraft propellers and helicopter rotor blades are also inversely proportional to the lengths of the blades, but for reasons that can't apply to birds. In the case of propellers and rotors, the speed of the tips of the blades should not approach the speed of sound, if energy losses due to miniature sonic booms are to be avoided, but birds are in very little danger of encountering this problem with their wings. The reasons why wing-beat frequency is inversely proportional to wing length over a large range of flying animals is not yet completely understood, although the British fluid dynamicist Sir James Lighthill has pointed out that, for geometrically similar animals, the requirement of constancy of skeletal stress generated by wing-root bending moments due to either the drag force on the wing or to the angular accelerations required at the extremes of the wing oscillations would explain the result (Lighthill, 1977). The reasoning is no different in principle from the arguments we applied to the rotational speed of reciprocating engines in Chapter 2 or the argument A. V. Hill applied to the flexion stroke of geometrically similar running animals, as explained earlier in this chapter.

**Wing Loading and Flying Speed.** The principles behind the diversity in flying behavior among birds are the same principles Ludwig Prandtl discussed more than 60 years ago as he wrote on the aerodynamics of aircraft. In fact, the aerodynamics of the 1920s is more suited to the description of animal flight than is the aerodynamics of the 1980s with its emphasis on supersonic airplanes and re-entering space shuttles.

An important concept in all that follows is wing loading, defined as the mass of the bird divided by the projected area of its wings (plan area as seen from above or below). In animals obeying isometry, in which length is proportional to $m^{1/3}$, the projected area of the wings is proportional to $m^{2/3}$, so the wing loading would increase as $m/m^{2/3} = m^{1/3}$. In the figure at the top of page 170, wing loading is shown as a function of body mass for the entire range of flying animals from the smallest flies and wasps to the largest shore birds. Also included (as a heavy broken line) is the wing loading of a hypothetical set of isometric animals. In this figure (which is from a paper by Crawford H. Greenewalt), if the broken line were continued four and one-half powers of 10 to the right, it would pass directly through the point representing the mass and wing loading of a Boeing 747. Along the way to the 747 point, we could expect to find hang gliders, Piper Cubs, and DC-3s.

Through dimensional analysis, we can discover the two dimensionless groups that can be formed from the bird's mass, its wing area, the air density,

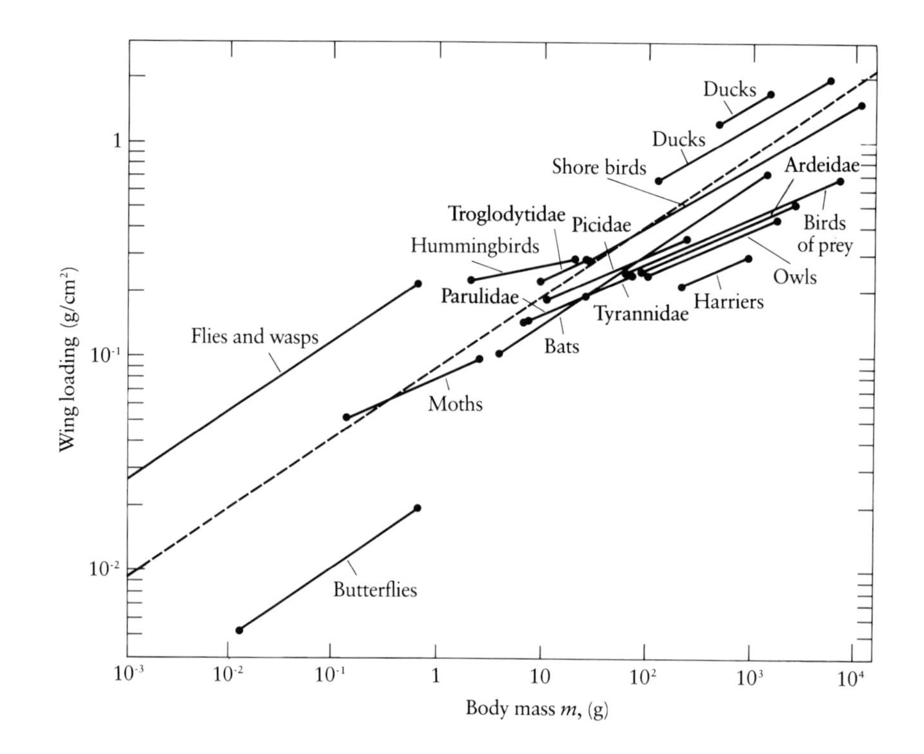


Wing loading of all fliers, birds and insects, plotted against body mass. The heavy broken line has a slope of 0.33 and, if extended, would pass through the point appropriate for a Boeing 747 (61 g/cm$^2$, $3.2 \times 10^8$ g or 705,000 lb. In this log–log plot, the vertical axis has been expanded by a factor of 2 for clarity.

the acceleration due to gravity, the speed of the bird through the air, and the angle at which the wing meets the air (its angle of attack). One of the two dimensionless quantities is the angle of attack. (Angles are always dimensionless, as we have mentioned before.) The other dimensionless quantity is the weight (body mass times gravity) divided by the product of air density, wing area, and the square of the speed. If these two dimensionless quantities are a complete set describing the problem, and if one of them is held constant (the angle of attack), then the other must also remain constant, no matter how large or small a bird we are considering. We can conclude that, for a set of isometric animals whose wings are flying at the same angle of attack, the wing loading (body mass / wing area) is proportional to the air density times the square of the speed. The conventional way of expressing this is to write

$$\begin{aligned}\text{wing loading} &= \text{body mass / wing area} \\ &= \tfrac{1}{2}\ \text{air density} \times \text{speed}^2 \times \text{lift coefficient / gravity} \quad (5.5)\end{aligned}$$

in which the lift coefficient is a dimensionless quantity, a number that changes with the angle of attack but remains fixed as long as the angle of attack is fixed. It may be helpful to think back to the submarine problem in Chapter 3 and remember that, when the Reynolds number was fixed, the drag coefficient

The speed for minimum power plotted against the span loading for birds of a range of sizes from hummingbirds to albatrosses. The span loading is defined as the bird's mass divided by the square of its wingspan; it is proportional to the wing loading for geometrically similar birds. The speed for minimum power was calculated from aerodynamic formulas using the bird's actual wingspan, wing area, and body mass. Since the slope of this log–log plot is 0.55, one may conclude that the speed for minimum power essentially is proportional to the square root of the wing loading over a wide range of sizes in birds.

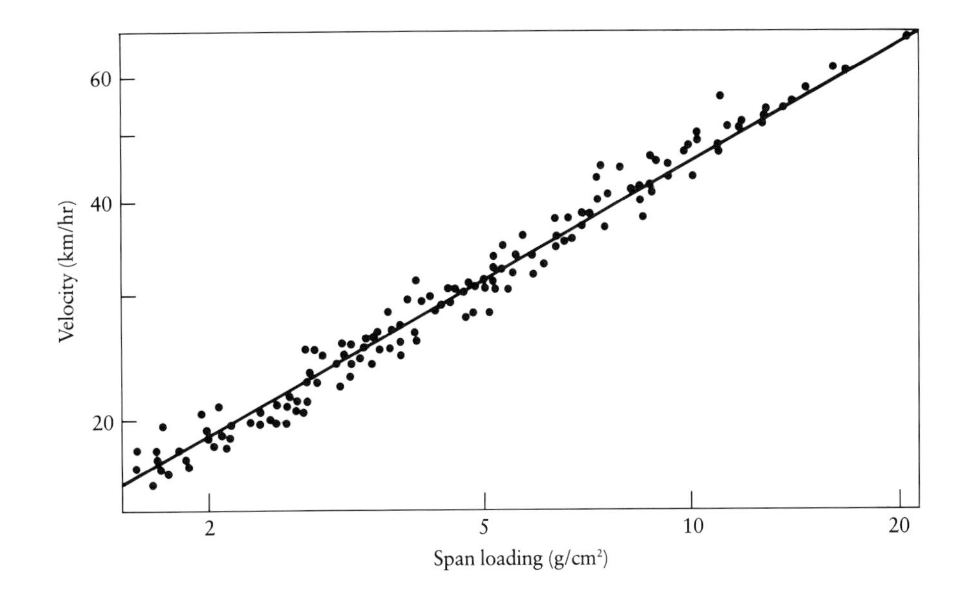


was also fixed. Notice that the drag coefficient there was defined in much the same terms as the lift coefficient here. Both are dimensionless ratios involving a force (either the drag force or the lift force) divided by the product of air density, wing area, and the square of the speed.

By the way, the reader who has been paying close attention may wonder why we have not considered the Reynolds number as another dimensionless variable in this problem. It turns out that the lift coefficient *does* vary with Reynolds number—but not by very much in the range of high Reynolds numbers ($10^3$ to $10^5$) that describe the flight of most birds.

Now we can understand why flying speed increases with animal size. According to the preceding result, cruising speed is proportional to the square root of wing loading. This prediction is well confirmed by the figure above, which shows the cruising speed for minimum power (corresponding to a particular lift coefficient) as a function of wing loading for a range of birds from hummingbirds to albatrosses. In this figure, wing loading has been estimated by dividing the bird's mass by the square of its wingspan (the wingspan is the distance between wingtips). Following the preceding arguments, we would expect this log–log plot to have a slope of ½. The straight line that best fits the data shows a slope of 0.55, which is very close to the expected value.

A simple conclusion from all of this is that, for geometrically similar flying animals, the characteristic airspeed should vary as the square root of the wing loading, or as the square root of $m^{1/3}$, which is $m^{1/6} = m^{0.167}$. For the large group labeled shore birds in the figure on page 170, Greenewalt (1975) finds the minimum-power airspeed to be proportional to $m^{0.15}$; for the largest group of ducks, it is proportional to $m^{0.16}$. Lighthill (1977) has noted that, when this same relation between speed and size is extrapolated down to an insect with a mass of 1 gram, we get a speed of about 4 meters per second, which is the speed at which many insects of that size actually fly.

**The Power Required to Fly.** The key to understanding why small flying animals hover, intermediate ones bound, and large ones soar is understanding the margin between the power required for flight and the power available for that purpose.

The drag force on a bird or an airplane consists of two components, the parasitic drag and the drag due to lift. The parasitic drag would be present even if the bird were supported by a thin wire trailing from a zeppelin, because it is caused by the fact that the bird is a solid obstacle in an air stream. The parasitic drag itself has two parts, called the skin-friction drag and the pressure drag. A flat plate aligned with the flow (like a weather vane) has a parasitic drag that is almost entirely due to the friction of the air scrubbing past. The same flat plate turned broadside to the oncoming airstream has a parasitic drag due almost entirely to the higher pressure on the front side and the lower pressure on the back side of the plate. The bird, having a body and wings with some thickness, has a parasitic drag that includes contributions from both skin-friction drag and pressure drag. Just as was true of the submerged submarine in Chapter 3, the parasitic drag is proportional to the surface area times the square of the speed, so the power required to overcome parasitic drag, the *parasitic power*, increases with the cube of the airspeed.

The second component of the total drag is the drag due to lift, called the induced drag. The induced drag is explained by the fact that the wings create a wake of descending air behind them, and creating this downward flow requires power (the *induced power*). (A side note: One way of making a wing that exhibits a minimum of induced drag is to make the shapes of the leading and trailing edges of the wing (as seen from above) elliptical. The famous Spitfire fighter flown by the Royal Air Force in the Battle of Britain had gorgeous elliptical wings and tail surfaces, giving the lowest practical induced drag for that particular wingspan.)

Supermarine Spitfire Mk. 1A. This famous British fighter from World War II had an elliptical wing shape (as seen from above) for low drag due to lift (induced drag).

Taking into account both the parasitic power and the induced power, one may calculate the total power required to maintain level flight. The results of these calculations, applied to a bird the size of a pigeon, are shown in the figure on page 173.

Let's try to make sense out of this figure. The parasitic power increases rapidly with airspeed, so that most of the power needed for high-speed flight goes to satisfy this requirement. The phenomenon of a sharply increasing power requirement with increasing speed is well known to bicyclists, who also expend most of their power overcoming air drag at high speeds.

At low speeds, the induced-power requirement dominates. The wings must generate their lift by changing the momentum of the passing air, and with the mass flow rate of the air moving past them diminished, they must create a

The power required for level flight for a bird about the size of a 400-gram pigeon. The total power required is the sum of a parasitic component plus a component needed to overcome the drag due to lift. The available power is indicated by the dotted horizontal line. It is assumed to be 8.0 watts, independent of speed. Ordinary forward flapping flight is possible under these assumptions when the available-power line lies above the total power required—that is, between about 22 and 66 kilometers per hour (14 and 41 miles per hour). The speed that requires the minimum power is indicated by the dot on the total-power curve; it is about 42 kilometers per hour (26 miles per hour). At speeds below 22 kilometers per hour (14 miles per hour), the bird can still fly for brief periods using special high-lift techniques not considered in these calculations.

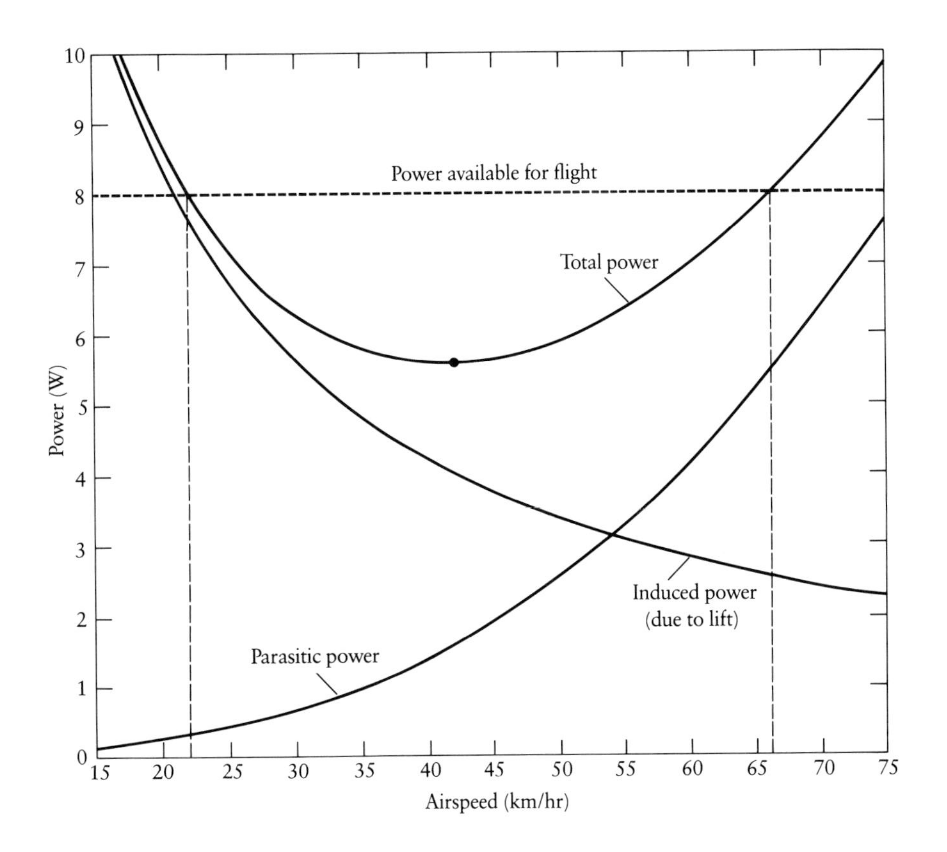


stronger downwash to generate that lift at low speed. The result is that the induced power is inversely proportional to the airspeed.

The total power required for flight has a minimum. The minimum-power speed, which is the most efficient speed to use for cruising, is about 42 kilometers per hour for the pigeon in the figure above. We have already seen how the minimum-power speed scales with the square root of span loading and increases with the size of the bird in the figure on page 171.

Supposing that the power available from the animal's muscles is not a function of the airspeed, we can draw a horizontal line in the figure representing the power available. The intersections of the available-power line with the curve specifying the total power required define the minimum and maximum airspeeds for level flight. In climbing flight, the separation between these two speeds would be narrower, because the total power required would now include an additional factor proportional to the bird's mass times the rate of its climb.

In the figure on page 174, the minimum power required for flight has been calculated for the whole size range of birds from hummingbirds to vultures. When the basal metabolic rate has been added to this minimum metabolic power, we have an estimate for the total power required to fly. C. H. Greenewalt, who made these calculations, points out that a hummingbird with a mass of 4 grams must increase its metabolic rate above the resting level by

For birds or aircraft flying at low speeds, the induced power dominates. The wings must create a strong downwash of air to keep the bird up. This is a swallow-tailed gull (*Larus furcatus*) in slow forward flight just before landing.

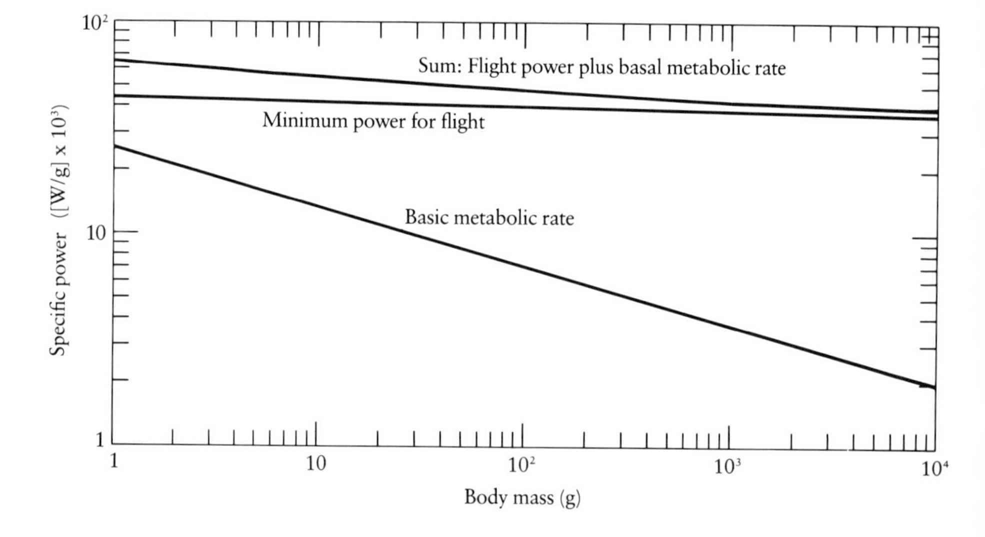


Specific metabolic power required for flight and basal metabolic rate. The specific metabolic power is given as watts per gram of body weight. The basal metabolic rate follows Kleiber's law, with the constants appropriate for birds at rest. The minimum metabolic power for flight is calculated using the same information that produced the total-power curve for the pigeon in the figure on p. 173, and then multiplying by a factor of 4 to take into account the fact that muscle is only approximately 25 percent efficient in converting chemical energy into mechanical work. The curve labeled *Sum* is the minimum specific metabolic power for flight plus the basal metabolic rate.

only a factor of about 3.3 to fly, while a 7.3-kilogram Griffon vulture has to raise its metabolic activity to a rate 20 times the resting level to stay airborne. Recalling that C. R. Taylor and his collaborators (1981) found that terrestrial animals can increase their metabolic rates to about 10 or 15 times the resting level, and taking into account the fact that birds have relatively larger hearts and lungs than mammals, we arrive at the conclusion that the largest birds are expending energy at rates close to the upper limits of their abilities while sustaining level flight in still air. The smaller birds, by comparison, have a surplus of power, and we shall see shortly what they do with it.

A high-performance sailplane, the Schempp-Hirth Nimbus 3, made in West Germany. The long, narrow wings give the sailplane a low induced drag so that it can have a long gliding range. The wingspan of the Nimbus 3 is 23 meters (75 feet). At its best gliding speed in still air, it moves 55 meters forward for every meter of altitude lost. This is an extraordinarily good gliding ratio.

**Soaring.** The fact that the largest birds are right at the ragged edge of not having enough power for sustained level flight leads them to search for free rides in ascending currents of air. Ascending air can be found at low altitudes on the windward sides of cliffs and mountains. It can also be found at intermediate altitudes in the thermal updrafts that occur when the sun heats the earth and at high altitudes in the standing waves of rising and falling air downwind of uneven topography.

In the period following World War I, when German aviation was severely restricted by the Treaty of Versailles, motorless flight, which had not been specifically prohibited by the Allies, became a popular recreation. Since its beginning in Germany and Switzerland, soaring has spread around the world. The Schempp-Hirth Nimbus 3 shown above is typical of modern high-performance sailplanes. The fuselage is very nicely smoothed and streamlined for low frictional drag, and the wings are long and narrow. A long, narrow wing has a lower induced drag than a short, stubby one, even though both may have the same lift. This is because a long wing can spread the downwash over a large area behind the aircraft, with the result that the downwash velocity is reduced substantially by comparison with a short wing. Within the constraints of structural safety, the best sailplane is the one with the longest wings.

By and large, the soaring birds also obey this rule. The albatross in the figure on page 166 looks somewhat like a sailplane. From a perch on a high cliff, it merely jumps out into an updraft when it wants to soar. The updraft is always there, provided a sea breeze is blowing, and the bird rarely has to flap its wings.

As the sun warms the ground in the morning, the intensity of vertical updrafts slowly increases, reaching a peak in midafternoon. The speed of the smallest vertical updraft necessary for soaring is a fixed fraction of the minimum-power airspeed for the bird in question and therefore is proportional to the square root of the bird's wing loading. Because wing loading increases with the size of the bird, the minimum conditions for thermal soaring are reached for smaller birds earlier in the morning. Kites soar earlier in the day than vultures. There is some natural fairness in this, because it gives the smallest soaring birds first crack at whatever food opportunities have accumulated during the preceding night.

Birds soaring over the Gulf of St. Lawrence.

The American basilisk (*Basiliscus basiliscus*) (Jesus Christ lizard) can run for short distances on the surface of the water, just as shore birds do when they take off. These lizards are as much as two feet (about 60 centimeters) long.

Andean condor.

**Flapping Flight.** Shore birds of intermediate size, including swans and geese, are excellent fliers and can cruise at relatively high speeds, but they also have high take-off speeds because of their high wing loadings. They can reach their take-off speeds by running along the water, using their big feet as paddles. (Even nonflying animals sometimes paddle themselves above the water this way. In the American tropics, there is a remarkable lizard, known as the basilisk, ("Jesus Christ" lizard) that can run for many yards across the water

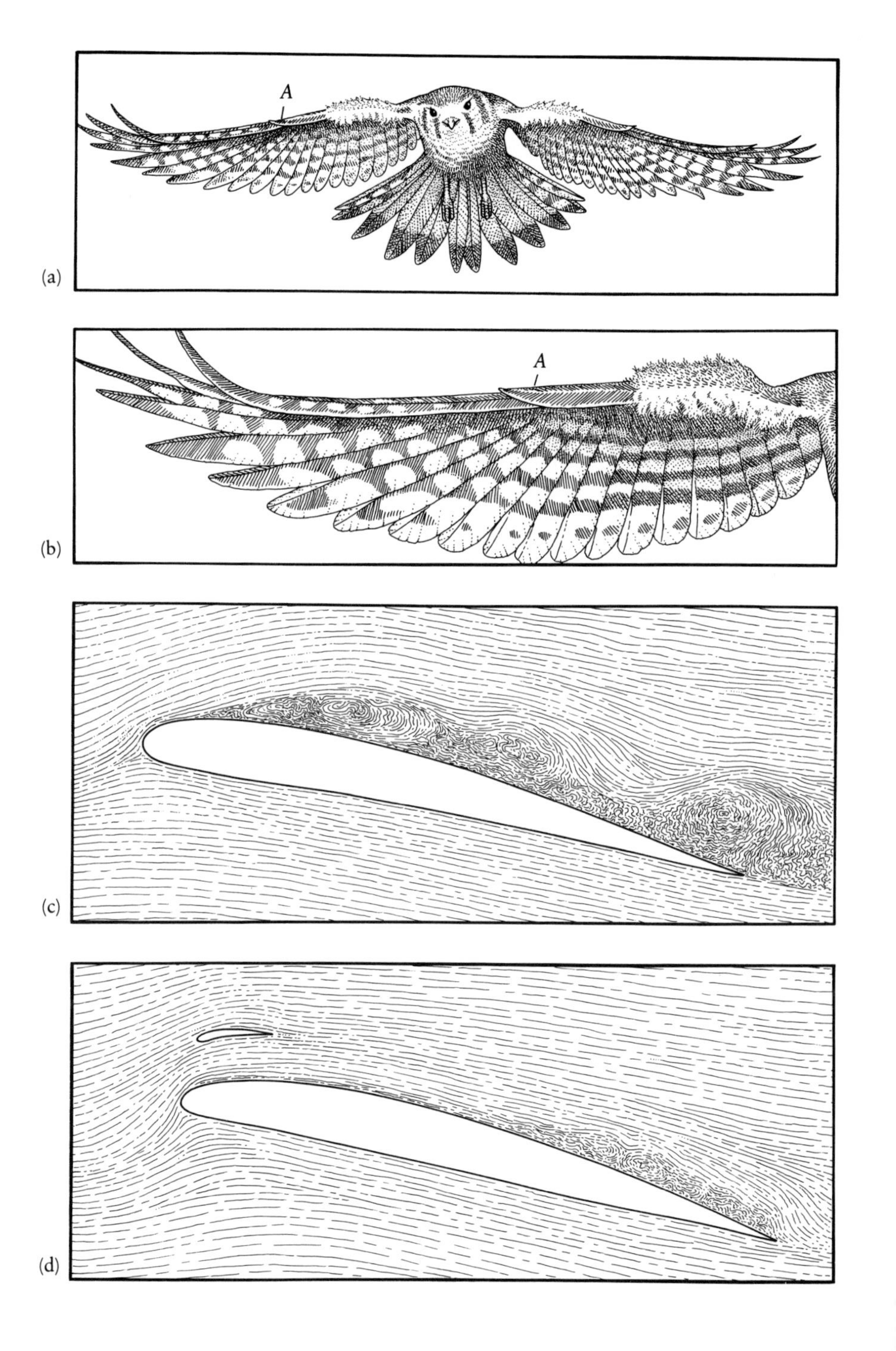


*a*, A sparrow hawk (*Accipiter nisus*) in low-speed flight. *b*, The alula, marked A, on the sparrow hawk's wing. *c*, A wing in a wind tunnel at a high angle of attack. This wing is stalled, which means that the streamline pattern separates from the top surface, reducing the lift. *d*, A small airfoil above a wing, called a leading-edge slat, accelerates the air over the top surface and prevents stalling. The alula on a bird's wing works as a leading-edge slat.

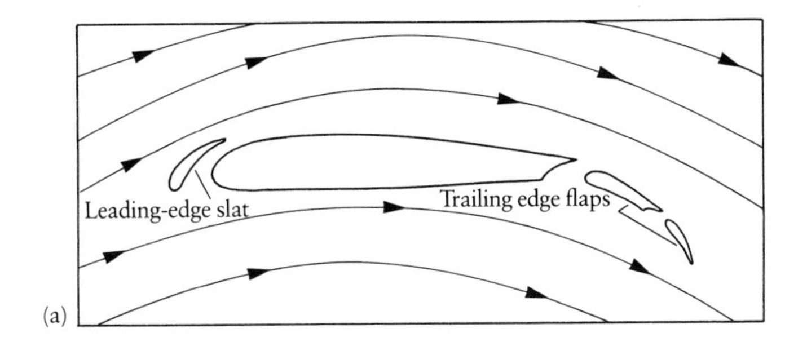


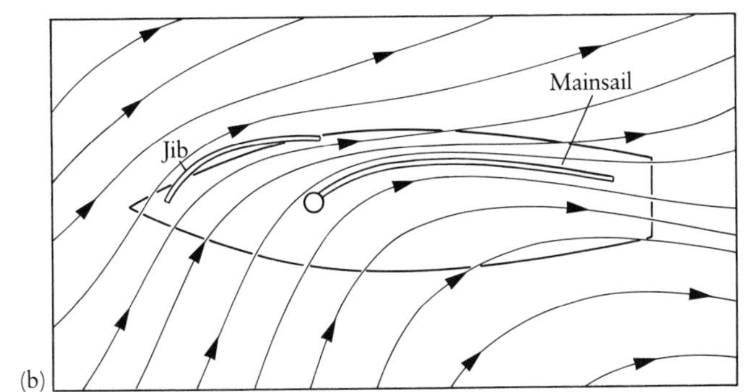

*a*, Aircraft use leading-edge slats and trailing-edge flaps as high-lift devices for slow flight. *b*, A sailboat uses a jib to improve lift on the mainsail by preventing stalling. (The jib is also a sail in its own right). *c*, Sailboat, showing the mainsail and the jib. The mast on this boat is high, giving the mainsail some of the characteristics of a long, narrow bird's wing. *d*, an airliner landing, showing the leading-edge slats and trailing-edge flaps used to obtain high lift at low speed.

surface using the same paddling foot motions these shore birds used.) The large birds of prey—hawks and eagles—use an even better technique than running on water to launch themselves. They jump off tall trees to get up flying speed.

When landing, large birds have several mechanisms for maintaining slow forward flight just before touching down. First, they can make use of what aeronautical engineers call "high-lift devices." The sparrow hawk in the figure on page 178 has spread the feathers of its wings and tail to obtain some of the same effects achieved by the trailing edge flaps shown on the aircraft wing in the figure above. Both the bird feathers and the aircraft flaps promote greater lift by increasing the wing area and the camber (the concave curvature of the wing). The hawk's wing also features a special device called an alula, a short "extra wing" formed by the thumb feathers. These alulae are maneuvered by the bird to form a converging channel between the thumb feathers and the top of the leading edge of the wing. This tends to keep the boundary layer attached and thus to prevent stalling. Aircraft wings have leading-edge slats and sailboats have jibs to accomplish the same thing.

Employing a mathematical model of flapping flight, Sir James Lighthill (1977) calculated that the power per unit of mass needed by a flying animal to overcome induced drag in slow forward flight just above the stalling speed is less than half the power required for the same animal to hover. Thus, birds that are much too large to hover for sustained periods are still able to land at very low forward speeds by an exceptional short burst of effort. This mode of flight is useful for landing in trees or alighting on narrow rock ledges.

**Bounding Flight.** Many small birds, including skylarks, swallows, and martins, use an intermittent mode of flight in which they alternately flap their wings and then fold them at their sides. Sir James Lighthill has also provided a simple analysis of this strategy. He has shown that, whenever the drag due to air friction acting on the outstretched wings exceeds the drag due to lift, there can be a savings in energy for covering a given distance when the wings are folded a part of the time. This can be true only of birds that fly at speeds well in excess of their minimum-drag speed (a speed somewhat greater than the minimum-power speed), and therefore it can be true only of small birds with plenty of power to spare. As an example, Lighthill considers a small bird traveling at an average airspeed 50 percent greater than its minimum-drag speed. If it flapped its wings half the time and kept them folded the rest of the time, the average drag (and, therefore, the energy consumption for traveling a given distance) could be reduced by 17 percent.

**Hovering.** The largest birds capable of hovering at a spot for the appreciable periods necessary for feeding are hummingbirds weighing about 20 grams. They do this standing on their tails in the air and moving their wings back and forth, like someone clapping his hands alternately in front of his chest and behind his back.

By creating a downward jet of air, the animal gives itself an upward thrust. The rate at which the downward airflow carries away momentum is equal to the animal's weight. Following these ideas, Lighthill (1977) showed that the power per unit mass required to generate lift is directly proportional to the

Helicopters, like birds, require less power for forward flight than for hovering. The engine must develop nearly maximum power for landing as well as for take-off. For several reasons, including the very great power requirements for hovering, helicopters did not appear until long after airplanes had already become successful.

square root of the mass and inversely proportional to the wingspan. The mass-specific power for hovering animals increases by almost a factor of 6 in going from the tiny cabbage butterflies to hummingbirds. Helicopters are another factor of 6 higher. Their high power requirements were among the reasons why practical helicopters were not developed until long after airplanes were in common use.

Lighthill applied this reasoning to a 20-gram hummingbird with a wingspan of 30 centimeters and calculated a specific power required of 13 watts per kilogram, which he says is uncomfortably close to the greatest sustainable muscular power for an animal of this size. Hence, we do not expect to find any larger hovering animals. In fact, Greenewalt has noted that even hummingbirds seem to be deliberately bending the rules, because the ratio of the bird's mass to the wing area is nearly constant over the whole range of hummingbird sizes. They accomplish this by increasing the wingspan more rapidly than the body length as the size of the bird increases. This regular distortion in shape with size preserves the mass-specific power at a constant value, no matter what the size of the hummingbird.

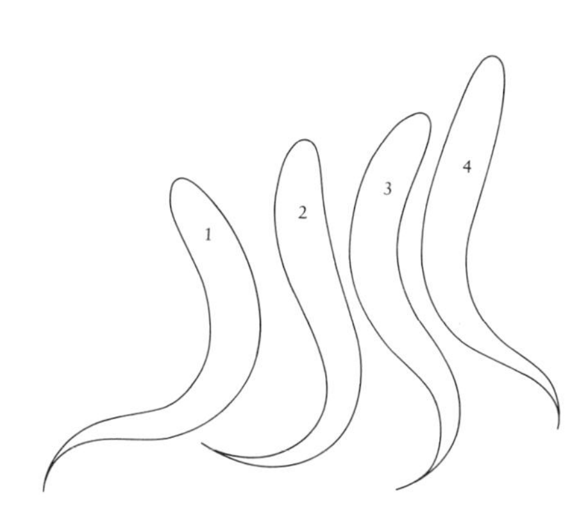


A rainbow trout (*Salmo gairdneri*) accelerating forward by sending a wave of lateral flexion posteriorly along its body.

## Swimming

Aquatic animals move through the water using a huge variety of swimming styles and techniques. Protozoans use flagellar beating and ciliary wave motions. Fish with long, thin tails move in a mode classified as anguilliform (eel-

Spotted dolphins swimming in the waters off Hawaii.

Swimming motions of a dolphin. Dolphins and whales bend their backbones up and down when they swim, whereas fishes bend their backs from side to side.

like). Those with tapering tails of medium length use a motion called carangiform (mackerel-like). The trout in the figure on page 181 is accelerating from a standstill using the carangiform mode. Although the fish's muscles are pushing its body from side to side, the lateral wave-shaped curve of the body is responsible for aiming part of the reaction force of the water forward to give the fish thrust. The lunate-tail swimming mode characteristic of sharks is similar in principle, but it makes a somewhat different use of the tail.

Dolphins and whales bend the other way. Perhaps reflecting the fact that their ancestors were land animals who may have galloped using the front-to-back flexures of the spinal column common among modern mammals, dolphins and whales beat their tails up and down instead of from side to side. It works just as well.

There are still other fascinating and diverse modes of swimming. Rays, turtles, and penguins "fly" through the water using the wing-beat motions of

Rays "fly" through the water using the wing-beat motions of birds.

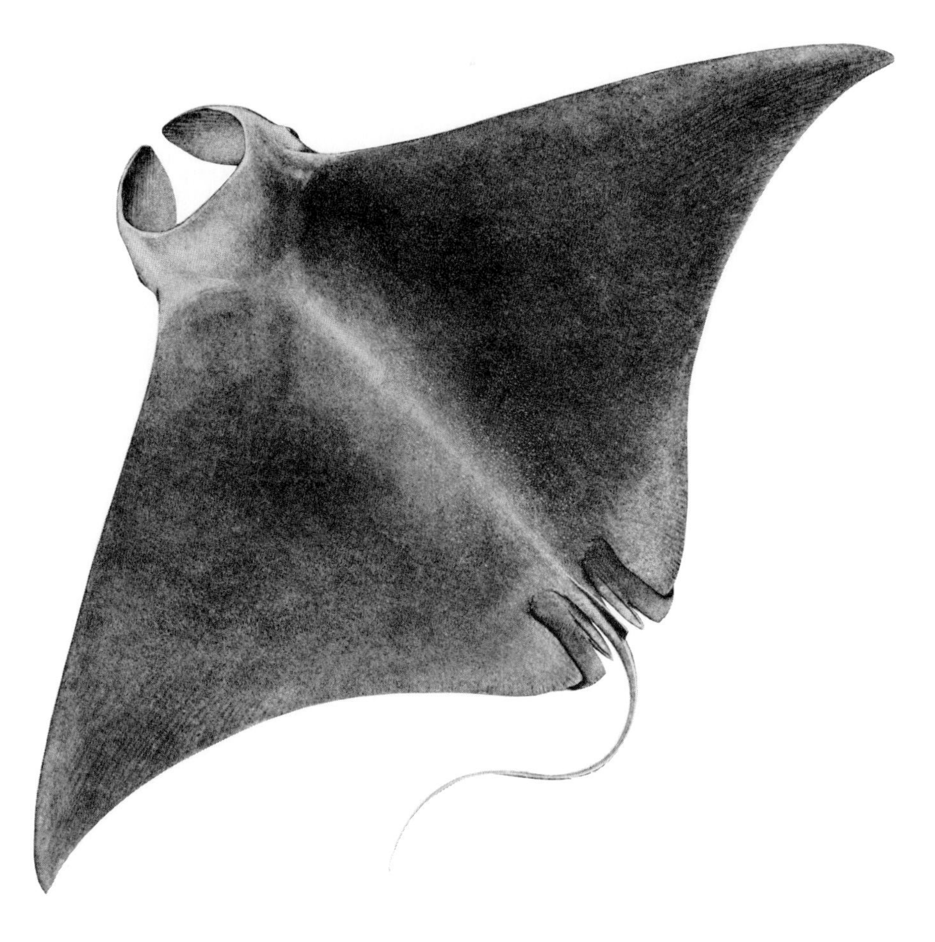

birds. Squids and other softies use jet propulsion. A lobster moves by leaping backwards with an energetic flick of its tail. Crabs use their feet as paddles, and the tiny water fleas of the genus *Daphnia* row through the water by beating two legs that resemble hairy paddles. Human beings use the breast stroke, crawl, and dog paddle. There are only a few different ways to walk and to fly; in the water, however, anything goes.

Because this chapter is about being large, most of what we shall have to say about swimming will apply to fishes and marine mammals. The microscopic swimmers will have their day in Chapter 6.

A shadowgraph shows where the water has been disturbed by the swimming motions of a fish. The wake (disturbed region) immediately behind the fish's tail is smooth and thin. The wake behind a ship would be much broader and more turbulent. This small tropical fish (a zebra danio, *Brachydanio verio*, about 3 centimeters long) has given a few pushes and is now coasting.

**Gray's Paradox.** Although the total range of Reynolds numbers for aquatic animals runs from $10^{-6}$ for such microscopic creatures as the paramecium to $10^8$ for the blue whale, the range for fishes and marine mammals swimming normally starts at about $10^4$ and ranges up to $10^8$. These high values for the Reynolds number mean that pressure gradients are balanced by inertial accelerations in the water around the animal's body, and the effects of viscosity are confined to a narrow boundary-layer region immediately adjacent to the animal's skin.

The streamlines toward the stern of a broad ship will usually show a phenomenon known as flow separation, in which fluid in the boundary layer near the ship's hull will slow down so much that it may actually come to rest with respect to the moving ship. This leads to separation of the boundary layer and the production of a wide, eddying wake. By contrast, the streamlined shapes of fishes and marine mammals combined with the pressure gradients due to their undulatory swimming motions produce only very thin wakes behind their bodies. The shadowgraph in the figure above shows the wake behind a small fish made visible by a technique invented by C. W. McCutchen, which uses a layer of cold water in a tank carefully overlain by layers of warm water. The fish swims through, disturbing this layering, and the disturbances can be picked up optically. The wake immediately behind the fish's tail is quite smooth and thin.

Dolphin swimming at high speed.

The boundary layer does not detach, as it would from a bluff shape, and the maximum thickness of the boundary layer near the fish's tail is not more than a small percentage of the maximum body thickness.

A. V. Hill (1950, p. 217) provided a relevant observation:

> Mr. G. A. Steven of the Marine Biological Laboratory, Plymouth, who spent much time during the war in a corvette in the Atlantic, tells me that at 15 knots he has frequently observed dolphins keeping up with the ship, apparently without effort, for long periods. At the season when the sea is alive with phosphorescent protozoa, which give out light on being stirred or touched, he has frequently noticed a clear thin line of light as a dolphin approached the ship. With dolphins there was no sign of a turbulent wake: with seals, on the other hand, a large amount of turbulence was easily visible. This beautiful observation, made many times, confirms the result of the calculations.

The calculations Hill speaks of in this quote were estimates of the power required to drive a dolphin-sized rigid body through the water at 15 knots, a speed Hill presumed a dolphin could sustain for long periods. When he assumed the existence of a laminar boundary layer around a dolphin's body, Hill calculated that about one-quarter horsepower would be required; and, assuming that the propulsion was about as efficient as a man walking up a hill, this power corresponds to the effort required for a human being of the same weight as the dolphin to climb 2,600 feet (about 790 meters) in an hour. Hill says that this is hard going but within the realm of possibility. By comparison, when turbulent flow is assumed to exist in the boundary layer, the power required rises to 1.7 horsepower, which corresponds to the effort required for a man to climb 17,680 feet (about 5,390 meters) in an hour, which is clearly beyond human ability.

For these reasons, Hill concluded that the dolphin somehow maintains not only an attached boundary layer (giving a thin wake) but also a laminar boundary layer over his skin, even though the Reynolds number is so high that the same flow over a smooth, rigid, full-scale model of the dolphin would be turbulent. The first discussion of the fact that dolphins apparently need more power to swim (assuming a turbulent boundary layer) than they actually have available was given by Sir James Gray in 1936. The apparent mismatch between the power required and the power available eventually received the name *Gray's paradox*.

Long after Gray, Hill, and many others, the U.S. Navy became interested in Gray's paradox. Photographs of dolphins during high-speed swimming showed circumferential wrinkles in their skin, probably a consequence of the deformation of the soft layer of fat that lies between the skin and the muscles of the dolphin's body. There was speculation that the dolphin had discovered a way of damping turbulent pressure fluctuations with its compliant skin, and some people visualized a brilliant sea of the future filled with ships whose exterior surfaces were as flabby as the hides of Barca-loungers.

None of it worked out the way the Navy had hoped. To date, tests have shown that torpedoes and ships with compliant skins would be no better than those with skins of ordinary steel. Furthermore, Navy studies with trained animals have indicated that dolphins and porpoises may not experience such unusually low drag after all (Lang, 1975). Dolphins trained to chase a lure through the open sea off Hawaii were photographed coasting to a stop after the lure was suddenly halted. The drag coefficient calculated from these measurements indicates that the flow in most of the boundary layer over the animals' skins could have been fully turbulent.

Finally, it is now clear that Gray's original assumption that dolphins and porpoises can sustain a speed of 10 meters per second is not quite true. The 91-kilogram porpoise Gray cited in his 1936 paper was clocked at a speed of 10.3 meters per second for 7 seconds. The extensive tests conducted by the Navy showed that dolphins and porpoises do indeed reach speeds of 10 and 11 meters per second for short bursts of about 7 seconds' duration, but their speed is reduced to about 80 percent of top speed when the duration is 60 seconds, and they maintain only about 50 percent of their top speed when swimming throughout a 24-hour day. Thus, a large part of Gray's paradox can be resolved by recognizing that dolphins cannot swim at 10 meters per second all day (unless perhaps by riding the waves of ships).

Fish, on the other hand, really do use a special strategy for maintaining a laminar boundary layer around their bodies, thus reducing drag. The mechanism depends on their slimy skins. Anyone who has ever caught a fish will recall the slippery feeling of the scales. Hundreds of small glands between the scales release a mucus consisting of long-chain molecules (polysaccharides and proteins) into the water. The mucus serves several purposes—it affords protection from bacteria and microscopic parasites, and its slippery property can be useful in escaping predators. Just as significantly, it has been well established that such long-chain molecules inhibit the transition to turbulence in a boundary layer. Tests show that as few as 6 parts per million of organic substance from the slime of the Pacific barracuda are responsible for a 45 percent reduction in the friction of seawater flowing in a pipe. In fact, firemen now use tiny amounts of the synthetic polymer poly(ethylene oxide), a mucuslike substance injected into their fire hoses, to obtain substantial reductions of fluid resistance. Moreover, the rulebooks of sailboat racing and crew racing now specifically prohibit the use of such substances in competition, because they are known to be effective in reducing the drag on hulls.

**The Shapes of the Fast Swimmers.** There is a strong tendency toward isometry in fishes and marine mammals of different sizes, presumably because they are relatively free from the influences of gravity, which produced the elastic similarity distortions we saw among terrestrial animals and the disproportionately large wings we noted among large hummingbirds. The dolphin, the blue whale, the tuna, and the Greenland shark in the figure on page 188 are all approximately the same shape, in the sense that the ratio of diameter to length of an equivalent streamlined body of revolution (shown shaded) is about the same.

The ratio of diameter to length is called the *profile thickness*, in hydrodynamic parlance. In the lower figure on page 188, the drag coefficient is shown as a function of the profile thickness for bodies of revolution of a variety of thicknesses. One way to establish the value of the drag coefficient experimentally is to make a small model of the body out of a material heavier than water and to drop the model into a deep tank. Because the weight of the model in water is known, this establishes the drag force at the terminal velocity, and thus the drag coefficient may be calculated (for one velocity and, therefore, one Reynolds number). One way of interpreting this figure is to imagine that the

The bodies of some marine mammals and large fishes fitted to equivalent bodies of revolution (shapes with circular cross sections). The ratio of diameter to length is confined mostly to the range 0.21–0.26.

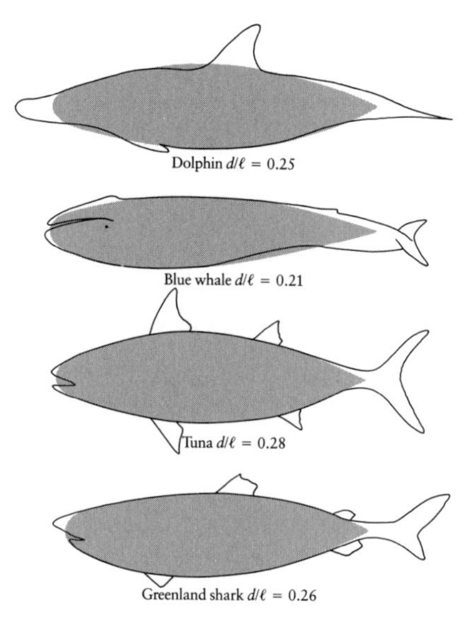


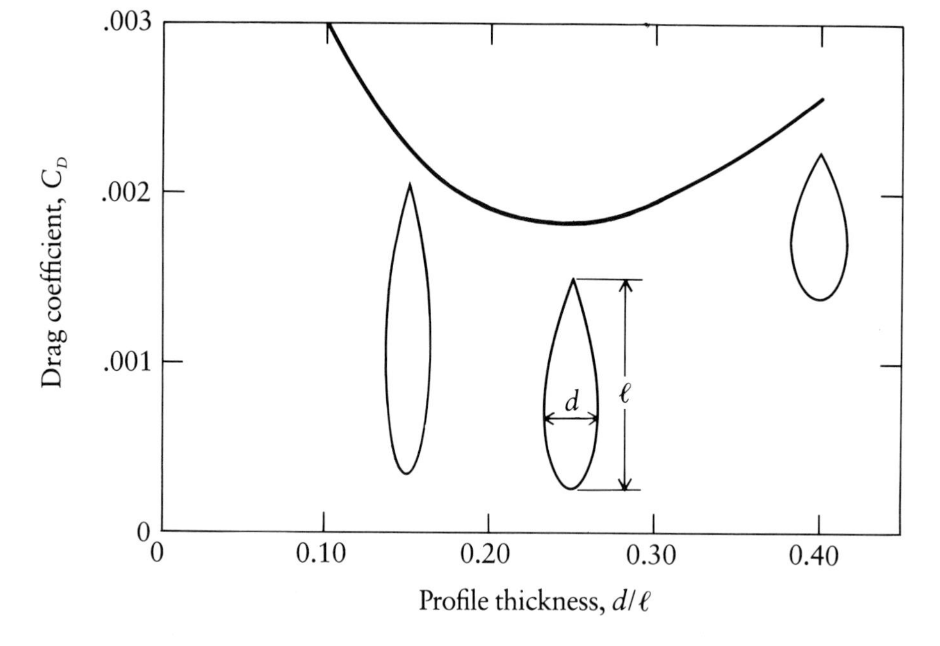


Drag coefficient vs. profile thickness for bodies of revolution, established by experiment. All these bodies have the same volume. The drag coefficient, $C_D = D/(\frac{1}{2}\rho v^2 V^{2/3})$ refers to an area calculated as the ⅔ power of the volume of the body, $V$.

models that were dropped in the tank all had the same volume but different shapes; the body with the lowest drag coefficient would have reached the bottom in the shortest time.

The figure shows that the drag is at a minimum when the profile thickness is close to 0.25. Heinrich Hertel, the German aeronautical engineer who made these observations, points out that many of the fast-swimming fishes and mammals have a profile thickness near the minimum-drag thickness. For narrower shapes, boundary-layer turbulence occurs too close to the nose of the model. Furthermore, the larger ratio of surface area to volume in very narrow shapes (by comparison with the optimum) would result in higher drag even if the boundary layer remained attached and laminar. With a thicker shape, the drag due to lowered pressure at the back of the body becomes more important than the skin-friction drag, and the body again has a total resistance higher than the resistance of the optimum shape.

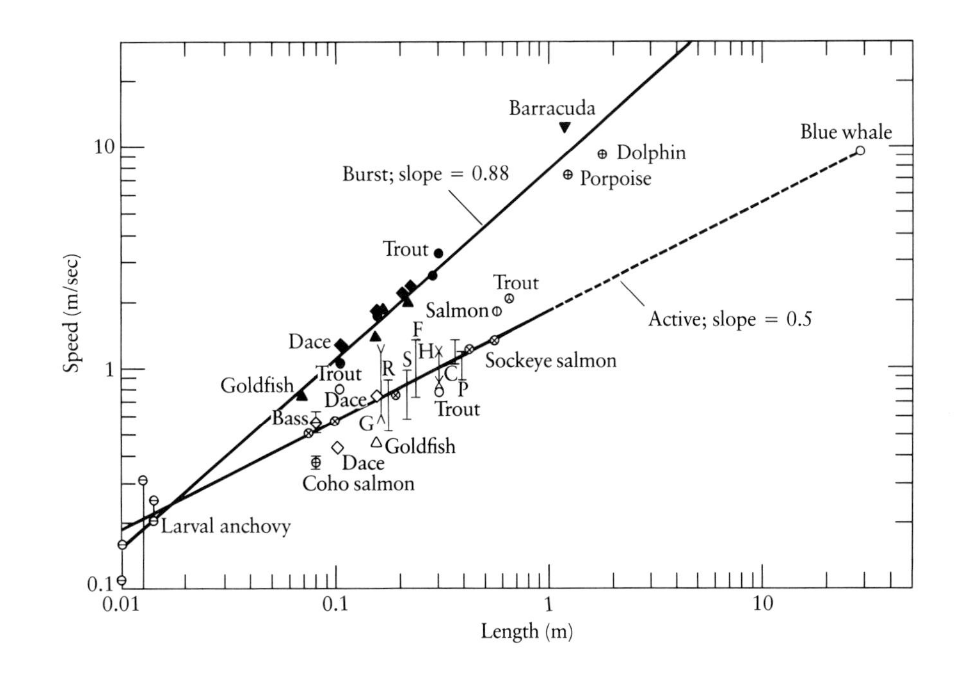


Speed plotted against length for swimmers of intermediate and large size. The slope of the line marked Active is 0.5, showing that the speed of sustained swimming is proportional to the square root of length. The slope of the broken line marked Burst is 0.88. Because the slopes of the two lines are very different, the larger fishes in this series are able to increase their swimming speed by a considerably larger factor than the smaller ones can. The vertical bars marked with letters correspond to: *C*, cod; *F*, flounder; *G*, goldfish; *H*, herring; *P*, pout; *R*, redfish; *S*, sculpin.

**Speed of Swimming.** It is quite difficult to obtain reliable measurements of the swimming speeds of fishes. Anecdotal accounts of observations in the wild tend to be fish stories, although some fairly reliable measurements of the swimming speeds of tunas, sharks, and barracudas have been obtained by clocking the time required for a given length of line to be pulled out after a fish has been hooked. Using a trough-shaped circular fish tank mounted on a device like a carousel, the English zoologist Richard Bainbridge was able to photograph the swimming movements of fishes with a stationary motion-picture camera. The fishes swam at a steady speed while the tank rotated in the opposite direction, so that the fishes appeared to be stationary with respect to the camera. This device, which Bainbridge called a fish wheel, had the advantage of eliminating most of the wall effects that usually complicate measurements made in a water tunnel. In the fish wheel, the water remains stationary with respect to the moving walls of the trough-shaped circular tank.

Through the use of fish wheels and other aids to observation, it is now known that fishes adopt two different patterns of swimming activity, as if they had gearboxes with only two forward speeds. The two speeds are called *active* and *burst*. As the names imply, the active speed can be maintained almost indefinitely. The test of whether a fish is swimming faster than its active speed is to see if it tires after minutes or hours at that speed. The burst speed, which is

Dolphins leaping. The high swimming speeds of dolphins, which can exceed 10 meters per second, allows them to leap as much as 5 meters straight up.

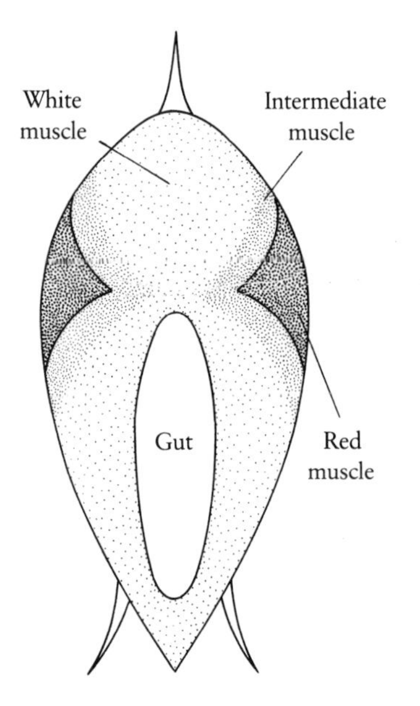


A cross section of a hypthetical fish, showing the types of muscle fibers. The red muscle fibers are used for sustained swimming; they have more mitochondrial enzymes and myoglobin and thus are better suited for continuous aerobic work. The white muscle fibers are recruited for burst swimming.

the top speed, attainable only with maximum effort, can typically be maintained for 30 seconds or less by fishes of intermediate size.

As shown in the figure on page 189, in which speed is plotted against length on a log–log graph, the burst speed increases with length raised to the power 0.88, whereas the active speed is proportional to length raised to the power 0.50. The largest fishes are able to increase their speed in going from the active to the burst mode by a much larger factor than smaller fishes are. In fact, fishes smaller than about 10 millimeters in length appear not to have a burst speed that differs from their active speed. The lines on the graph intersect just above this size.

Several fish researchers have said that the burst speed of a streamlined fish is about 10 lengths per second, and inspection of the figure shows that this is approximately true, although no one knows quite why. The slope of the active line has received a better explanation. Bainbridge (1961) and Webb (1977), who have presented analyses based on equating the power required to the power available for swimming, have calculated that the speed should be proportional to length raised to the power 0.56 if the boundary layer were laminar and proportional to length raised to the power 0.39 if the boundary layer were turbulent.

The high swimming speeds of large swimmers make it possible for them to jump far out of the water. Dolphins and sailfish can leap as much as 5 meters straight up, more than twice the height reached by the best terrestrial jumpers.

**Different Muscles for Different Speeds.** The skeletal muscles of the vertebrates are composed of three distinct populations of fibers: red, intermediate, and white. The red fibers make up 5 to 20 percent of the total muscle mass in most fish. They are found in a discrete red band running along the side of the body, as shown in the bottom figure on this page.

The red fibers are darker because they contain more mitochondrial enzymes and myoglobin than the surrounding white fibers, which suits them better to continuous exercise. In fact, Johnston and Goldspink (1973) have shown that the red muscles alone are used during the active mode of swimming, with the white muscles joining them only during burst swimming. The intermediate fibers, with properties that fall between those of the red and the white, are recruited at fast cruising speeds.

The white muscle is intrinsically capable of greater speeds of contraction than the red muscle, as is appropriate to the higher tail-beat frequencies that characterize the burst mode of swimming. Thus, the two gears that we spoke of in explaining active and burst swimming are accounted for by two separate

A mother humpback whale supporting her calf on her back. When a whale calf is young, its body is soft and its thoracic cage is not yet fully developed. The mother must therefore provide assistance to keep the young whale from sinking.

muscle masses. Both muscle masses are used in burst swimming, but the white muscles are turned off to save energy during sustained (active) swimming.

**How Fast Do Whales Swim?** All of this brings up an obvious question: Can the blue whale, with its body length of 25 to 30 meters (roughly 80 to 100 feet) really swim at 120 meters per second (270 miles per hour) in bursts, as one would predict by continuing the burst line of the figure on page 189 to the right? The answer would have to be that if they can, no one has ever seen them do it.

Following the pattern we remarked on earlier in the chapter, whales seem to have given up the burst swimming mode in parallel with the way that elephants have given up galloping and condors have given up fast flapping flight. As shown in the figure on page 189, an extrapolation of the active line to whale lengths produces the perfectly plausible sustained swimming speed of 9 meters per second, or 20 miles per hour. This fits with the observation made by Hill (1950) and others that whales and dolphins seem to swim at about the same maximum speed. At 9 meters per second, whales are in low gear (active mode), while dolphins are in high gear (burst mode), but both are swimming at maximum speed.

Perhaps there are ecological as well as physiological reasons why whales don't sprint. Many large whales don't have to catch anything except the tiny krill which they sieve through the great baskets of whalebone in their mouths. Some others, such as sperm whales, are known to eat large prey, especially giant squid. No one has clocked them at the great depths at which they hunt, and perhaps they swim much faster down there, but then how fast do they need to swim in order to catch a squid? Probably 9 meters per second would be adequate to the job. The truth may be that there is absolutely no advantage to being a super-fast whale, and perhaps that's why they don't bother.

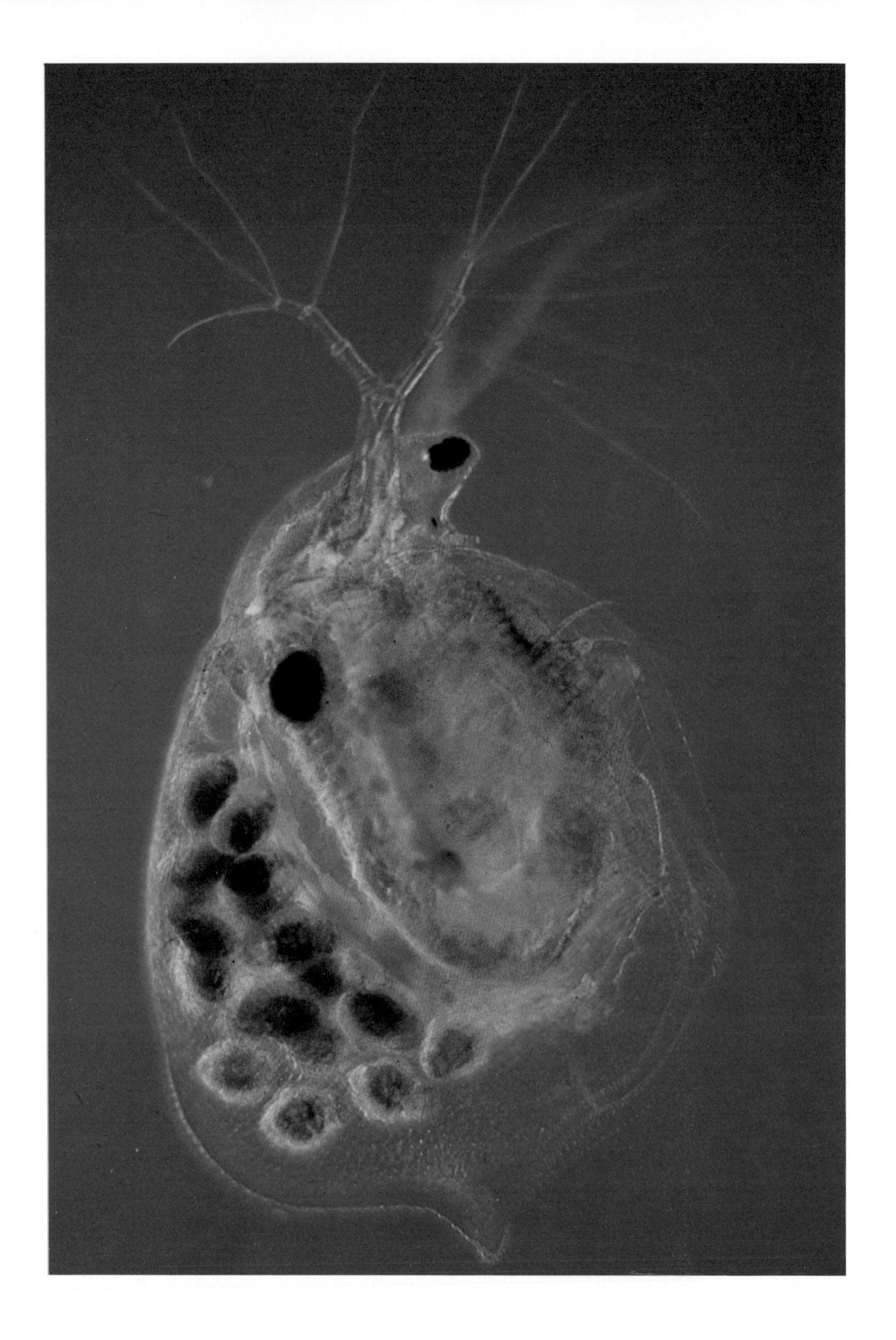

# Chapter 6

## On Being Small

By being small, an organism avoids most problems related to gravity and strength, but there are two new problems that appear with decreasing size. Both are of great interest, and their consequences are far reaching. One has to do with Reynolds numbers and the other with the forces of molecular cohesion, including surface tension. But before we go into each of these intriguing problems, let us first briefly consider the question of why it is to their advantage to move.

When we considered similar questions for larger organisms, we argued that, because speed and size are related, increased size would give advantage to prey and predator alike. One advantage of small size in a world of large animals is the ability to hide from predators in small places. But certainly we cannot explain the small size of bacteria (and their corresponding slow rate of locomotion) as an adaptation to escape from predators. Among other reasons, bacteria seem to be the first cells that appeared on the surface of the earth; they appeared long before any larger predators. Presumably, therefore, the initial explanation for their being small is that they were a first, early invention, and they have retained their small size because of their great ecological success.

Size is not always associated with motility. At one end of the spectrum, we have giant trees firmly anchored into the ground; at the other, we have nonmotile bacteria that are just as common as the motile ones. It seems reasonable that the presence or absence of motility is due entirely to adaptation. For a tree, whose energy is acquired by photosynthesis, immobility is clearly advantageous. It is not simply that there is no need to move in order to catch the sun's rays; it is also that the amount of additional energy that could be acquired in this way would not be sufficient to permit all sorts of energy-consuming rushing about. It is possible to imagine that, for some bacteria, the ability to move towards richer sources of food might be desirable; but, for those bacteria that have landed by accident in a great patch of food, it is more efficient simply to eat and multiply—there is simply no need to move. What we find among bacteria is that there are many species that have both motile and nonmotile strains, and, depending on the immediate environment, one or the other may be favored.

But this still does not say why an expensive (in terms of energy consumption) locomotory apparatus seems to be present not only in bacteria but in other small organisms as well. Let us consider the specific adaptive advantages of such a locomotory apparatus. There are two obvious ones: we have already mentioned the pursuit of food, and the other is attraction of the sex cells (gametes) to one another. In both these cases, minute organisms (or cells) can somehow orient in chemical gradients, and this emphasizes an extremely im-

A female *Daphnia*. Her parthenogenetic eggs are visible in her dorsal brood chamber.

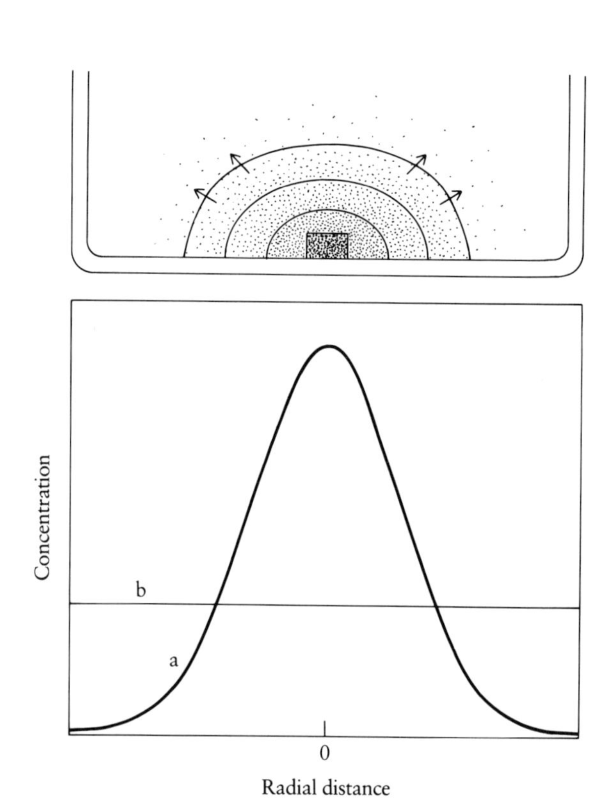


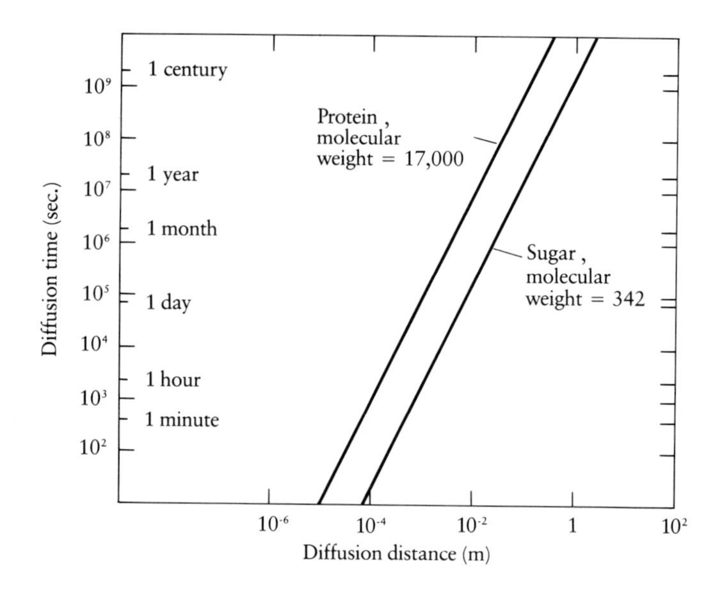


*Top left*, a sugar crystal dissolving in the bottom of a glass. *Bottom left*, the concentration is highest near the crystal shortly after it has dissolved (*a*), but eventually the sugar diffuses evenly throughout the water, leading to a uniform concentration (*b*).

*Right*, diffusion time vs. distance on a log–log plot. A protein molecule, being much larger than a sugar molecule, requires more time to diffuse over the same distance.

portant point for microscopic life: Gradients promote diffusion, and diffusion is a dominant physical phenomenon in the activities of microbes.

Diffusion concerns the random motions of molecules, which rush about at great speed until they collide and then change their direction. If one examines a liquid containing barely visible particles, one can see the particles move with irregular jerks as they are struck by moving molecules, a phenomenon known as Brownian motion. If one puts a sugar crystal in the bottom of a glass of water and allows it to sit undisturbed, it will soon dissolve. At first, there will be a high concentration of sugar molecules around the place where the crystal lay before it dissolved and a lower one farther away (see figure *top left*). Given enough time, the sugar molecules will eventually become evenly distributed in the water; this is the inevitable result of their random motion. The rate at which diffusion occurs depends on various physical factors, such as temperature, but there are two factors that play a most significant role. One is the steepness of the concentration gradient (the steeper it is, the more rapid the rate of diffusion from the region of high concentration). The other is the size of the molecule (see figure *top right*).

If bacteria or other motile unicellular organisms (including amoebae) lie in a gradient of some substance to which they are attracted (or by which they are repelled), they have the means of sensing the gradient and will move toward the greater concentration of the substance (or away from it, if they are repelled). Of course, larger animals can also orient themselves in a chemical gradient. We do this ourselves when we try to locate the source of an odor. The remarkable thing is that something as small as a bacterium can do it, too. The bacterial mechanism that serves this purpose has some similarities to our own in that there are specific receptor proteins that detect the key molecules.

To summarize, we could say that bacteria and eukaryotic microbes are small primarily because they were originally constructed that way and were offered no inducement to become larger. But, because they are so enormously widespread over the surface of the earth, we can only assume that they continue as extraordinarily successful organisms. In some instances, this success is associated with, among other things, the ability to respond to, and to move in, concentration gradients of such things as nutrients and toxic substances. But microbial movement cannot be understood in terms of our own problems of locomotion. To think about the problems bacteria experience in moving, we must be able to imagine their world, an environment of low Reynolds numbers.

## A World Governed by Low Reynolds Numbers

As we said earlier, the Reynolds number is the ratio of the inertial forces to the viscous forces acting on a fluid particle. By definition, then, viscous forces are more important than inertial forces at low Reynolds numbers.

**Stopping Distance.** One way to visualize the greater importance of viscous forces at lower Reynolds numbers is to imagine what would happen if a fly were to try to fish like a pelican. Anyone who has ever been near a fish market knows that flies enjoy eating fish. Suppose that a fly decided one day to catch his own. Nothing would prevent his flying over the water, spotting a fish, and diving after it. But once through the surface of the water, the fly would be brought to rest by viscosity within a few centimeters, because he would not have the pelican's large inertia to carry him down to make the catch.

This same effect becomes even more important for microscopic swimmers. For spermatozoa or motile bacteria (which have Reynolds numbers on the order of $10^{-3}$ and $10^{-6}$, respectively), the distance the cell glides after turning

Two brown pelicans (*Pelecanus occidentalis*) diving for fish.

off its propulsive machinery (called the stopping distance) can be very small indeed. The stopping distance, expressed as a fraction of the cell's diameter, is one-eighteenth of its Reynolds number based on its diameter and its initial speed. For a bacterium 2 micrometers long swimming at a Reynolds number of $10^{-6}$, the stopping distance is about $10^{-7}$ micrometers, much less than the diameter of a single atom. Clearly, these cells don't move with the thrust-and-glide motions characteristic of fishes, some of which glide more than five body lengths between propulsive strokes. In order to make any progress, an organism swimming at a low Reynolds number must have its engine on all the time.

**Reversibility.** Another important thing about swimming at low Reynolds numbers is its reversibility. A very small organism makes progress through the water the way a corkscrew makes progress through a cork or the way a pinion gear makes progress over a rack. In theory (ignoring the effects of thermal diffusion), if a ciliated protozoan makes a certain number of motions with its cilia and then reverses those motions exactly, not only will it return to its starting place, but all the water molecules surrounding it will return to their starting places as well. This means that, again ignoring the minor effects of thermal diffusion, a little blob of dye mixed into the water by the swimming

An experiment demonstrating the reversibility of a low Reynolds-number flow. The narrow gap between an outer cylinder and an inner cylinder is filled with glycerine. A blob of dyed glycerine is introduced (via a hypodermic needle) into the space between the cylinders (*left*). When the inner cylinder is turned four turns in one direction, the dyed fluid appears to mix with the clear fluid (*middle*). When the inner cylinder is then turned four turns in the opposite direction, the particles of the dyed fluid return to positions close to their starting points (*right*). (The effects of thermal diffusion have caused the blob to appear somewhat fuzzier at the end of the experiment than at the beginning.) The reversibility of the flow demonstrated here depends on the very viscous property of the glycerine. The same experiment performed in the same amount of time but using water instead of glycerine reveals no such reversibility.

motions of a microscopic organism would be unmixed by an equal number of swimming motions carried out in reverse. This stands in sharp distinction to what would happen if a blob of dye were placed next to a fish swimming in a tank. In that case, the inertial motions of the water would carry the dye particles everywhere in the tank, and carrying the fish's motions out in reverse would only disperse the dye further.

A sailboat can be propelled forward, even in the complete absence of wind, by swinging the rudder back and forth. This wouldn't work at all in a suitably miniaturized sailboat, because the fluid motions generated by pulling the rudder to the right (or the left) would be exactly reversed by returning the rudder to the center.

Similarly, an organism with two rigid "oars" sticking out on either side could swim at high Reynolds numbers by stroking rapidly backward and slowly forward, using the inertial properties of the water to obtain greater drag on its oars on the backward (thrust) stroke. At low Reynolds numbers, it would go forward on the thrust stroke but backward an equal distance on the return stroke.

The American physicist Edward Purcell (1977) has given a vivid illustration of how swimming at a low Reynolds number would feel:

> It helps to imagine under what conditions a man would be swimming at, say, the same Reynolds number as his own sperm. Well, you put him in a swimming pool that is full of molasses, and then you forbid him to move any part of his body faster

than one centimeter per minute. Now imagine yourself in that condition: you're in the swimming pool in molasses, and now you can only move like the hands of a clock. If under those ground rules you were able to move a few meters in a couple of weeks, you may qualify as a low Reynolds number swimmer.

**Ciliary Propulsion.** The problem for a microscopic swimmer is to come up with a paddle or oar that will actively produce forward motion. If one were in a rowboat in Purcell's molasses-filled swimming pool and were not allowed to feather the oars or to take them out of the liquid, then, as we have said, no matter how one tried to vary the rate of the pull with the return push for each stroke cycle, the boat would remain in virtually the same place because of the reversible nature of flow at low Reynolds numbers. But Purcell points out that the way to get around this is to have a flexible oar, so that the shape of the oar varies with the push and the pull. With a flexible oar, the blade and shaft can be held out straight on the power stroke, providing an effective forward thrust. Then the oar can be allowed to fall limp on the return stroke, so that the blade

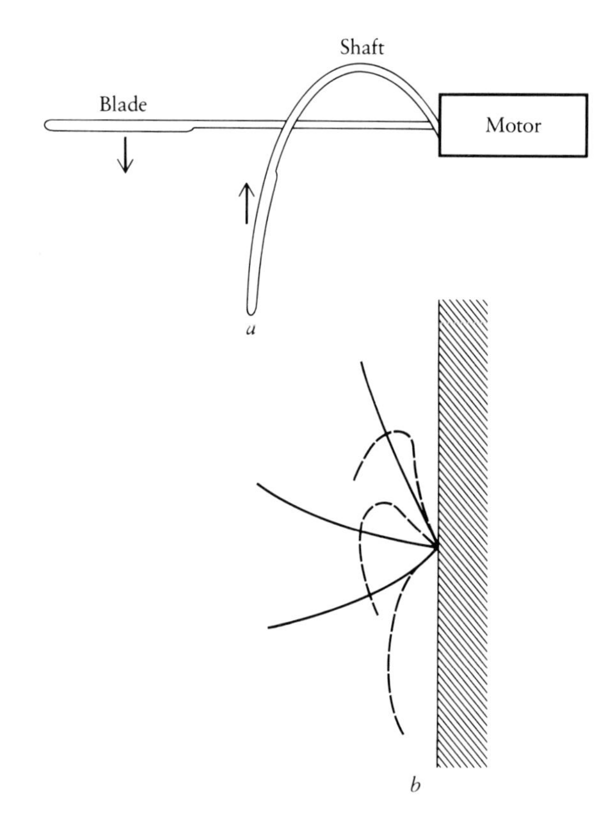


A flexible oar (*a*) for rowing at low Reynolds numbers remains reasonably rigid on the thrust stroke (moving downward) but bends significantly on the return stroke. Ciliary beating (*b*) uses the flexible-oar principle. The solid curves show successive positions of the cilium beating downwards; the broken curves show the cilium on the return stroke.

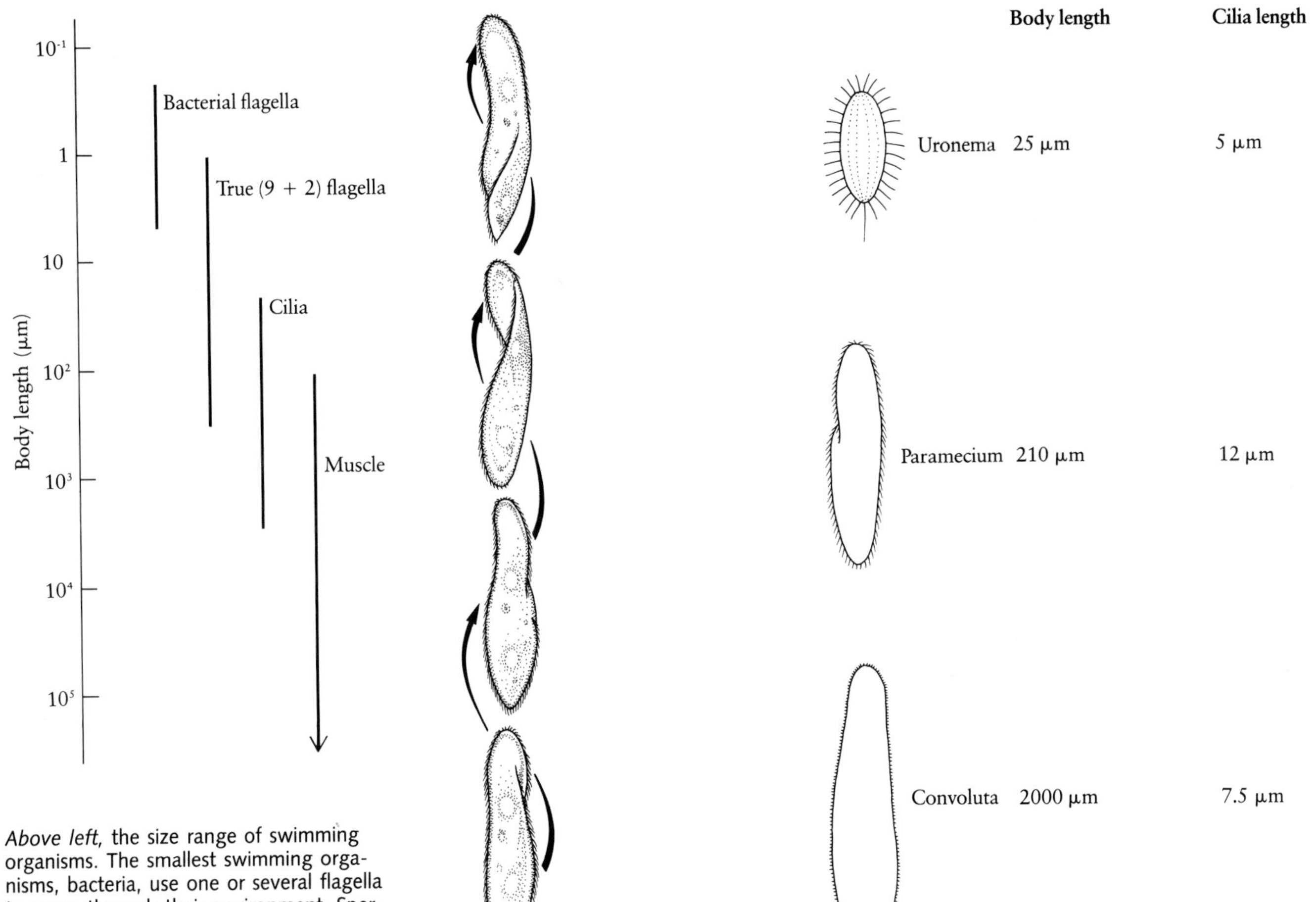


*Above left,* the size range of swimming organisms. The smallest swimming organisms, bacteria, use one or several flagella to move through their environment. Spermatozoa are the best-known examples of cells with "true" (9 + 2) flagella, the next size class. Larger life forms use cilia for propulsion, and even larger forms move about using muscle. The range of ciliates considered does not include organisms with comb plates.

*Above right,* ciliated organisms of different sizes, showing that the length of the cilia remains roughly constant as body size increases.

*Center,* a *Paramecium* propels itself forward at a speed of 2–3 millimeters per second by beating the hairlike cilia covering its body. These protozoans rotate as they swim. They reverse their ciliary motion and go backwards after encountering a solid object.

and the end of the shaft are drawn through the water roughly parallel to their own long axis, which is a configuration of relatively low drag, even at low Reynolds numbers. This is precisely what occurs with the hairlike cilia covering the surfaces of many protozoans (or covering the epithelium of multicellular organisms, such as that lining the gills of clams and other filter feeders).

Organisms that get about using ciliary propulsion range in size from about 20 micrometers to 2 millimeters. At their largest, they overlap the size range of animals that use muscles for movement—for example, the smallest species of the water-flea genus *Daphnia*, crustaceans that swim using paddlelike limbs. At their smallest, ciliated organisms overlap the size range of those organisms that use flagella for propulsion (see figure *top left*). The main difference between cilia and flagella is that flagella function by means of propagated waves that move along their length opposite to the direction of swimming, while cilia beat with a flexible-oar motion, as explained above.

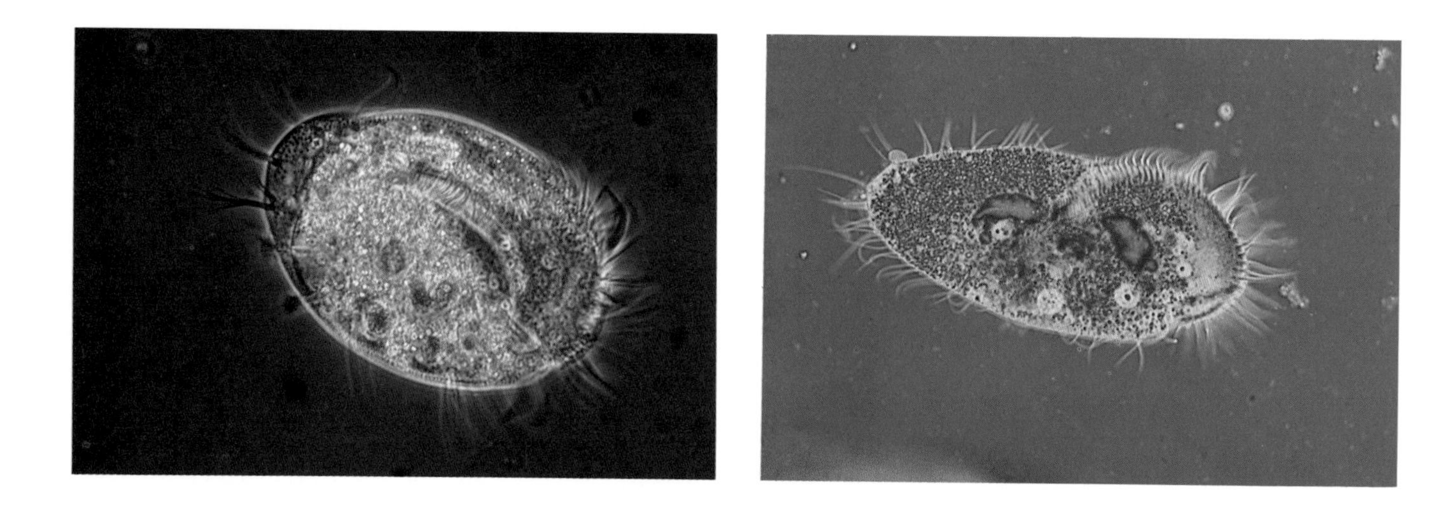

Two ciliated protozoans: *(left) Euplotes patella; (right) Stylouchia pulsata.*

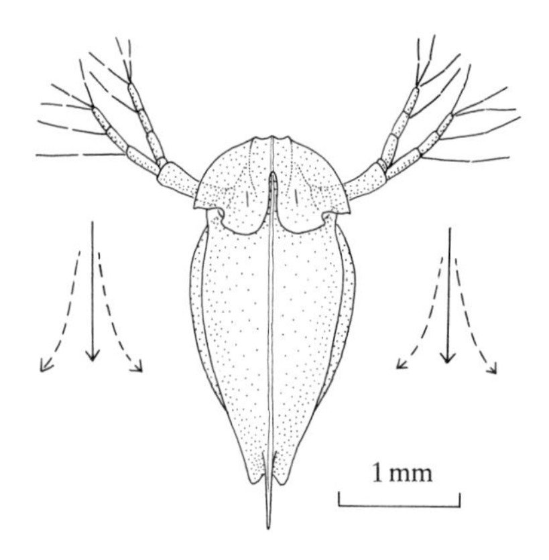


A tiny water flea of the genus *Daphnia* (actually not a flea but a crustacean) moves forward by rowing with its two oarlike limbs.

Although the lengths of ciliated organisms range over two orders of magnitude, the lengths of their cilia (excluding those of organisms with compound cilia) are confined to a narrow range between 5 and 12 micrometers (*right* figure, page 199). The cilia in mammalian lungs are also generally in this range of lengths. Most cilia beat in the frequency range of 10 to 20 cycles per second. This remarkable consistency in ciliary length and frequency of beating is responsible for the fact that most ciliated organisms swim at the same speed, about 1 millimeter per second. All these observations imply that there is some very strong constraint on ciliary design that prevents any cilium from growing longer or beating faster than any other one. The constancy of swimming speed of ciliated swimmers means that the largest ones are very slow by comparison with muscular swimmers of the same size. For example, the ciliated flatworm *Convoluta*, which is 2 millimeters long, swims at 0.6 millimeters per second, whereas 2-millimeter-long water fleas of the genus *Daphnia*, which move their limbs by muscular action, swim more than 10 times faster at 7 millimeters per second (Lochhead, 1977). The comparison is fair because the water fleas swim using a rowing motion of their limbs that is similar in some respects to ciliary rowing, although their limbs are much longer and thicker than cilia.

**Flagellar Propulsion.** Below the ciliated organisms on the size scale are those unicellular swimmers that use flagellar propulsion, including spermatozoa and flagellated bacteria. The sperm cells shown at the *top left* of page 201 swim forward by sending a backward wave along their flagella, which may be 0.2 to 1 micrometers in diameter and 10 to 500 micrometers in length. The flagellar

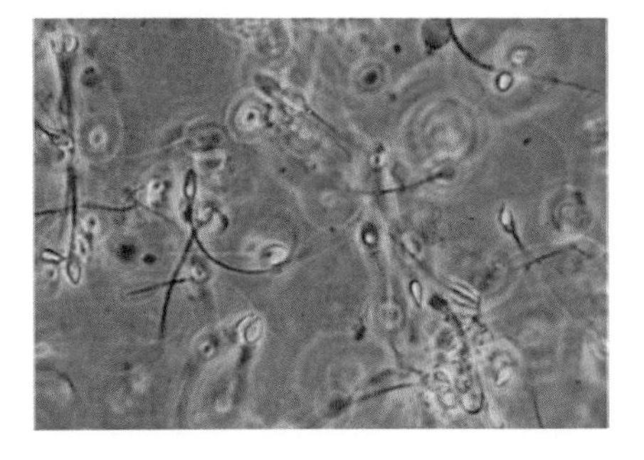

*Above,* human sperm cells. Each consists of a head attached to a whiplike tail, the flagellum. They swim by beating their tails from side to side; part of the viscous force on the tail pushes the sperm ahead.

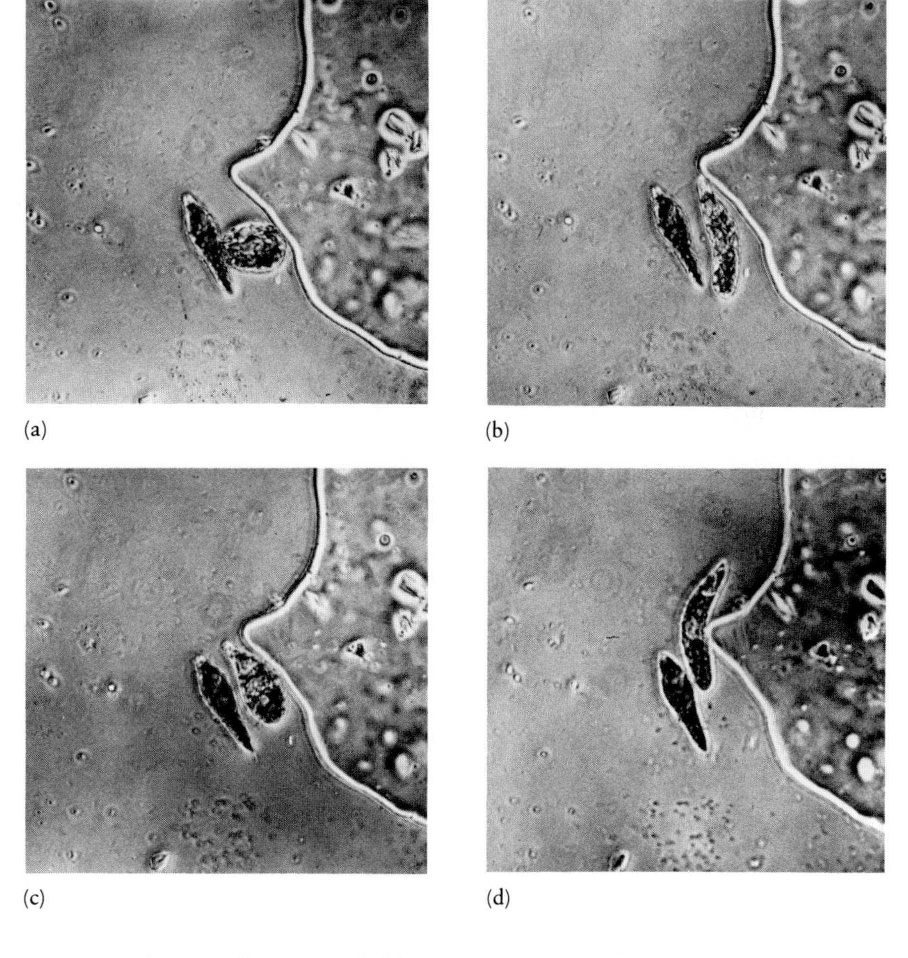


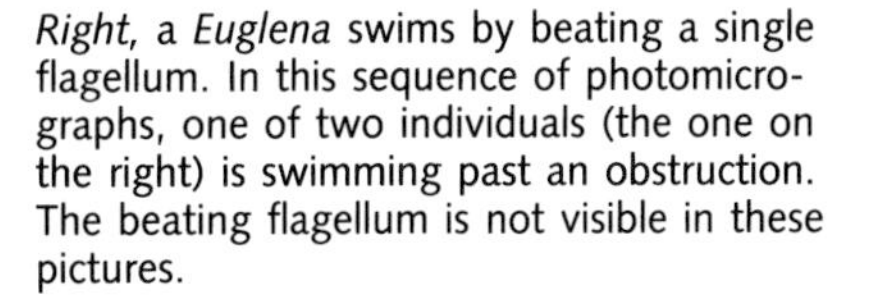

*Right,* a *Euglena* swims by beating a single flagellum. In this sequence of photomicrographs, one of two individuals (the one on the right) is swimming past an obstruction. The beating flagellum is not visible in these pictures.

wave may be two-dimensional (like a wavy line) or three-dimensional (like a corkscrew). The wave of bending requires energy and utilizes adenosine triphosphate (ATP), just as skeletal muscle employs ATP as a high-energy fuel. A cross section of the flagellum (which looks much the same as a cross section of a cilium) reveals its characteristic structure of nine outer double hollow tubules surrounding a set of two inner tubules. The active wave of bending is now thought to involve relative sliding of these tubules in a manner comparable to the relative sliding of thick and thin filaments in striated muscle fibers.

Flagellates all tend to swim at the same speed, regardless of their size. The swimming speed of the flagellates, approximately 0.1 to 0.2 millimeters per second, is only about one-tenth the swimming speed of the ciliates (which also

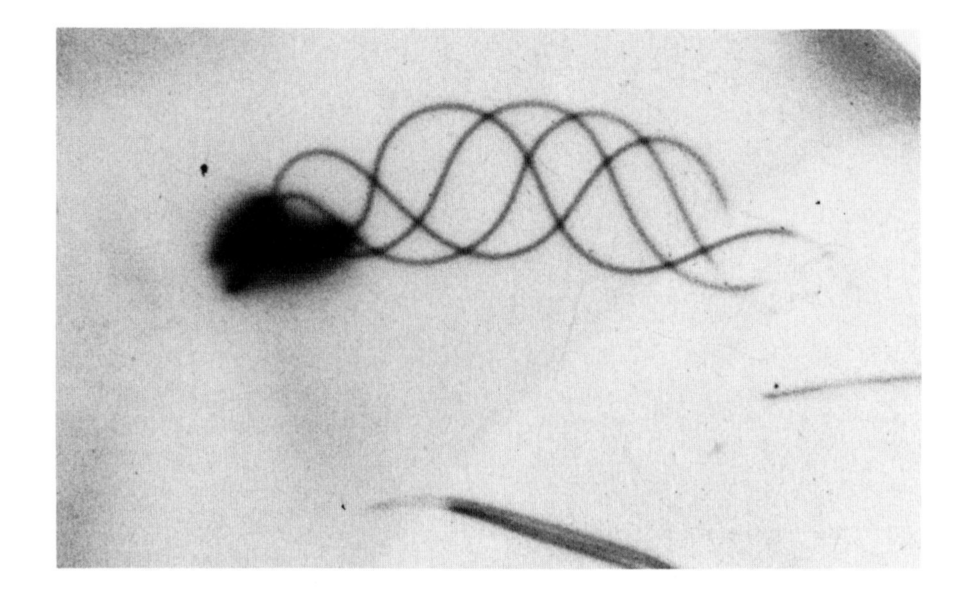

A multiple-exposure photograph using four light flashes 10 milliseconds apart shows the swimming motions of this spermatozoon from the sea urchin *Lytechinus pictus*.

tend to swim at one speed, no matter what their size, as was mentioned earlier). Quite clearly, there is a large speed advantage in being a ciliate rather than a flagellate in the range of sizes in which both types of swimming are observed.

**Flagellated Bacteria.** Some types of bacteria use one or several flagella for swimming. The remarkable part of this story is that it was assumed for many years that motile flagellated bacteria use a sliding-filament system like the one used by eukaryotic flagellated cells (such as spermatozoa). Yet it was known that the bacterial flagellum is much smaller and simpler in structure than the flagellum of a eukaryotic cell. Then H. C. Berg and R. A. Anderson (1973) made what seemed like a wild suggestion: that bacterial flagella rotate about their own axes, employing rotary joints that connect them to the cell body. Shortly afterwards, M. Silverman and M. Simon (1974) provided evidence in favor of this idea. When the end of the flagellum of a bacterium was attached to a glass slide using a suitable antibody as a kind of glue, then the bacterium would rotate about its anchored flagellum. This mechanism is so different from any other mechanism known to be used in cell locomotion that its proposal was quite unexpected. If indeed there is a rotary joint between the flagellum and the cell body, no one knows how it works.

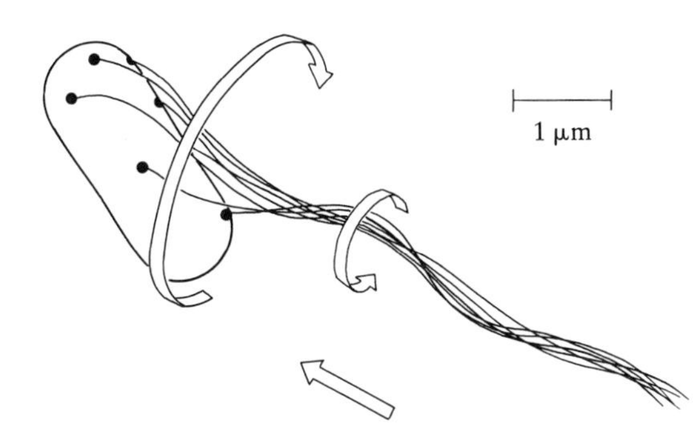


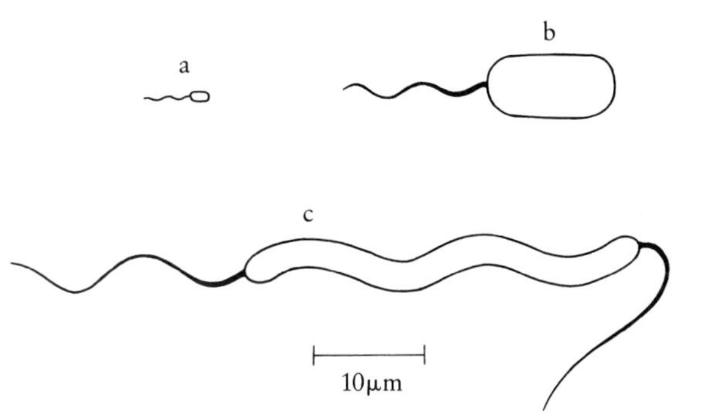


*Left,* a typical flagellated bacterium. One or several long, helical filaments (flagella) project from the surface of the cell. When the bacterium is swimming steadily, the separate filaments are united in a flagellar bundle, which pushes the cell along. The arrows show that the cell body and the flagellar bundle rotate in opposite directions during steady swimming. Periods of steady swimming are interrupted by periods of erratic motions in which the bundle flies apart.

*Right,* scale drawings of flagellated bacteria show several representative species. In *Escherichia coli* (*a*), approximately six filaments originating on the sides of the cell form a flagellar bundle. In *Chromatium okenii* (*b*), about 40 filaments arise at one pole. In *Spirillum volutans* (*c*), about 25 filaments leave the cell at either end. This cell is shown swimming from left to right.

It should be added that motile bacteria can change direction by momentarily reversing the direction of flagellar rotation, causing the flagellar bundle to splay outward and resulting in a reorientation of the cell. Then the normal direction of rotation resumes, and the bacterium swims off on a new course.

Another curiosity found among the bacteria is that some species that live in mud are able to orient themselves with respect to weak magnetic fields. They contain crystals of magnetite (lodestone) of just the right size for optimum orientation in magnetic fields (Frankel et al., 1979). It is hardly certain why this should be so, but a reasonable hypothesis has been put forward. Because bacteria are too small to have conventional gravitation sensors (large animals use otoliths—literally "rocks in the ears"—mounted on sensitive hair cells to sense the direction of gravity), and if it is important for bacteria to move downwards into the mud for nutrients, then magnetic sensing would accomplish this goal, because the earth's magnetic field has a component that points up and down as well as one that points north and south. The use of magnetic sensing is not confined to microbes, however; there is increasing evidence that it is involved in bird and bee behavior as well.

**Insect Flight.** The most important thing about insect flight is that it is carried on exclusively at small scale—the largest flying insect is smaller than the smallest mouse. The flight muscles of insects are among the most active animal tissues known, but a remarkable fact is that the insect's blood is not the mechanism for transport of oxygen to the muscles. Instead, air-filled tubes, or tracheae, carry oxygen from the surface of the body directly inward to the muscles and other organs. Currents of air are forced through these tubes to make them work more efficiently. It is possible that the upper limit on the

A dragonfly of the genus *Aeshna* in flight.

The wings of a light aircraft and a dragonfly contrasted: *a*, the shape of the wing of a Piper J-3 Cub at mid-span; *b*, cross section of the fore wing of the dragonfly *Aeshna interrupta*; *c*, plan view of a dragonfly wing.

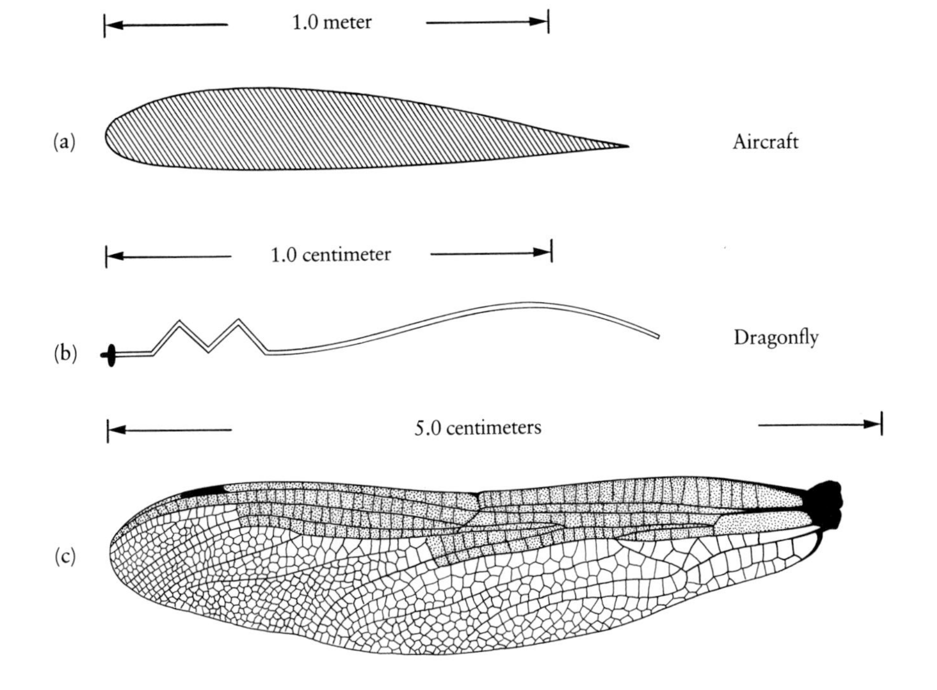


range of insect sizes is dictated by the rate at which oxygen can be supplied through this tracheal system, but Knut Schmidt-Nielsen (1977) has argued that the upper limit on the power output of insect muscle is more likely to be a property of the muscle itself and not constrained by the rate at which oxygen arrives. The fossil remains of certain giant dragonflies of the Carboniferous period had wingspans of nearly 750 centimeters, but these must have been exclusively gliders. The smallest fly is shown on page 3 in Chapter 1; it is about the length of a human ovum and much smaller than the largest ciliate protozoan. The smallest flying insects weigh as little as 1 microgram and have wingspans of only 0.2 millimeter. The Reynolds numbers of hovering insects range from about 1 to $10^4$. Although this range does not include Reynolds numbers as low as those we have just considered for swimming microorganisms, it does consist entirely of Reynolds numbers sufficiently low that wings of a design conventional for small airplanes will not work effectively. An airfoil like the one pictured on page 205, part *a*, will show an abrupt increase in drag and a reduction in lift below a Reynolds number near $5 \times 10^4$. This increase in drag is caused by a separation of the laminar boundary layer from the top surface. At low Reynolds numbers, flat or curved plates work better than the thicker shapes appropriate for airplanes.

The dragonfly wing shown on page 205 has two big horrible-looking pleats in it just aft of the leading edge. Although it is clear that these pleats contribute substantially to the mechanical stiffness of the wing in bending along the wingspan (just as the corrugations in a piece of cardboard give it strength), intuition suggests that this irregular shape must exhibit very poor aerodynamic properties.

In this case, intuition is wrong, as B. G. Newman and his collaborators have shown in an ingenious way. Wind tunnels inevitably have turbulent flow in them, and, although the turbulence can be reduced to low levels through proper care, even low levels of turbulence have a profound effect on the flow of air past wings at low Reynolds numbers. Because of this, Newman not only flew dragonfly wings in a wind tunnel but also supplemented his wind-tunnel studies with the free flights of model gliders whose wings had the dragonfly-wing cross section. One model had a wingspan of 1 meter and was designed to fly at a Reynolds number near $3 \times 10^4$; the other was half this size and flew at

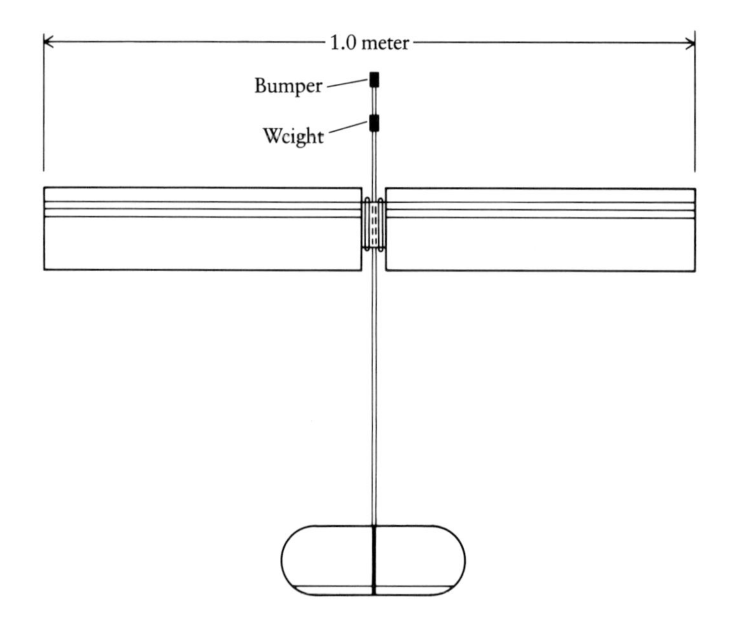


A model glider with wings like those of a dragonfly was used to study dragonfly aerodynamics.

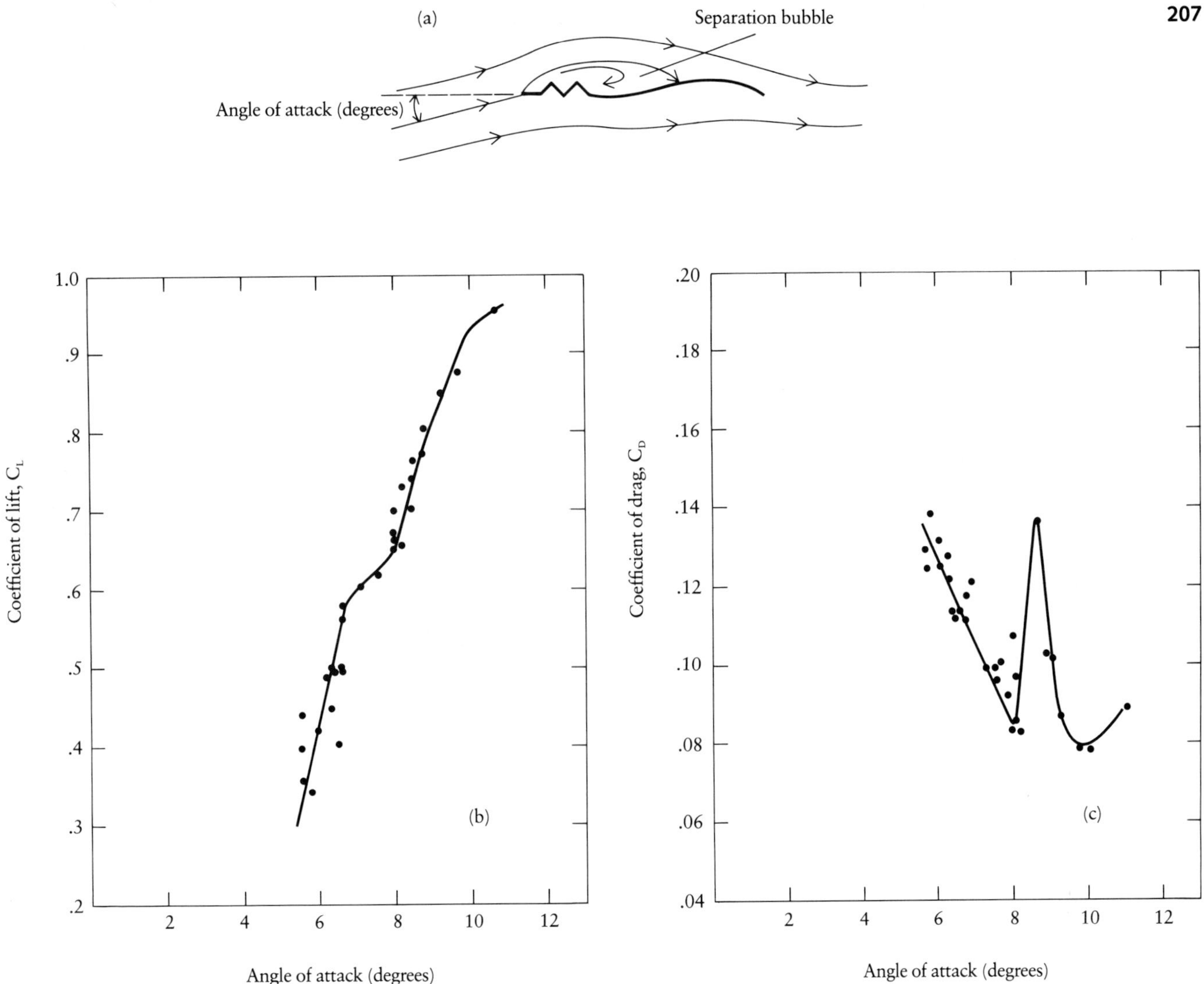


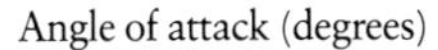


Results of experiments with dragonfly wings. Flow visualization using smoke in a wind tunnel (*a*) shows a large separation bubble above the pleats in the wing. The flow reattaches aft of the bubble. Lift coefficient (*b*) and drag coefficient (*c*) are shown plotted against angle of attack. The lift and drag coefficients were obtained by analyzing the flights of model gliders. At an angle of attack near 10 degrees, the ratio of $C_L/C_D = 0.95/0.08 = 11.9$.

a Reynolds number near $1 \times 10^4$. Although the models were larger than dragonflies, these Reynolds numbers are still typical of the values appropriate for forward flight in the large dragonfly *Aeshna cyanea*. By flying his models indoors in a gymnasium, Newman was able to reduce the level of turbulence in the incident air nearly to zero.

Newman's models were trimmed to glide at a constant speed along a straight path by moving a weight near the nose either forward or back. Stroboscopic photographs were analyzed to give the flying speed and the angle of glide. Using this information, the investigators calculated the lift coefficient and the drag coefficient as a function of the angle of attack, as is shown above. For a given wing, the ratio of the lift coefficient to the drag coefficient is at a maximum at a particular angle of attack. This ratio, called the maximum

*Left,* serrations and small hairs protruding from the leading edge of the hind wing on a large dragonfly, *Aeshna cyanea*. The leading edge of the wing runs from the upper left to the lower right of the photograph; the wing extends to the left from the leading edge. The small hairs are about 35 micrometers long. The precise function of these serrations and hairs is not known. It is possible that the hairs serve a sensory function, detecting the direction of the airflow. B. G. Newman notes that roughness elements, including the hairs, may also promote a transition to turbulence in the boundary layer, allowing the wing to achieve a higher performance than it would with a smooth leading edge.

*Right,* a very small insect, the white plume moth. Its wings are so loosely joined together that their design would serve a bird or large insect quite poorly, but they work fine at this small scale.

lift-to-drag ratio, reveals (among other things) the greatest distance the wing could move forward per unit distance of vertical descent when the wing is used for gliding. Conventional aircraft wings, like the one shown in part *a* of the figure on page 205, have a maximum lift-to-drag ratio greater than 30 at the high Reynolds numbers at which they were designed to operate, but the lift-to-drag ratio is typically below 5, and may fall as low as 2, for Reynolds numbers in the dragonfly realm. By contrast, the figure on page 207 shows that, when the angle of attack is near 10 degrees, the dragonfly wing has a lift coefficient of 0.95 and a drag coefficient of 0.08, giving a lift-to-drag ratio of 11.9.

Newman and his collaborators found that the comparatively good performance of the dragonfly wing is aided by the longitudinal pleats. Partly because of the presence of the pleats, a large bubble of separated flow exists on the upper surface of the wing. If this separated flow continued past the trailing edge, the wing would be stalled—and the lift would be low and the drag would be high, as they always are on a stalled airfoil. Instead, the flow in this configuration reattaches to the top surface aft of the bubble and remains smooth over the rest of the wing (part *a* of the figure on page 207). The bubble acts as if it were part of the wing, guiding the air around it in the proper way to maintain lift.

In addition to the pleats, a dragonfly wing has minute hairs and saw teeth

Among the smallest insects are the several species of fairyflies, which are actually tiny wasps. Their body lengths can be less than 0.5 millimeter (about 0.02 inch). The wings are slender stalks, the margins of which are fringed with hairs (setae). The wings can be used for swimming underwater as well as for flying.

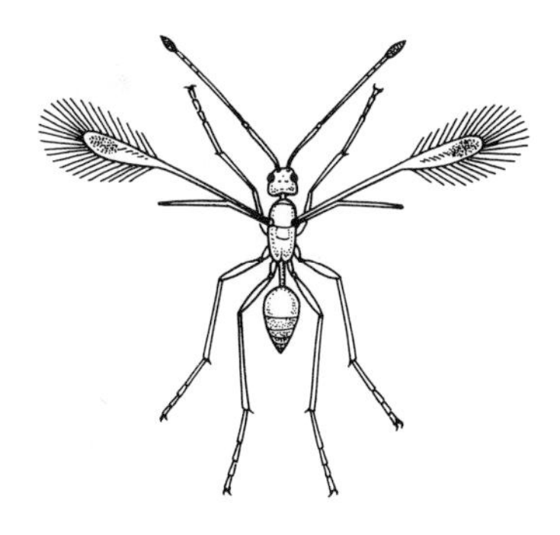

(serrations) along the leading edge, which, Newman found, improve the performance still further by promoting the transition to turbulence in the separated layer of air above the bubble. As a consequence, the boundary layer can reattach in a turbulent (and therefore more stable) state. He comments that model aircraft flying at the same range of Reynolds numbers often use equivalent leading-edge serrations to obtain the same benefits. These leading-edge serrations play the same role as the trip wires mentioned at the end of Chapter 3.

These tests investigated forward flying only and did not address hovering, but it is well known that the cross-sectional shape of an airfoil is less important in hovering than it is in fast flapping flight. Small insects with all sorts of bizarre airfoils can clap them together front and back and somehow get enough lift to hover. In fact, the wings of the various species that constitute a family of tiny wasps known as fairyflies look like feather dusters—they are not membranes at all but fringed paddles—and yet, at their low Reynolds numbers, these insubstantial wings can be as effective as solid ones.

**Jumping.** When jumping height is compared to body length, all the champion jumpers are small animals. Fleas, who jump for a living, must jump high enough to reach the dog or cat as it walks past, a distance that can be more than 100 times their own length. They do this not by using their muscles

A rhododendron leafhopper. Leafhoppers, locusts, and many other insects use their muscles to wind up some elastic element in their bodies before they jump, and then they depart in an explosion of energy as the elastic element is released.

directly in the jump but by using their muscles beforehand to store energy in a pad of protein called resilin. The resilin is placed in shear between two stiff skeletal elements, so that the whole mechanism behaves like a cocked spring. This energy is released abruptly like the snap of a slingshot when the animal wishes to ascend. In locusts, the jumping energy is stored in the bending of certain portions of the hind leg. Although kangaroos and other large jumping animals cannot jump using quite the same principles as insects, they do store energy in their massive Achilles tendons, as we have seen in the hopping experiments described in Chapter 5.

Zebras and wildebeests jumping.

A woman jumping from stone to stone. This is plate 170 from *Animal Locomotion*, which was published by Eadweard Muybridge in 1887. An ordinary jump like this usually involves a vertical rise of less than 20 centimeters.

A jumping animal begins from rest with a certain quantity of stored energy and uses that energy to bound upward. A. V. Hill (1950)—and D'Arcy Thompson (1917) before him—pointed out that, if the stored energy is assumed to be proportional to the jumping animal's mass, then, because the gravitational potential energy at the greatest height of the jump is given by the product of the animal's mass times gravity times height, the law of conserva-

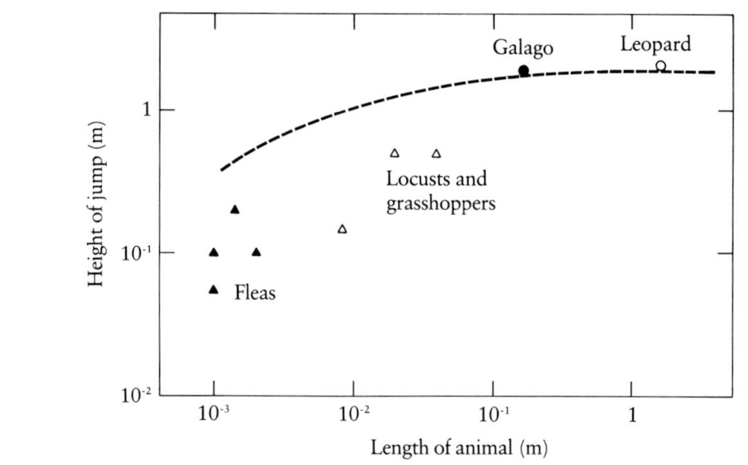


Jump heights for various jumping animals. The broken line indicates the theoretical height reached assuming a specific energy of 20 joules per kilogram, taking air resistance into account.

tion of energy would lead us to expect that the height of the jump would be independent of body size. Indeed, the maximum height reached by antelope, leopards, dogs, cats, and even humans is in the neighborhood of 2 to 2.5 meters and is practically independent of size.

What, then, prevents fleas from reaching 2 meters in height instead of a few centimeters? First, there is air resistance, which H. C. Bennet-Clark (1977) has estimated absorbs only 10 percent of the energy in a vertical jump of 1 meter by an animal 10 centimeters long but consumes 50 percent of the energy in a 10-centimeter jump by an animal 1 millimeter long. The broken curve in the figure above represents the maximum jumping height versus animal length assuming a specific energy of 20 joules per kilogram, taking air resistance into account.

As is apparent from the figure, air resistance can't be the whole reason why fleas can't jump as high as leopards. Another very important matter concerns acceleration. As size decreases, the peak acceleration rises rapidly, as do, therefore, the peak stresses borne by muscles and skeletal elements. For geometrically similar animals jumping to the same absolute height, because the vertical velocity on leaving the ground is independent of size, the time of thrust should be proportional to the animal's characteristic length, and consequently the vertical acceleration should be inversely proportional to that length. In fact, the peak vertical acceleration of a leopard jumping is only about three times the acceleration due to gravity, but the peak acceleration of a flea jumping one-tenth as high is more than 200 times that of gravity, and the vertical acceleration of jumping click beetles may exceed 400 times that of gravity.

Bennet-Clark points out that all sorts of design problems come up when

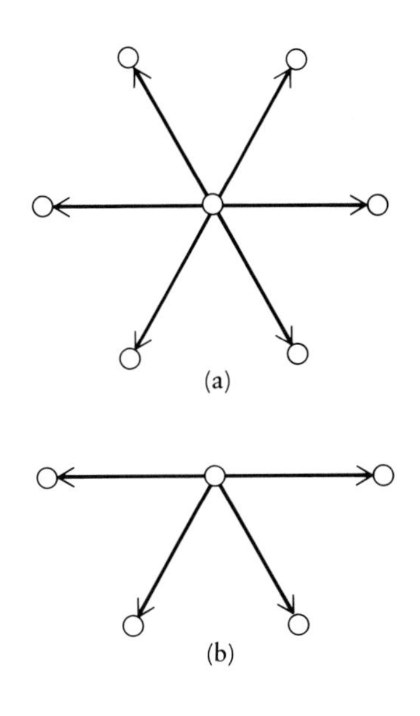


Surface tension. A molecule in a liquid that is far from any surface (*a*) is equally attracted in all directions by other molecules. A molecule at the surface (*b*) is attracted only by molecules at the sides and below and thus has a net tendency to be pulled away from the surface.

animals must experience such high accelerations. Not only must the skeleton be built strongly enough to withstand the acceleration, but the internal organs must be well enough anchored that they stay firmly in place. An animal with the internal skeleton and the haphazard attachments of internal organs characteristic of large vertebrates could not stand an acceleration of 400 times that of gravity without beginning to come apart. Perhaps this is the most important reason fleas don't jump 10 times higher than they do. It could be that they have already reached the limits of acceleration that even their rugged bodies can stand.

## A World Governed by Molecular Cohesion

Just as the world of large organisms is ruled by the forces required to overcome inertia and gravity, the world of minute organisms is ruled by the forces of attraction between molecules. These forces are manifested in a variety of phenomena, including surface tension.

**Surface Tension.** In any liquid, the component molecules will attract one another. Any molecule that is in the middle of a body of liquid will be pulled by the molecules surrounding it equally on all sides (part *a* of the figure above). A molecule lying on the surface, however, will have neighboring molecules below and to the sides only, and therefore it will tend to be pulled away from the surface (part *b* of the figure). As a consequence, the molecules at the surface have higher energy than those in the body of the liquid. The difference between the energy per unit area of a sheet of molecules at the surface and the energy per unit area of a sheet of molecules in the body of the liquid is called the *surface tension*. Notice that, because surface tension is measured in units of energy per unit area, its dimensions may also be expressed as force per unit length.

A practical method for measuring the surface tension between a liquid and a gas is shown at the top of page 214. A round wire hoop has been brought in contact with the surface of a dish of liquid. The vertical force (less the weight of the hoop) is measured as the hoop is lifted slowly upward. The force is recorded just before the hoop breaks free from the surface-tension film, and the surface tension is estimated to be one-half of this force divided by the perimeter of the hoop. (The factor ½ is necessary to account for the two gas–liquid interfaces, one continuous with the surface inside the hoop, the other with the surface outside.)

A method for measuring the surface tension between a liquid and a gas. *Left,* a wire hoop is lifted from the surface of the liquid in a pan. *Right,* the force lifting the hoop is created by twisting one end of a wire under tension. A lightweight arm supporting the surface-tension hoop is attached to the center of the wire. The force required to break the hoop free of the surface tension of the liquid in the pan can be read from the calibrated dial at the right in this view of the instrument.

**Surface Tension and Gravity.** In Chapter 3, we used dimensional analysis to show that there is an inverse relation between the pressure and the radius in soap bubbles. The same is true of liquid drops, because they are also bounded by a surface-tension skin. A liquid drop suspended in space or floating at neutral buoyancy in an emulsion is spherical, but, when such a drop is placed on a solid surface in air, it adopts an oblate shape (see *top left* figure on page 215). Gravity causes the pressure to be higher at the bottom than at the top, and, because the surface tension is the same everywhere, the surface must have a higher curvature near the bottom than at the top.

The liquid-storage tank shown in part *b* of the figure looks like such a drop. When it has this shape, the steel skin bounding the liquid volume is not required to endure bending stresses, only tensile stresses. Furthermore, the skin is placed under the same tension per length everywhere, and, therefore, no one region of the tank is more likely to rupture than another.*

In the drop problem, the interaction between surface tension and gravity is controlled by the surface-to-volume ratio, with the result that very small droplets sitting on a nonwetting solid surface may be almost spherical, whereas very

*This is strictly true only for a full tank.

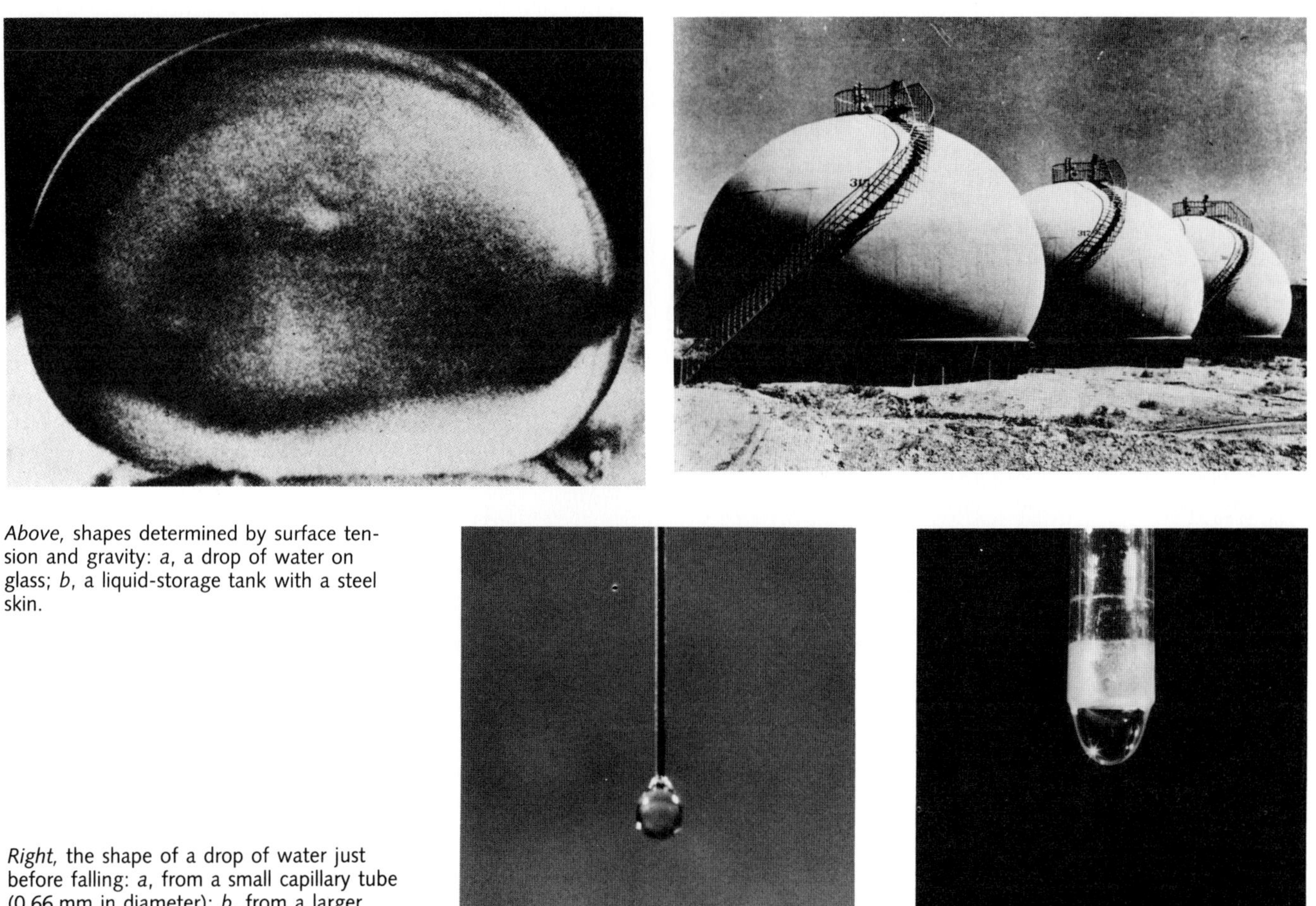

*Above,* shapes determined by surface tension and gravity: *a*, a drop of water on glass; *b*, a liquid-storage tank with a steel skin.

*Right,* the shape of a drop of water just before falling: *a*, from a small capillary tube (0.66 mm in diameter); *b*, from a larger tube (7.9 mm in diameter).

large drops will have much flatter shapes. In an alternative view of the same phenomonon, the balance between surface tension and gravity depends on the ratio of a perimeter to a volume—as, for example, in the case of a drop of water hanging from the mouth of a tube. If the faucet supplying the tube is leaking slightly so that the hanging water drop is growing slowly, the drop will grow until its weight exceeds a critical force proportional to the surface tension times the perimeter of the opening of the tube, at which time the surface-tension membrane will break and the drop will fall. A water drop hanging from a

A water strider on the surface of a pond. The dimples created in the surface of the water under the legs would hold a weight of water, if filled in, equal to the weight of the insect.

small capillary tube has a bulbous shape, but the largest drop that can hang from a big tube looks quite different.

Whether an animal can stand on water also depends on the ratio of a perimeter to a volume. In this case, the perimeter is the total perimeter of the leg surface wetted by the water. The water strider shown above has long, thin legs that serve to increase this perimeter. Fine, velvety hairs help to protect the legs against wetting. Water striders occasionally do break through the surface tension and fall in. When this happens, they swim back to the surface, aided by the buoyancy of air bubbles trapped among the hairs of the body. If their tarsal hairs are wetted, they must crawl out onto some dry surface until their legs are dry again.

**Surfactants.** When a small amount of household detergent is added to water, the surface tension is markedly reduced. For example, the surface tension of water at 30 degrees Celsius is 72 dynes per centimeter; but, when a drop of detergent is added to a cup of water, the surface tension may fall to as little as 30 dynes per centimeter. Substances that have this property are called *surfactants,* and there are naturally occurring biological ones as well as synthetic ones. In the mammalian lung, a natural surfactant, dipalmitoyl lecithin, reduces the negative pressure within the chest required to keep the lung's tiny air sacs (alveoli) open against the forces of surface tension, which tend to close them.

Surfactants modifying surface-tension forces can also be used for locomotion. A tiny chip of camphor placed on a water surface will race all about in

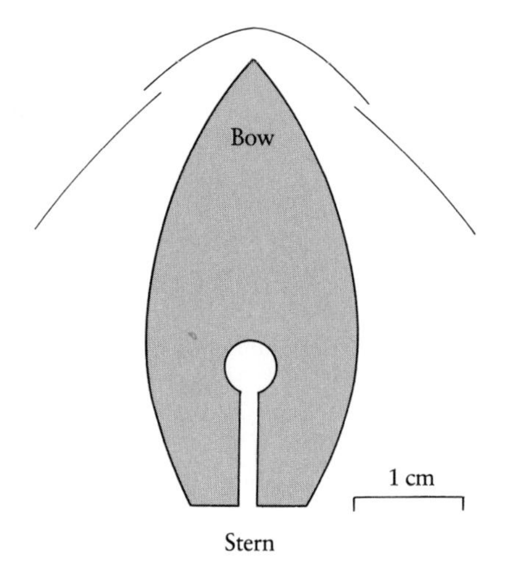


A boat propelled by a surfactant. Placing a chip of camphor (or a drop of detergent) in the hole allows the surfactant to escape at the stern. The resulting imbalance of surface-tension forces propels the boat forward.

random directions. This is because the camphor, which is a surfactant, dissolves unevenly, which means that the water on one side of the chip or the other will contain more dissolved surfactant, with the net result that the greater surface tension on the opposite side will pull the camphor chip in that direction. A more directed motion can be obtained with the small boat shown on the left, which is nothing more than a sheet of wood with a keyhole-shaped cutout. A chip of camphor or a drop of detergent placed in the hole will send a surfactant stream flowing out the stern. The boat will then go forward because the surface-tension forces on the stern are less than those on the bow.

Given that biological surfactants exist, it is amazing that small organisms do not make more extensive use of them. No water striders or whirligig beetles yet described propel themselves with surfactants. No known predators drown their prey by spraying detergents on them, although there is no reason why this should not work. Perhaps a careful analysis would show that surfactant propulsion and surfactant warfare are always slower, or less effective, or more hazardous than other methods of doing the same thing.

**Coming Out of the Water.** Consider the *micro*conquest of land. It is often assumed that, when one uses the term "conquest of land," one is referring to the evolution of algae into the early land plants and the subsequent rise of angiosperms or to the transition from fish to amphibians and thence to the reptiles and higher vertebrates. These are celebrated examples of the conquest of land, but they hardly constitute the only examples. Many microorganisms, even bacteria, have had their own conquests of land. Among the bacteria, the cells of myxobacteria (with the help of chemotaxis) come together in swarms, and these swarms build small fruiting bodies, each consisting of a stalk with terminal spore-filled cysts, that stick up into the air about 1 millimeter (see part *a* of figure on page 218). These fruiting bodies serve as an effective method of cyst dispersal, and the cysts are carried off to begin a new generation elsewhere. It is known that one finds essentially the same phenomenon in slime molds (part *b*), and there is even one example among the ciliate protozoans (part *c*). Finally, many small fungi, usually referred to as molds and mildews, produce similar structures made out of stiff filaments (part *d*).

Each of these organisms must, during the course of its life cycle, break through the water film and stick up into the moist air. Pure water has a very high surface tension, as mentioned earlier, and one can imagine that this might be a phenomenon requiring considerable strength. Unfortunately, this has not been studied, and we do not know whether these lowly beasts develop a strong

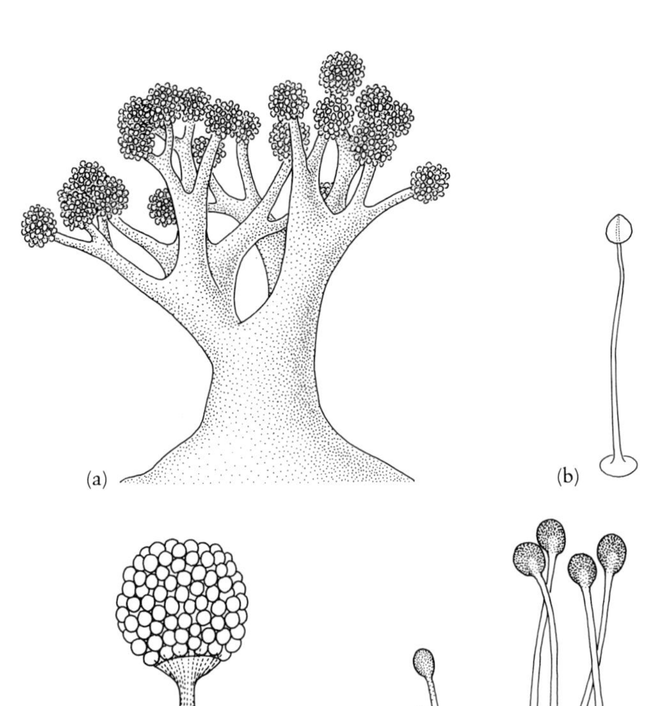


Fruiting bodies of (*a*) myxobacteria (height = 1 mm), (*b*) cellular slime molds (height = 2 mm), (*c*) ciliate protozoans (height = 0.5 mm) and (*d*) small fungi, (height = 10 mm).

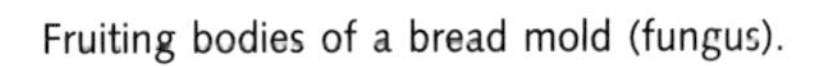

Fruiting bodies of a bread mold (fungus).

force for their upward push or whether they use a more subtle approach and release some surfactant that makes the task easy. One can see that, if an organism needs to separate its location for feeding (in liquid) from its location for dispersal (in air) during the course of its life history, then one could well imagine that special mechanisms to fight through the barrier imposed by the surface tension of water might well have evolved through natural selection.

**Surface Tension, Adhesion, and Cells.** Here we must discuss specifically what cohesive forces mean for cells. Cells are very different from drops of liquid. A droplet of water can generate enormous internal pressures if it becomes very small for the simple reason discussed earlier: that the internal pressure is high when the radius of curvature is small. Does this mean that a single cell is under great pressure, and that a small bacterium is under even greater pressure? The answer is *no*, and the reasons are most interesting.

For one thing, most cells, unlike water drops or soap bubbles, are not bounded by an air–liquid interface; they are bounded by elastic membranes. One can demonstrate the difference between a surface-tension membrane and an elastic membrane easily using the apparatus below. Suppose there is a soap

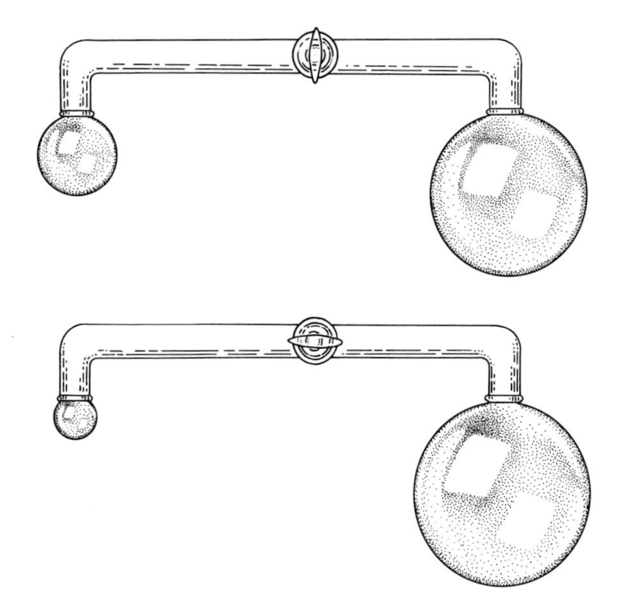

Two soap bubbles of unequal size connected by a tube containing a closed valve (*top*). When the valve is opened (*bottom*) the smaller bubble, with its higher pressure, empties its volume into the larger bubble. This process continues until the smaller bubble practically disappears.

bubble at each end of the tube. If one bubble is smaller than the other, then, when the connecting valve is opened, air will pass from the small bubble into the large one. If they were substantially inflated rubber balloons instead of soap bubbles, the reverse would be true and the two balloons would come to equilibrium containing equal volumes of air.

Not only are cells more like balloons than water droplets or soap bubbles, but, some years ago, E. N. Harvey (1954) and others measured the "tension at the surface" (as they called it) of various cells and found the values invariably to be exceedingly low—too low to explain their shapes. The term "tension at the surface" was used to include a combination of the tension of the membrane as well as the surface tension of any liquid lying on the membrane. Today, we know that cell shape is more likely the result of internal structural proteins, such as microtubules and different kinds of filaments, and that the tension at the surface plays a small role in cell shape.

This does not mean that the attractive forces between molecules fail to play a significant role in cells; it is simply not in the form of tension at the surface. Cells in general, and specifically cells that make up multicellular organisms, have adhesive surfaces. As a result, cells stick to one another and to other surfaces. The sticking, which is due to molecular attraction, is known to play a major role in the ordering of cells within a developing multicellular animal. This ordering is the direct result of simple physical principles, as has been shown in the elegant work of M. S. Steinberg (1978). He isolated two tissues from the embryo of a chick—for example, liver tissue and heart muscle—and separated them into their constituent cells. He then made a mixture of 50 percent heart and 50 percent liver cells. After a period of time, during which the cells moved about and bumped into one another, they formed a solid ball of cells consisting of a central sphere of heart cells surrounded by a layer of liver cells. Steinberg showed that this configuration could be explained by making certain assumptions about the strengths of adhesion between the different cells—that is, between heart and heart, between heart and liver, and between liver and liver cells. If the force of adhesion for heart–heart is greater than that for liver–liver, and heart–liver is equal to or slightly greater than liver–liver, then one would predict precisely what Steinberg found: a central ball of heart cells surrounded by liver. Steinberg and his colleagues have measured these forces of adhesion for a variety of different cell types besides heart and liver, and, in each case, the equilibrium configuration of the two tissues corresponds to the predictions.

These findings support the assumption that molecular forces of attraction play a key role in animal development. (The reason for stressing *animal* is that only animal embryonic cells are motile, and, in order for this tissue ordering to work, the cells must move.) Furthermore, it is important to stress that this system of cell sorting by virtue of differences in forces of adhesion can only work because cells are minute. The forces we are discussing are so small that they would be unable to hold larger bodies together.

**The Fly on the Wall.** Let us now move from the specific case of the adhesion of cell surfaces to a more general consideration of the role of cohesion in living organisms. First, it may be helpful to compare cohesion forces with those of gravity. For this purpose, we will borrow an argument and an illustration from the distinguished plant physiologist Frits W. Went (1968). As can be seen from the figure below, there are two lines with different slopes placed on a log–log plot. The vertical axis measures force in arbitrary units. The horizontal axis measures length of the organism. One line is the familiar one of mass or

Forces due to mass (proportional to length cubed) and molecular cohesion (roughly proportional to length) are shown on the same graph so they can be compared. The lines are arbitrarily allowed to intersect at a length of 1 mm.

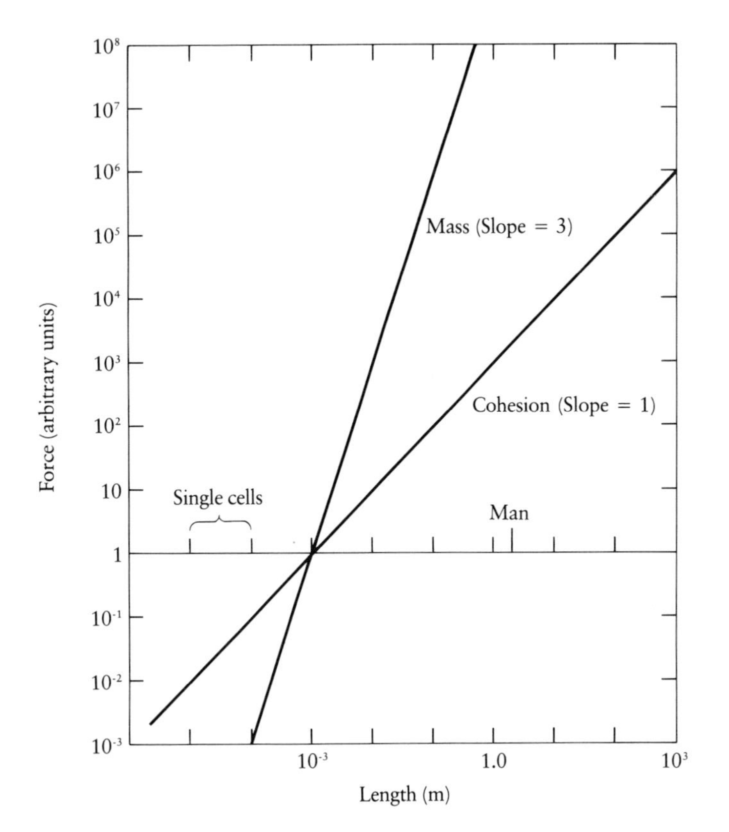

weight, which is much affected by gravity and rises as the linear dimension cubed ($\ell^3$). The other is the line for cohesion, which is considered to rise approximately as the linear dimension ($\ell$).

Why are cohesive forces roughly proportional to length? As we have been discussing in this chapter, the forces due to molecular cohesion are of two kinds: those appearing as surface tension at the interface between two or more immiscible fluids and those due to molecular adhesion between solids. The forces due to surface tension for a bug standing on the water are proportional to the perimeter of the wetted area of the legs where they touch the water and, therefore, to length in a series of geometrically similar bugs or other animals. Molecular adhesive force increases in proportion to length squared, in which length is the linear dimension of the adhesive surface. But the strength of the attraction between the molecules (or surfaces) that are attracting each other falls off as the distance between them increases. As bodies increase in size and the chances that they can come close together decrease (for intimate contact is greatly aided by small size), the two above factors will average out, making adhesive forces proportional to length seem a reasonable compromise. Thus, both surface-tension forces and surface-adhesion forces are roughly proportional to length.

Note that the $\ell^3$ line and the $\ell$ line in the figure on page 221 are made arbitrarily to cross at 1 millimeter on the horizontal axis. This comes from the empirical observation that organisms approximately that long seem to be at the point at which the importance of cohesive forces (due to surface tension, or molecular adhesion, or both) and gravitational forces are roughly equal. Now let us examine the consequences of this graph by considering how a fly can walk on the wall. Our task is to explain the fact that a fly or any other small insect can walk up a wall and across a ceiling, something that neither you nor your cat can manage. The best aid to sticking to things in general, and to walking on the wall in particular, is to be of small size. Beetles have foot pads made up of many small protuberances, and these act to increase the intimacy of their contact with the surface of the wall, even if it is a very irregular surface. Thus, the principle on which the protuberances function is molecular adhesion, and, returning to the figure on page 221, we can see that small insects and other wall-walking animals must operate below a critical size at which one line proportional to body weight ($\ell^3$) on a log–log plot intersects another line proportional to cohesive force ($\ell$). If we could redesign a beetle's feet, we might be able to raise its cohesive-force line somewhat, but this does not change the fact that any one design has an absolute limitation on size established by the intersection.

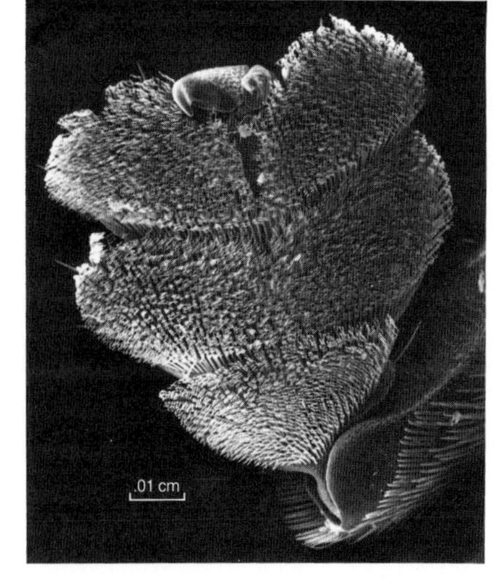


A high-power micrograph of the foot of a chrysomelid beetle, showing the hairlike protuberances that allow close contact with rough surfaces.

A tokay gecko (*Gekko gecko*) clinging to a vertical surface. Many small lizard species are capable of climbing walls that appear smooth to the naked eye.

The toe pad of a gecko (*Gekko vittatus*) at three magnifications. (*a*) The entire toe pad. The magnification (*b*) shows that the pad is constructed of a series of waves. Each wave is made up of tufts, like the tufts of a carpet. At the higher magnification (*c*), the individual tufts are shown; these look like miniature stalks of cauliflower. The branched endings of the tufts present a large surface area for clinging to minute irregularities on apparently smooth surfaces.

(a)

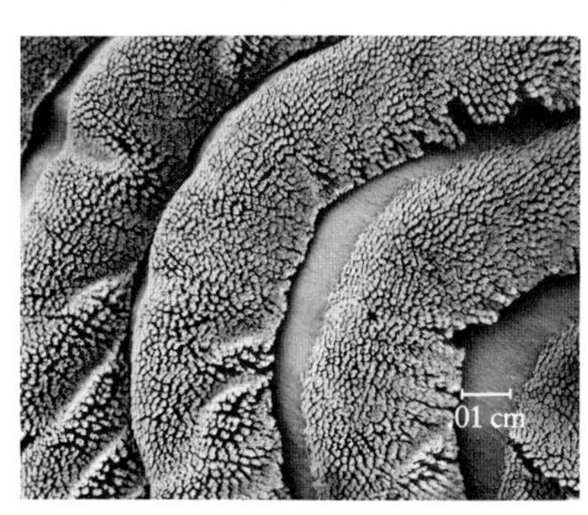


(b)

(c)

Geckos, those small, elegant lizards of warm climates that are so clever in climbing walls, have pads on their toes that use the same principle. Apparently, a gecko on a smooth wall must be much closer to the point of falling off than a beetle is—simply because it is much larger.

## A Further Note on Diffusion

We made the point earlier that diffusion plays a particularly significant role in the lives of small organisms. The rate of diffusion is severely impaired by increasing distance, and, because of the vital importance of the diffusion of gases and the assimilation of food, the surface of a large organism must be relatively extensive in order to allow for the diffusion of sufficient nutrients and gases into the large volume it contains. Similarly, a small organism can get away with relatively less surface area. This means that the guts of small animals tend to be simple, short, straight tubes. Many small animals do without circulatory systems because they are so compact that the necessary exchange of gases and food in solution can take place entirely by diffusion.

The larvae of a culicine mosquito hanging from the water surface by their breathing tubes. A special nonwetting structure at the end of the breathing tube breaks the surface and supports the weight of the larva by surface tension.

**Air-Breathing Larvae.** This kind of modification can be seen especially clearly in small organisms that have descended from larger ancestors. We have already discussed the small rotifer whose gut has disappeared and whose digestive system consists mainly of food vacuoles that resemble those of unicellular protozoans. Another example may be found among the larvae of aquatic insects. Most such larvae have breathing systems similar to those of terrestrial forms. They have many tubes (tracheae) that bring air directly to the tissues. The larva of a mosquito lives beneath the surface of a pond or a shallow puddle. It gets its air through a breathing tube that works like a snorkel. This snorkel, which protrudes from its tail, has a set of flaps at the end that are kept closed when the larva is under water. When fresh air is needed, the larva swims to the surface, allowing a special nonwetting structure at the tip of the snorkel to penetrate the surface film of the water so that air can go directly into the tracheal system. Surface-tension forces keep the opening of the snorkel floating at the surface; an oily secretion that repels water keeps the snorkel dry. Mosquito larvae may be killed by spreading a thin film of petroleum over the surface. The petroleum penetrates the snorkel's water-repelling agent and causes the larva to suffocate.

The aquatic larvae of some other very small insects seem to have no air openings at all, yet one can plainly see that there is air in their tracheae, and the larvae are fully motile and expending energy in their submerged habitat (see figure on page 225). How do such larvae manage without any intake of fresh air? The air in their tracheae contains both oxygen and nitrogen. As the oxygen is consumed by the tissues, fresh oxygen diffuses into the tracheae from the water (the oxygen in the water is in equilibrium with the oxygen in the air), and the normal oxygen–nitrogen ratio of the air is thus restored. This can

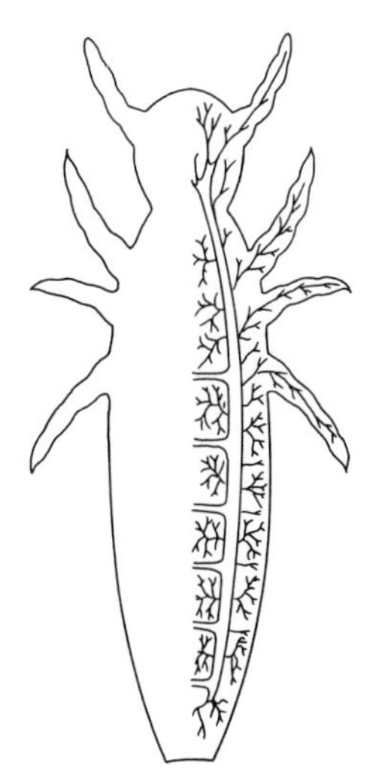

Diagram of the tracheal system of a very small aquatic insect such as the larva of the wasp *Polynema*. (Only half of the tracheal system is shown.) Note that, unlike the tracheal systems of other insects, this one has no opening to the outside. Oxygen penetrates into the tracheal system entirely by diffusion through the cuticle.

happen only because the small larva uses so little total oxygen that it can be resupplied easily by diffusion. A large larva (for example, the mosquito larva) would overtax such a diffusion system.

**Decreasing Complexity.** If an aquatic animal is sufficiently small, then even a tracheal system is not necessary because the gases can diffuse directly into and out of the tissues. Thus, an evolutionary trend toward decreasing size may lead to a disappearance of ancestral complexity. Complexity that is retained may be used, but complexity that is not needed may slowly disappear over evolutionary time.

Because natural selection has caused an overall size decrease in these instances, there has been a corresponding decrease in complexity. In some ways, it would appear to be an "evolution in reverse," because organisms are becoming less complex, more "primitive." But there are limitations to this kind of evolution. It is not an evolution by invention; it is merely an evolution by attrition. There is no evidence that, for instance, the corkscrew or flexible-oar modes of locomotion have been reinvented by the small descendants of large ancestors. There has been much more innovation associated with the evolution of size increases than with that of size decreases.

# Chapter 7

## The Ecology of Size

Tiger and prey. Mosaic from the Musei Capitolini in Rome.

Thus far, we have seen a great variety of mechanical or engineering consequences of size in different kinds of organisms, and we have periodically shown by many examples in varied contexts that, because living organisms evolved by natural selection, there has been a selection for better mechanical design for organisms of different sizes. Sometimes the shape of the organism has been subject to natural selection and the size has been affected secondarily, but often it is the reverse: size has been the direct object of selection, and the individual design has been appropriately modified, by means of genetic changes, to maintain optimal mechanical efficiency.

There is a further consequence of the evolution of size that involves an even larger picture. We will no longer confine our discussion to the properties of individual organisms but will be concerned with organism size at the population level—that is, the distribution of sizes within groups of organisms and how that distribution is affected by the presence of other kinds of organisms, the climate, and even the reproductive strategies of any particular species.

Let us begin at the simplest level. In any one habitat, be it tropical or temperate, aquatic or terrestrial, there will be a distribution of the sizes of organisms. For example, the size range in a local woods will be from the smallest bacteria to the largest deer. Using a logarithmic scale, one can rank the sizes in such a range into roughly seven levels, each a power of 10 (or 1 log unit) apart (see *left* figure on page 228; also figure on page 4). Bacteria range in length from 1 micrometer to 10 micrometers, whereas large mammals range in length from 1 meter to 10 meters. One could do the same for aerial forms, going from the smallest fly (about 100 micrometers in length) to the largest bird, such as a bald eagle or a swan, which will approach 1 meter in length. In ponds or in the oceans, the story would be the same. All these examples stress animals, but the same can be done for plants, from the smallest fungi and algae to giant beeches and tulip trees.

If any one community is made up of animals and plants having such a huge range of sizes, can one say anything about the abundance of the organisms of different sizes? As a rough approximation, it is possible to compare the population density of animals of different sizes (*left* figure on page 228), and it is clear that, as their size decreases, their abundance increases. The same is true of plants: we expect grasses in a field to be very close together, while great oaks will stand far apart.

Recently, such data were gathered for mammals from a large number of different habitats by J. Damuth (1981). He plotted the number of animals per

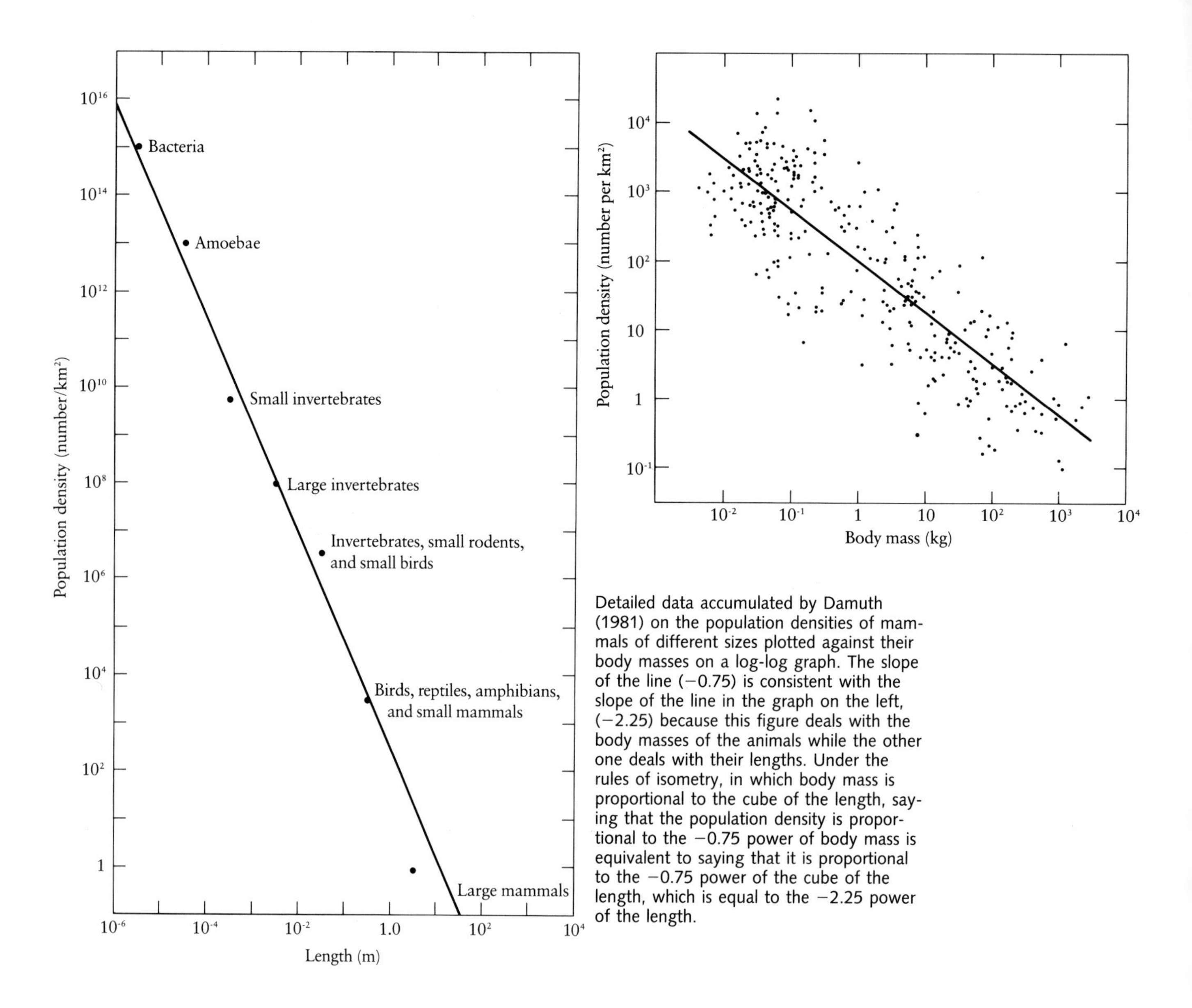


Detailed data accumulated by Damuth (1981) on the population densities of mammals of different sizes plotted against their body masses on a log-log graph. The slope of the line (−0.75) is consistent with the slope of the line in the graph on the left, (−2.25) because this figure deals with the body masses of the animals while the other one deals with their lengths. Under the rules of isometry, in which body mass is proportional to the cube of the length, saying that the population density is proportional to the −0.75 power of body mass is equivalent to saying that it is proportional to the −0.75 power of the cube of the length, which is equal to the −2.25 power of the length.

The number of individual organisms per square kilometer (population density) plotted against the length of the organism, from bacteria to large mammals, on a log–log graph. The points are estimated from a variety of sources. The slope of the line is −2.25.

square kilometer (population density) against the body mass of the individual animals and found an allometric relation (*right* figure above) in which

$$\text{population density} \propto (\text{body mass})^{-0.75}.$$

Notice that the exponent of body mass in Damuth's rule is −0.75. If you recall from Chapter 4 that Kleiber's law stated that an individual animal's resting metabolic rate (measured, for example, in the amount of food consumed per day) is proportional to body mass raised to the power 0.75, you will see that an alternative statement of Damuth's finding is possible:

$$\text{population density} \propto 1 \,/\, \text{food consumed per animal per day}.$$

The larger the organism, the greater the mean distance between individuals. Here a herd of barren-ground caribou (*Rangifer articus*) fords a shallow river in Alaska. Herds containing several thousand caribou make seasonal migrations each year between their summer and winter feeding grounds, covering a vast area.

Expressed in this way, Damuth's rule has a simple interpretation. Multiplying both sides by the food consumed per animal per day, we obtain the statement that the food harvested and eaten per square kilometer per day is a constant, independent of the body size of the animals that do the harvesting. This statement evokes the food pyramid originally described by C. S. Elton, the pioneer British ecologist. If one examines an animal food chain, the bottom organism is usually a small herbivore, which is eaten by a carnivore, which in turn is eaten by a larger carnivore, and so on. The interpretation we have given Damuth's rule says that the flux of biomass through each level of body size, from herbivore to small carnivore to larger carnivore, is a certain constant that depends on, among other things, the productivity of the land.

The rule that large carnivores eat small ones has many obvious exceptions. When a hyena kills a zebra or a wolf kills a moose, the prey is larger than the predator. Small size in carnivores can be compensated for in two ways: one is by aggressive skill and the other is by cooperative hunting. In the latter, the hunter is really two or more hyenas or wolves functioning as a unit. Elephants

Predator larger than prey: *a*, false vampire with mouse; *b*, giant praying mantis (10 cm long) feeding on a dragonfly; *c*, red fox with rabbit; *d*, smoky jungle frog eating a puddle frog.

(a)

(b)

(c)

(d)

Predator smaller than prey: *a;* hyenas killing a wildebeest; *b*, black widow spider eating a scorpion; *c*, lion killing a buffalo; *d*, eagle in pursuit of a wild dog.

Sparsely growing oaks.

are so large that they have no predators—they have outgrown them all (although *Homo sapiens,* by virtue of the cleverness made possible by a big brain, may manage to do them in after all). But, exceptions aside, one would expect there to be fewer larger animals on the basis of energetics alone.

For trees and other photosynthetic organisms, size is the result of another kind of competition. In any given place, the largest tree has the biggest spread of leaves and therefore literally takes the largest share of available energy and, in so doing, deprives the lesser trees. This means that it competes not only with other species in the forest but even with its own growing offspring. Here, the reason for the decreased population density among larger organisms is even more obvious than it is among animals. Each tree will have a radius of shade, and the larger the tree, the larger the radius. Therefore, low population density among large trees is a direct result of size competition among individual trees in the population.

There is another aspect of size differences that applies to closely related species. It is sometimes called the Hutchinson ratio, after the noted ecologist G. E. Hutchinson of Yale University, who showed that the ratio of the body lengths of two closely related species will often be a certain constant, whether two large species or two small ones are compared. This may apply not only to body lengths but also to the lengths of parts of an animal, such as the beaks of birds. Note that, if the ratio of body masses of adjacent species is found to be a constant, then the ratio of their body lengths (approximately the cube root of the ratio of the masses) is also constant.

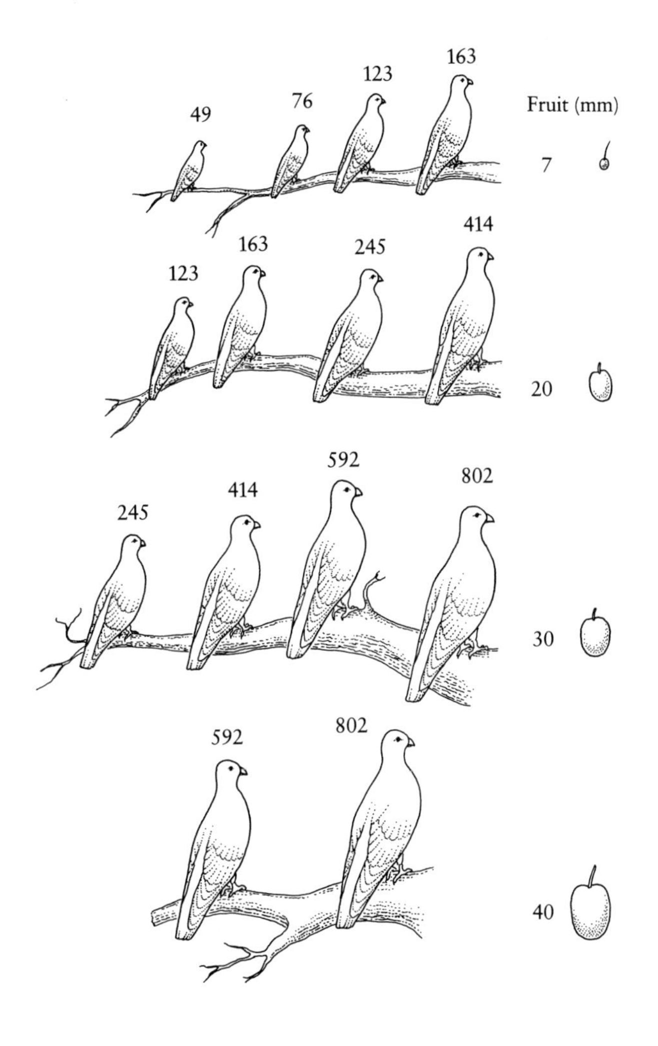


Schematic representation of niche relations among the eight species of fruit pigeons of the genera *Ptilinopus* and *Ducula* in the lowland rain forest of New Guinea. On the right is a fruit of a certain diameter (in millimeters) and on the left are pigeons of different masses (in grams) arranged along a branch. The mass of each pigeon is approximately 1.5 times the mass of the next. Each fruit tree attracts up to four consecutive members of this size sequence. Different species of trees with increasingly large fruits attract increasingly larger pigeons. In a given tree, the smaller pigeons are preferentially distributed on the smaller, more peripheral branches. The pigeons having the masses indicated are: 49 grams, *Ptilinopus nanus;* 76 grams, *P. pulchellus;* 123 grams, *P. superbus;* 163 grams, *P. ornatus;* 245 grams, *P. perlatus;* 414 grams, *Ducula rufigaster;* 592 grams, *D. zoeae;* and 802 grams, *D. pinon.*

To give an illustration of how this principle works, J. M. Diamond (1973) describes eight different species of fruit pigeons that live in New Guinea. Because their body masses are marked on the figure, one can see that the mass of each species is approximately 1.5 times greater than the next, implying that the length ratio between closest species is 1.15 to 1. The figure is helpful also in that it shows that these size differences are associated with the fruit sizes that the birds can eat (also shown) and the branch sizes that will support them, the smaller birds being able to perch on the slenderest branches without danger of bending them very much.

The presumption is that these eight size classes are the result of a competition for food and a partitioning of the food resources. But one must apply some caution here, as Henry S. Horn and Robert M. May (1977) of Princeton University have shown. They examined various families of musical instruments and even bicycles of different sizes for children and adults only to discover that those objects also followed the Hutchinson ratio. In spite of this caution, it seems fair to say that competition for food does play an important role in producing size classes in many groups of organisms. This can be demonstrated by the fact that size divergence between two species will be greater when the two occur together on an island than it will when they live on separate islands. This has been shown very clearly with Darwin's finches on the Galapagos Islands (Lack, 1947).

Besides the overall competition for energy, there are other kinds of relations between organisms that will affect size. These amount to special cases of the general proposition already outlined, and they are in themselves of particular interest. The first two examples we will discuss involve the relation of reproductive strategies to size. The other two are the special problem of brain size and the very general problem of species diversity and species size.

## Reproductive Rates and Size

An important idea, crystallized by Robert MacArthur and E. O. Wilson (1967), is that, on a continuous scale of reproductive strategies, there are two extreme kinds. One is called "$r$ selection" ($r$ stands for rate of increase by reproduction), in which an individual has many offspring either by having a few offspring at frequent intervals or by having large numbers of offspring at one time. One characteristic of organisms that exhibit $r$ selection is that they are small. A good example would be a bacterium that divides at a very rapid rate to produce masses of offspring in a very short period of time. If one were to compare different mammals, one would surely conclude that a mouse exhibits $r$ selection but that an elephant does not. A mouse will produce many generations in a short time, while an elephant has few young and each generation takes more than 10 years. The elephant, being at the other extreme from the mouse, would be said to exhibit "$K$ selection" ($K$ stands for the carrying capacity of the environment, which becomes limiting for these organisms). Two characteristics of organisms that exhibit $K$ selection are large size and an accompanying long generation time, clearly exemplified by the elephant.

The length of an organism at the time of reproduction in relation to the generation time, plotted on a logarithmic scale. The first four organisms at the bottom are bacteria. The next five going up the graph are protozoans. *Daphnia* is a water flea, *Drosophila* is the fruit fly made famous by experimental geneticists, and all other organisms are easily recognized by their common names.

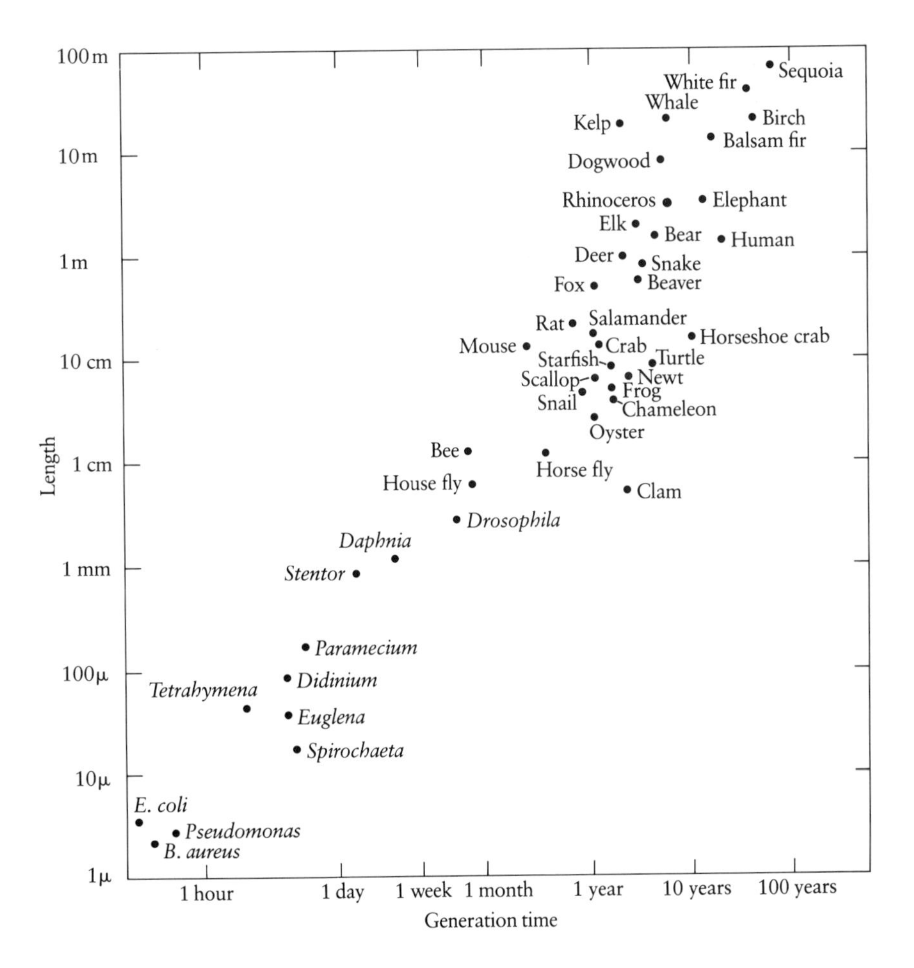


*K* selection and *r* selection are the two extremes of a range of strategies for reproductive increase. *K* selection is especially suited to stable climates in which the full resources of the environment can be exploited safely. The tropics are a good example. Not only are there many *K*-selection organisms in the tropics, there is also an enormous number of species. Therefore, *K* selection goes with environments with large species diversity, a point to which we will soon return. By contrast, *r* selection is best suited to unpredictable environments, such as temperate and subpolar regions where the production of large numbers of offspring is insurance against environmental catastrophe. Bacteria are very vulnerable to what we might see as a slight change of local conditions, but for them it may be a major flood or drought. In their unstable microworld, the best reproductive strategy for bacteria is to reproduce as rapidly as possible and whenever possible so that, through massive dispersal of offspring, they will continue to survive even though whole local populations may suddenly be wiped out.

Sexual dimorphism, female larger than male: *a*, praying mantis mating pair; *b*, golden silk spider, female (larger) and male (smaller).

## Sexual Dimorphism

One effect of reproductive strategies on size in animals is sexual dimorphism. Largely through competition between males, and sometimes through choice of mates by the females, there arises a difference in appearance between the males and the females of a species. This phenomenon was first described and discussed in detail by Charles Darwin in his *Descent of Man* (1871). He called it *sexual selection.* In its more extreme forms, it accounts for the extraordinarily elaborate coloration of the males of some species of birds in contrast to the cryptic coloration of the females. The peacock and the peahen provide an example, as do the birds of paradise, the males of which have incredibly elaborate and fantastic plumage. But here we are concerned primarily with size differences between the sexes.

In some species, the female is larger than the male. This is especially true of many raptorial birds. The female peregrine falcon is a third again as large as the male, and the same is true for many species of owls. In these cases, we do not know what selective forces produced the size differences, although it has often been suggested that the larger female investment in reproduction (the necessity of producing large eggs) favors large size. The only difficulty with this argument is that, in many other species of birds, the sexes are equal in size, and in some, like the capercaillie, a huge European grouse, the male is considerably larger than the female.

There are far more extreme examples among invertebrates. Female spiders may be huge compared to the males of the same species, and, because they are carnivorous and have voracious appetites, the spindly male has a delicate time trying to persuade the female which appetite to satisfy first. Even greater size differences are found in a variety of invertebrates, most notably among those in which the male becomes a small parasite attached to the female, as is the case in some crabs. In these extreme cases, the male may be less than one-tenth the size of the female.

Among many vertebrates, including human beings, the male is larger. This can be seen among carnivores (such as lions), many herbivores (such as cattle and elephants), and such primates as baboons and gorillas. Some of the most

Sexual dimorphism, male larger than female: *a*, bison; *b*, fur seals breeding.

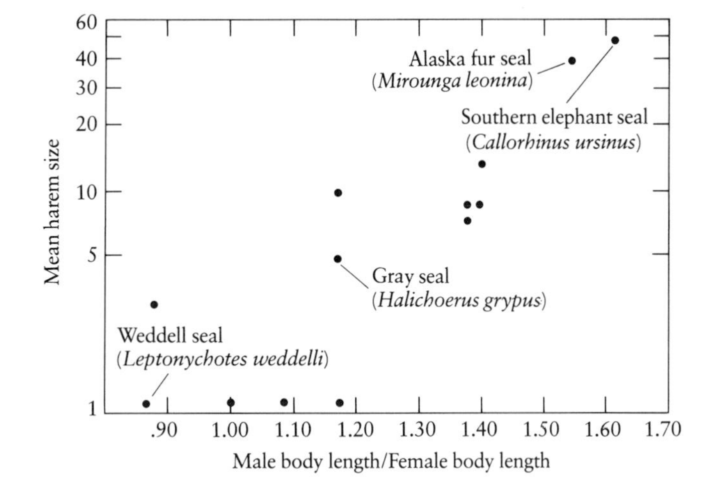


Harem size plotted against sexual dimorphism (as size differences) in thirteen species of seal. Note that the species with the relatively largest males have the biggest harems.

striking cases are found among seals: A bull Alaska fur seal will weigh more than 225 kilograms (roughly 500 pounds), while a cow will weigh roughly 30 kilograms (about 70 pounds), again a difference of almost an order of magnitude. This means that, as in the invertebrate examples mentioned above, the size differences between the sexes can be sufficiently large to make them comparable to two species belonging to entirely separate size classes.

The reasons for these extremes in sexual dimorphism vary, but somehow they always involve reproductive strategies. In the case of the small males of marine invertebrates, M. T. Ghiselin (1974) has argued that small size increases their chances of dispersal and outbreeding in an ocean environment with currents. Among mammals, such size differences are associated with polygamy. If one plots the harem size of various species of seals against the degree of sexual dimorphism they exhibit, there is an excellent positive correlation (see graph above).

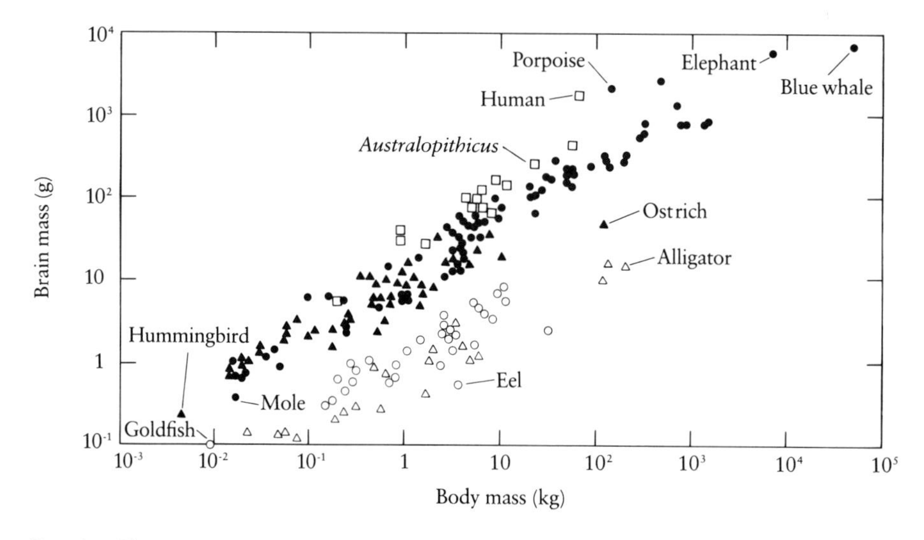


Brain size of 200 species of vertebrates plotted against body size on a log–log graph. Primates are open squares; other mammals are solid dots, birds are solid triangles, bony fishes are open circles, and reptiles are open triangles.

## Brain Size

The whole matter of brain size has a special fascination for us. If one plots brain size against body size, one can see that there are allometric relations (see figure above). Three things in the figure are of special interest. One is that the slope of the line for any one major group of organisms is approximately ⅔, which means that, for larger organisms, there is relatively less brain. It may be presumed that this is because, in order to govern a larger machine, more controls are needed; but the differences between the small and large machines are not sufficient to warrant a corresponding increase in brain size to control the working of the larger machine. There are those who have argued—especially B. Rensch (1956), the distinguished evolutionary biologist from Germany—that the absolute size of the brains should nevertheless result in smarter animals. Rensch did show that large breeds of chickens can learn more and retain the information longer than small ones, and he did some elaborate studies on elephants to show that they never forget. He went on to argue that the superiority of a larger brain may have been a leading reason why there is a selection pressure for larger animals—that is, he offered an explanation for Cope's rule. This is certainly a theoretical possibility, although it is not a very convincing one.

The second lesson to learn from the figure above is that, even though the slopes are roughly the same for the various major groups of animals, the value of $b$ in the allometric formula is different for each group ($b$ is the value of $y$ when $x = 1$ in the formula $y = bx^a$). Furthermore, the values of $b$ follow some sort of a phylogenetic sequence, because primates have the largest $b$ value

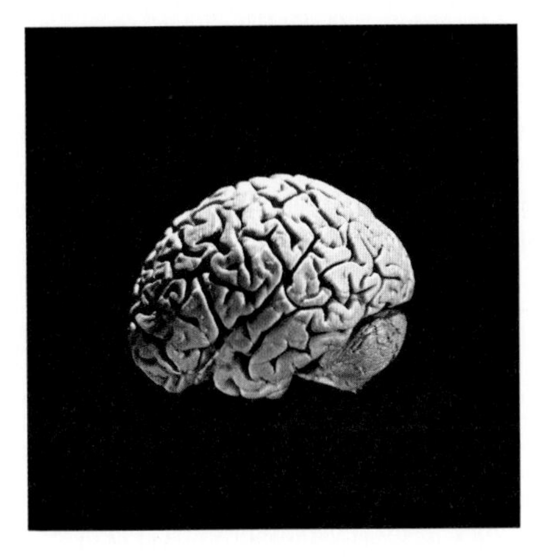

A model of the right half of the human brain shows the extensive folding and convolution of the surfaces. The cerebral cortex, in which speech, vision, and many motor functions are centered, is at the top and left of the picture. The cerebellum, which coordinates motor activities below the level of consciousness, is under the cortex at the right.

Approximate average encephalization quotient (EQ) for three groups of mammals.

| | Average EQ |
|---|---|
| Mammals (excluding primates) | 1.0 |
| Primates (excluding *Homo sapiens*) | 2.1 |
| *Homo sapiens* | 7.6 |

while bony fish and reptiles have the smallest. This fact has been used by H. J. Jerison (1973) to provide an effective way of judging the brain capacity of any individual species—that is, to answer the question, Are there some mammals or some reptiles that are brainier than others? He devised what he called an "encephalization quotient," or EQ for short, which is the ratio of the value of $b$ for a particular animal over the value of $b$ for all members of its group, be they primates, mammals, birds, reptiles, amphibians, or fishes.

Jerison's method has made it possible to examine questions concerning relative brain sizes within groups. For instance, it has been shown that there is a high correlation between the kind of food an animal eats (and its method of catching it) and the size of its brain. Among primates, mammals, and dinosaurs, carnivores have higher EQs than herbivores. In order to eat by catching prey, one has to be smarter and no doubt better coordinated than an animal that eats grass or leaves. There are great subtleties in diet and methods of food catching, and they seem to be better correlated with brain size than with other factors, such as the degree of social organization of the species.

The third remarkable aspect seen in the figure on page 238 is that *Homo sapiens* is above all other animals. Clearly, because of its brain size, *Homo sapiens* has taken a great leap forward, as is reflected in its enormous EQ. It would appear in this case, which is even more striking than the herbivore–carnivore comparison, that there has been a selection specifically for increased brain size in the evolution of hominids. And this relative increase in brain size is again quite independent of body size. How this hominid selection for increased brain size worked in our early history is a matter of great interest and considerable speculation (none of it, at the moment, very satisfactory). Among other lingering problems, porpoises and dolphins seem to have taken an equally impressive turn towards large brains and high EQs, and why that should be so is puzzling. Is there some special mechanism of prey catching among dolphins that requires the development of a complex brain, or (as so many people would like to believe) are they just as clever as we are, but without our being able to prove it simply because we have not yet cracked their language code? Unfortunately, the latter suggestion seems to be wishful thinking, but we still do not understand the reasons for their large brains.

## Diversity and Sizes

The question of why there are so many species was posed by G. E. Hutchinson (1959) in a famous essay, and some of the greatest insights into this problem came from his student Robert MacArthur (1972). Together they were the first to ask if there was a relation between species diversity and species size (Hutchinson and MacArthur, 1959). They concluded that there were, even within one size level in a food pyramid, fewer species of large animals than of small animals. This is in agreement with intuition. One assumes that any large organism requires a larger niche and is, therefore, automatically limited. It makes sense to propose that this limitation would apply to species as well as to individuals. Hutchinson and MacArthur presented a complex model that predicted that the number of species, $S$, will vary as the inverse square of the linear dimension, $\ell$:

$$S \propto \ell^{-2}$$

As Robert M. May (1978) shows in an excellent review article, there are many sets of data for different groups of animals that support the above relation (including those of van Valen, 1973), and May has summed a large num-

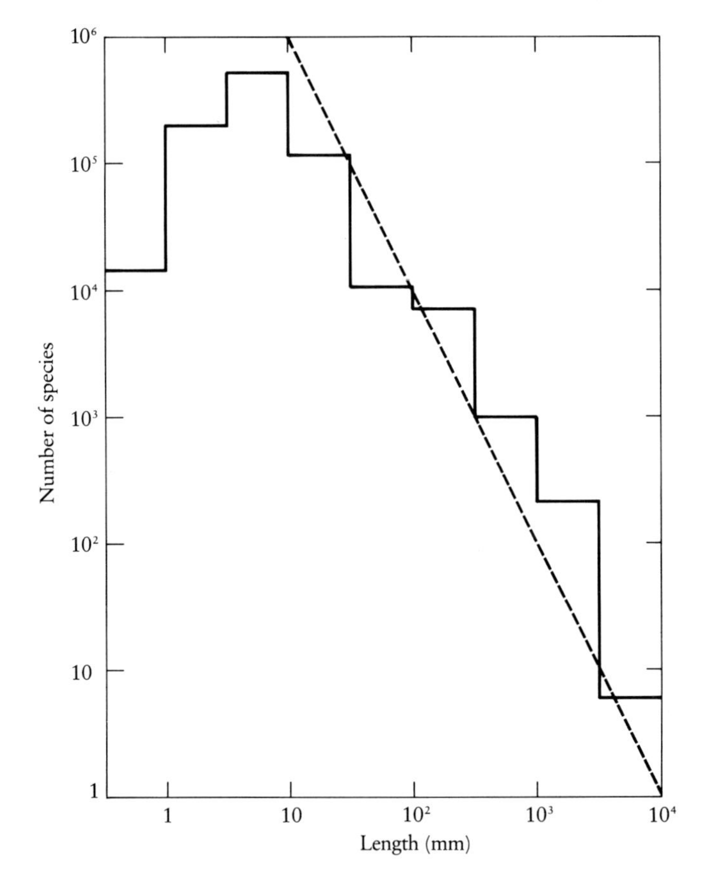


The number of species of all terrestrial animals classified according to lengths. The numbers used are very rough estimates. The dashed line illustrates one case in which the number of species is inversely proportional to the square of the length.

ber of such data for terrestrial animals to show the empirical evidence for the relation (see figure on page 238). This relation means that, if there is a tenfold reduction in length (for example, from 1 centimeter to 1 millimeter), there will be a hundredfold increase in the number of species. The curve always drops off at the lower end of the size range and departs from the theoretical prediction. Certainly, for very small organisms, such as bacteria, it could be that our notion of what constitutes a species is very fuzzy, but there are other reasonable explanations for the drop-off in the number of species of small insects, such as the possibility that we simply have not sampled small insects as well as the larger ones.

In any event, there is a relation between the number of species and the lengths of individuals of those species in the natural environment. In general, there are more small species than large ones. As May (1978) cautions, it is not at all clear how seriously we should take the value of −2 for the length exponent, because this has only been determined for restricted groups of terrestrial organisms, and even then it is not precisely −2. There are really two problems here for the future. One is the gathering of more data for other types of organisms. But far more important is the need to understand why there should be a precise relation between the adult size of a species and the number of individuals of that species per unit area in a particular environment.

· · ·

We have in this book examined an enormous range of subjects connected the problem of size in living organisms. We began with a little natural history, which quickly brought us to an appreciation of the fact that size is related to shape, and that both have, during the course of evolution, been deeply influenced and guided by natural selection. Then we looked into the matter of how one measures shape so that we could be more precise in our statements about the relation of size to shape. This was, at best, a compromise, because changes of shape introduce so many features to measure that it is far easier to compare only two of those features at one time—for example, the length of a particular bone as a function of body mass.

Armed with an understanding of the scope of our problem, we then took a close look at what the engineer does to understand size changes in the physical world. There we saw the power that can be gained from using dimensionless numbers; they provide a penetrating way of looking at the effects of scale. It turns out that size has an effect not only on shape but also on all the activities of an organism, including its ability to move and the way it is influenced by physical forces, such as surface tension, friction, and inertia. Through dimensional analysis, one can see the effects of size in a wonderfully compact yet panoramic way.

What one learns from bicycles, soap bubbles, raindrops, stars, submarines, atomic bombs, musical instruments, shaking bridges, ship collisions, and model airplanes is directly applicable to an understanding of the hovering and soaring flight of birds, the sprinting of fish, the proportions of trees, and the activities of water striders, mosquito larvae, flying insects, and bacteria. It is also possible to examine specifically the problems and consequences of being very large or very small and to determine how these extremes of size affect all the physiological and structural characteristics of the organisms listed in the previous sentence, and others as well.

Then we returned to the big question of evolution and its relation to size, approaching it from an ecological point of view. Not only has there been a general increase in the sizes of organisms over the course of evolution, but all the different size levels between the smallest and largest exist today in harmony. Living organisms, from bacteria to large mammals and trees, range across at least eight orders of magnitude in length. The smaller the organism, the more abundant it will be in any one environment. This size range and the abundance of each size class is partially, but not wholly, explained by the food pyramid and the question of who eats whom or what. And then there is the related fact that smaller size classes have more species than larger ones, a phenomenon we have yet to explain satisfactorily.

In fact, although we have reached the end of this book, there are still plenty of mysteries. An important one has to do with the way in which living cells seem to sense the size of the space they have to live in. Living cells are separate; in many circumstances, individual cells can carry on with their normal functioning in isolation. The amphibian heart still beats, even when cut into smaller and smaller pieces. Single nerve cells will give action potentials and single muscle cells will contract when they are stimulated appropriately in a bath containing the correct ions and metabolites. Even though this is true, the individual cells of the body always seem to know how large an animal they are in. The cell in a large animal is more specialized than one in a small animal; it has a lower metabolic rate, a lower rate of enzymatic activity, and it lives longer. Because it has joined forces with a vast society of other cells, it moves through its environment (by walking, swimming, or flying) faster and, on the whole, more safely than any single-celled organism. The cells in an animal's body move through the world together not like a flock of birds but like the crew members and paying passengers on a steamship. We can visualize a liver cell standing at the deck rail, contemplating life outside, watching the world pass-

The cells in an animal's body move through the world together not like a flock of birds but more like the crew and paying passengers on a steamship.

ing by like the waves. It may be thinking that one of its kind would last alone out there for only a few minutes; it is too specialized to make it on his own.

Something tells the organism how much living space it has. Goldfish remain small in a bowl but grow to relatively large dimensions in a lake. Something quite parallel, we assume, tells the liver or muscle cell the size of the organism it lives in so that it can adjust its metabolic rate and other biological activities accordingly, so that it can fit in with the other cells and do its job.

It may be a physical truth that space is unbounded, but it does not appear to be a biological truth. The fact that boundaries exist and may be perceived through unconscious mechanisms—the boundaries of the cell membrane, of the skin, of the fish tank—means that, as far as biology is concerned, space is bounded and finite. The body is subject to the dumb perceptions of boundedness and the limitations of size—the laws of scale—and this will never change. Among the special faculties of life, it appears that only imagination is unbounded.

# Credits

*© by the American Association for the Advancement of Science.

Cover image
Scala/Art resource

### Chapter 1

Facing page 1
Walters Art Gallery, Baltimore

pages 2 and 3
Adapted from Wells, H.G., Huxley, J.S., and Wells, G.P. (1931); Baluchitherium drawing adapted from Gregory (1951).

page 4
Data from Schmidt-Nielsen, K. (1975).

page 6 (above)
Courtesy professor E.S. Barghoorn, Harvard University.

page 6 (below)
Barghoorn and Tyler (1965)*

page 7 (top left)
Photo Researchers

page 7 (top right)
Fritz Goro

page 7 (below left and right)
Photo Researchers

page 8
Photo Researchers

page 9 (left and right)
Peter Arnold

page 10 (left)
Bruce Coleman

page 10 (middle)
Tom Stack and Assoc.

page 10 (right)
Bruce Coleman

page 11 (left and right)
Courtesy Dr. Marion I. Barnhart

page 12
The Bettmann Archives

page 13 (above and below)
Bruce Coleman

page 14
Bruce Coleman

page 15 (above and below)
Bruce Coleman

page 16
Rosenwald Collection, Library of Congress

page 17
Redrawn from *Evolution*. Life Nature Library. 1964.

page 19 (top)
Animals, Animals

page 19 (bottom)
A. Gunther. *The Reptiles of British India*. Ray Society. 1864. Plate 13.

page 20 (left, middle, right)
Bruce Coleman

page 21
Redrawn from Pennak (1953).

page 22
From Bonner (1968).

page 23
Daniel Adams

### Chapter 2

page 24
Photoatelier Jörg P. Anders, Berlin, West Germany

page 26
Kummer (1951)

page 27
Scala/Art Resource

page 32 (above)
Medawar (1945)

page 33
Library of Congress

page 36
Bonner (1965)

page 37 (left and right)
Bruce Coleman

page 39 (above)
Bruce Coleman

page 39 (below)
Peter Renz

page 40
Jell Foott

page 41 (left and right)
Bruce Coleman

page 42
National Library of Medicine, Bethesda, Maryland

page 43 (above)
© 1982 King Features Syndicate, Inc. Reprinted with permission.

page 43 (below)
(a) Tom Stack Assoc. (b) Animals, Animals (c) Bruce Coleman (d) Tom Stack Assoc.

page 44 (left)
McMahon (1971)*

page 44 (photo)
Stock Boston

page 45 (3 photos)
Photo Researchers

page 47
McMahon (1971)*

page 50 (above left)
Douglas Faulkner

page 50 (above right)
Sally Faulkner

page 50 (below)
Chrysanthemum photo before alteration courtesy Prof. Robert E. Lyons, Virginia Polytechnic Institute and State University College of Agriculture and Life Sciences, Blacksburg, Virginia.

page 52
Adapted and redrawn from Scientific American *Vertebrate Structure and Function*, Freeman, 1974.

page 53
Prange (1977)

page 54
Wide World Photos

page 55
Ultsch (1973)

page 56 (top left)
Wilkie (1977)

page 56 (below left)
Lietzke (1956)*

page 56 (right)
Jerry Cooke for *Sports Illustrated,*
© Time Inc.

page 58
Courtesy United States Steel.

page 59
Data from U.S. Steel.

page 61
Courtesy Technopower, Inc., Oak Park, Illinois.

page 65 (top)
Schmidt-Nielsen (1970)

page 65 (below)
Kleiber (1932)

page 66
Mickey Pfleger

## Chapter 3

page 68
Courtesy Bill Palmeroy, NASA.

page 70 (left)
N.C.F.M.F. Book of Film Notes, 1972; MIT Press with Education Development Center, Inc., Newton, Mass.

page 70 (right)
The Museum of Modern Art/Film Stills Archive.

page 72
Joseph Daniel

page 79 (above)
The Bettmann Archive

page 79 (below)
Photo by Gerald Martineau, The Washington Post.

page 82
Fritz Goro

page 83
Redrawn from C.V. Boys, *Soap Bubbles,* Dover Press, 1959.

page 84
Edward Arnold Ltd.

page 86 (above)
Los Alamos National Laboratory

page 86 (below) and 87
Photo by Mack in Taylor (1950, II).

page 91 (left and right)
Department of the Navy

page 93 (a,b,c,d and left)
N.C.F.M.F. Book of Film Notes, 1972; MIT Press with Education Development Center, Inc., Newton, Mass.

pages 94 and 95 (a,b,c)
N.C.F.M.F. Book of Film Notes, 1972; MIT Press with Education Development Center, Inc., Newton, Mass.

page 96
N.C.F.M.F. Book of Film Notes, 1972; MIT Press with Education Development Center, Inc., Newton, Mass.

page 97
Courtesy Ms. J.R. Naidish

page 98
Courtesy Bill Palmeroy, NASA

page 99
The Bettmann Archive

page 101 (left)
Gabriel Moulin Studios

page 101 (right)
Colour Library International

page 102
Wide World Photos

page 103 (left and right)
Department of the Navy

page 105
British Crown Copyright. Science Museum, London.

page 106
Department of the Navy

page 107
Courtesy Professor Milton Van Dyke (1983).

## Chapter 4

page 110
Library of Congress

page 112 (table)
Groups 1,2,3,7,9, and 10 Stahl (1962); 4 and 5, Calder (1981); group 6, Prosser (1973); group 8, Sacher and Staffelt (1974); group 11, Dedrick and Bischoff (1980).

page 113
Stevens (1974)

page 115 (above)
Data from Prosser (1973).

page 115 (below)
Modified from Noordergraaf et al. (1979). © Academic Press, London.

page 118 (above)
Courtesy Stuart Rabinowitz M.D.

page 118 (below)
Andreas Feininger, LIFE Magazine © 1952, Time Inc.

page 120
Schultz (1969)

page 121 (above)
Halpert (1982)

page 121 (below)
Courtesy Harvard Museum of Comparative Zoology.

page 122 (left)
Peter Johnson, Natural History Photographic Agency, Sussex, England.

page 122 (right)
Animals, Animals

page 123 (left)
Animals, Animals

page 123 (right)
Scientific American *Vertebrate Structure and Function,* Freeman, 1974.

page 125
Adapted from McMahon (1975c)
© University of Chicago Press.

page 126
Drawings by Gregory Paul

page 127
Scientific American *Vertebrate Structure and Function,* Freeman, 1974.

page 129
McMahon (1973). Adapted from Stahl and Gummerson (1967). Tracheal diameter from Tenney and Bartlett (1967).

page 130
McMahon (1973) adapted from Hemmingsen (1960).

page 131
Modified from Taylor et al. (1981).

## Chapter 5

page 136
Peter Johnson, Natural History Photography Agency, Sussex, England.

page 139 (left and right)
David Muench Photography, Inc.

page 140
Tom Stack and Assoc.

page 141
McMahon (1975a). *Scientific American* 233:93-102.

page 142
McMahon (1973)

page 143
Horn and Kiltie (unpublished results)

page 144
Thomas A. McMahon

page 146
McMahon (1975a). *Scientific American*.

page 147
Ralph Morse

page 148
W. Kopachik

page 150 (left)
McMahon (1975a). *Scientific American*.

page 150 (right)
Martin Zimmermann

page 150 (below)
Bonner (1965)

page 153 (below)
Adapted from Greenewalt (1977).

page 155 (below)
Adapted from Alexander (1977).

pages 156 and 157
International Museum of Photography at George Eastman House, Rochester, New York.

page 158
Courtesy Dr. Harold E. Edgerton

page 159
Hill (1950)

page 160
McMahon (1975b)

page 161 (above)
Heglund et al. (1974)*

page 161 (below)
Photo Researchers

page 163 (above)
Adapted from Halpert (1982)

page 163 (below)
Adapted from McMahon (1975b)

page 164
Taylor et al. (1972)*

page 165
Fritz Goro

page 167 (left)
Crawford H. Greenwalt

page 167 (right)
Library of Congress

page 168 (left)
Bruce Coleman

page 168 (right)
Lynn M. Stone

page 168 (below)
Modified from Greenewalt (1975).

page 170
Greenewalt (1975)

page 171
Greenewalt (1975)

page 172
Kenneth Sommerfield

page 173
Adapted from Greenewalt (1975)

page 174 (above)
Stock Boston

page 174 (below)
Greenewalt (1975)

page 175
Cambridge Aero Instruments, Inc.

page 176
Animals, Animals

page 177 (above)
Redrawn from a drawing by René Marten (Life Nature Library).

page 177 (below)
Photo Researchers

page 178
Adapted from Hertel (1963).

page 179 (below left)
Photo Researchers

page 179 (below right)
Stock Boston

page 181 (above)
John Long

page 181 (below)
Adapted from Hertel (1963).

page 182 (above)
Bruce Coleman

page 182 (below)
Hertel (1963)

page 183 (top)
From Memoirs of the M.C.Z. *36*, 1913, plate 38.

page 184
C.W. McCutchen (1976)

page 185
Bruce Coleman

page 188 (above)
Hertel (1963)

page 188 (below)
Hertel (1963)

page 189
Adapted from Wu (1977).

page 190 (above)
Tom Stack and Assoc.

page 190 (below)
Goldspink (1977)

page 191
Photo courtesy of The Cousteau Society, Inc., 930 West 21st Street, Norfolk, Virginia 23517, a membership supported environmental organization.

## Chapter 6

page 192
Photo Researchers

page 194 (right)
From Went (1968).

page 196 (right)
Bruce Coleman

page 197 (3 photos)
N.C.F.M.F. Book of Film Notes, 1972; MIT Press with Educational Development Center, Inc., Newton, Mass.

page 199 (left)
Sleigh and Blake (1977)

page 199 (middle)
Adapted from Sleigh and Blake (1977).

page 200 (left)
Photo Researchers

page 200 (right)
Peter Arnold

page 201 (left)
Photo Researchers

page 201 (right a,b,c,d)
Life Science Library/*The Cell.* Photograph by Sol Mednick. Time-Life Books, Inc. Publisher. © 1964 Time Inc.

page 202
Charles J. Brokaw (1972)*

page 203 (left)
Anderson (1975)

page 203 (right)
H.C. Berg (1975)

American Museum of Natural History

page 204
Stephen Dalton, Natural History Photography Agency, Sussex, England.

page 205 (b)
From Newman et al. (1977)

page 206
Newman et al. (1977)

page 207
Newman et al. (1977)

page 208 (left)
Stuart B. Savage

page 208 (right)
Photo Researchers

page 209
John Shaw Photography

page 210
Stephen Dalton, Natural History Photography Agency, Sussex, England.

page 211 (above left and right)
Peter Arnold

page 211 (below)
International Museum of Photography at George Eastman House, Rochester, New York.

page 212
Bennet-Clark (1977)

page 214
Thomas A. McMahon

page 215 (top right)
Courtesy Chicago Bridge and Iron Co., Oak Ridge, Ill.

page 215 (below right and left)
Thomas A. McMahon

page 216
Bruce Coleman

page 218 (below)
Photo Researchers

page 221
Went (1968)

page 222
Thomas Eisner

page 223 (above)
A. Gunther (1864)

page 223 (below a,b,c)
The low and intermediate magnifications are unpublished micrographs by Jane Peterson. The high magnification micrograph is from Williams and Peterson (1982).

page 224
J.R. Eyerman, LIFE Magazine © 1950 Time Inc.

page 225
Redrawn from Wigglesworth (1972).

## Chapter 7

page 226
Scala/Art Resources

page 228 (right)
Data accumulated by Damuth (1981).

page 229
Charles Ott (National Audubon Society) Photo Researchers

page 230
(a) Peter Arnold (b) Animals, Animals
(c) Peter Arnold (d) Photo Researchers

page 231
(a) Animals, Animals (b) Animals, Animals
(c) Animals, Animals (d) Natural History Photography Agency, Sussex, England

page 232 (left)
Dennis Brokaw Photography

page 232 (right)
Jeff Gnass, West Stock Inc.

page 233
Diamond (1973)*

page 235
Bonner (1965)

page 236 (below)
Natural History Photography Agency, Sussex, England.

page 236 (above)
Lynn M. Stone

page 237 (above left)
Leonard Rue Enterprises

page 237 (above right)
Lynn M. Stone

page 237 (below)
Alexander et al. (1979)

page 238
Redrawn from H.J. Jerison.

page 239 (above)
Manfred Kage, Peter Arnold

page 239 (table)
After Jerison (1973)

page 240
May (1978)

page 243
Ellis Herwig, Stock Boston

# References

ALEXANDER, R. MCN. 1981. Factors of safety in the structure of animals. *Sci. Prog., Oxf.* 67:109-130.

ALEXANDER, R. D.; HOOGLAND, J. L.; HOWARD, R. D.; NOONAN, K. M.; and SHERMAN, P. W. 1979. Sexual dimorphisms and breeding systems in Pinnipeds, Ungulates, Primates and Humans. In: *Evolutionary Biology and Human Social Behavior,* N. A. Chagnon and W. Irons, eds. Duxbury Press, North Scituate, Mass., pp. 402-435.

ALEXANDER, R. M. 1977. Mechanics and scaling of terrestrial locomotion. In: *Scale effects in animal locomotion* (T. J. Pedley, ed.), New York/London: Academic Press, pp. 93-110.

ANDERSON, R. A. 1975. Formation of the bacterial flagellar bundle. In: *Swimming and flying in nature* (T.Y.-T. Wu, C. J. Brokaw, and C. Brennen, eds.), vol. 1. New York: Plenum Press, pp. 45-46.

BAINBRIDGE, R. 1961. Problems of fish locomotion. *Symp. Zool. Soc. London.* 5:13-32.

BARGHOORN, E. S. and TYLER, S. A. 1965. Microorganisms from the Gunflint Chert. *Science* 147:563-577.

BENNET-CLARK, H. C. 1977. Scale effects in jumping animals. In: *Scale effects in animal locomotion* (T. J. Pedley, ed.), New York/London: Academic Press, pp. 185-201.

BERG, H. C. 1975. Bacterial movement. In: *Swimming and flying in nature* (T.Y.-T. Wu, C. J. Brokaw, and C. Brennen, eds.), vol. 1. New York: Plenum Press, pp. 1-11.

BERG, H. C. and ANDERSON, R. A. 1973. Bacteria swim by rotating their flagellar filaments. *Nature* 245:380-382.

BONNER, J. T. 1965. *Size and Cycle.* Princeton University Press, Princeton, New Jersey.

BONNER, J. T. 1968. Size change in development and evolution. *J. Paleontology* 42:(Part II)1-15.

BONNER, J. T. 1982. Evolutionary strategies and developmental constraints in the cellular slime molds. *American Naturalist* 119:530-552.

BROKAW, C. J. 1972. Flagellar movement: a sliding filament model. *Science* 178:4060, pp. 455-462.

BROKAW, C. J. and GIBBONS, I. R. 1975. Mechanisms of movement in flagella and cilia. In: *Swimming and flying in nature* (T.Y.-T. Wu, C. J. Brokaw, and C. Brennen, eds.), vol. 1. New York: Plenum Press, pp. 89-126.

BUCKINGHAM, E. 1914. On physically similar systems; Illustrations of the use of dimensional equations. *Phys. Rev.* 4:345-376.

CALDER, W. A. 1981. Scaling of physiological processes in homeothermic animals. *Ann. Rev. Physiol.* 43:301-322.

COPE, E. D. 1885. On the evolution of the vertebrata, progressive and retrogressive. *American Naturalist* 19: 140-148, 234-247, 341-353.

CURREY, J. D. 1977. Problems of scaling in the skeleton. In: *Scale effects in animal locomotion* (T. J. Pedley, ed.), New York/London: Academic Press, pp. 153-167.

DAMUTH, J. 1981. Population density and body size in mammals. *Nature* 290:699-700.

DARWIN, C. 1859. *On the origin of species.* A facsimile of the first edition. Harvard University Press, Cambridge, Mass. (1975).

DARWIN, C. 1871. *The Descent of Man and Selection in Relation to Sex.* A facsimile of the first edition. Princeton University Press (1981).

DEDRICK, R. L. and BISCHOFF, K. B. 1980. Species similarities in pharmacokinetics. *Federation Proc.* 39:54-59.

DIAMOND, J. M. 1973. Distributional ecology of New Guinea birds. *Science* 179:759-769.

ELTON, C. S. 1927. *Animal Ecology.* London: Sidgwick & Jackson.

FAIRCHILD, L. 1981. Mate selection and behavioral thermoregulation in Fowler's toad. *Science* 212:950-951.

FELDMAN, H. A. and MCMAHON, T. A. 1983. The ¾ mass exponent for energy metabolism is not a statistical artifact. *Resp. Physiol.* 52:149-163.

FRANKEL, R. B.; BLAKEMORE, R. P.; and WOLFE, R. S. 1979. Magnetite in fresh-water magnetotactic bacteria. *Science* 203:1355-1356.

GHISELIN, M. T. 1974. *The economy of nature and the evolution of sex*. Berkeley: University of California Press.

GOLDSPINK, G. 1977. Mechanics and energetics of muscle in animals of different sizes, with particular reference to the muscle fibre composition of vertebrate muscle. In: *Scale effects in animal locomotion* (T. J. Pedley, ed.), New York/London: Academic Press, pp. 37-55.

GOULD, S. J. 1975. Allometry in primates, with emphasis on scaling and evolution of the brain. In: *Approaches to Primate Paleobiology* (Contrib. Primate. vol. 5). Basel, Karger (Switzerland), pp. 244-292.

GRAY, J. 1936. Studies in animal locomotion. VI. The propulsive powers of the dolphin. *J. Exp. Biol.* 13:192-199.

GREENEWALT, C. H. 1975. The flight of birds. *Trans. Am. Philosophical Society,* vol. 65, part 4.

GREENEWALT, C. H. 1977. The energetics of locomotion—is small size really disadvantageous? *Proceedings of the American Philosophical Society,* Vol. 121, No. 2, pp. 100-106.

GREENHILL, A. G. 1881. Determination of the greatest height consistent with stability that a vertical pole or mast can be made, and of the greatest height to which a tree of given proportions can grow. *Proc. Cambridge Philosoph. Soc.* 4:65-73.

GREGORY, W. K. 1951. *Evolution emerging. A survey of changing patterns from primeval life to man*. Macmillan Co., New York. Volume 1, 736 pp.; Volume 2, 1013 pp.

GUNTHER, A. 1864. The reptiles of British India. Ray Society.

HALDANE, J. B. S. 1928. On being the right size. In: *Possible Worlds*. New York: Harper. pp. 20-28.

HALPERT, A. P. 1982. Structure and scaling of the lumbar vertebrae of African bovids (Mammalia: Artiodactyla). B. A. thesis, Dept. of Biology, Harvard University.

HARVEY, E. N. 1954. Tension at the cell surface. In: *Protoplasmatologia* (L. V. Heilbrunn and F. Weber, eds.) Band 2, E5, 30 pp. Wein: Springer-Verlag.

HEGLUND, N. C.; TAYLOR, C. R.; and MCMAHON, T. A. 1974. Scaling stride frequency and gait to animal size: mice to horses. *Science* 186:1112-1113.

HEMMINGSEN, A. 1960. *Energy Metabolism as Related to Body Size and Respiratory Surfaces, and Its Evolution*. Copenhagen: C. Hamburgers.

HERTEL, H. 1963. *Structure, form, movement*. New York: Reinhold Publishing Co.

HEUSNER, A. A. 1982. Energy metabolism and body size. I. Is the 0.75 mass exponent of Kleiber's equation a statistical artifact? *Respir. Physiol.* 48:1-12.

HILL, A. V. 1950. The dimensions of animals and their muscular dynamics. *Science Progress* 38:209-230.

HORN, H. S. and MAY, R. M. 1977. Limits to similarity among coexisting competitors. *Nature* 270:660-661.

HUTCHINSON, G. E. 1959. Homage to Santa Rosalia, or why there are so many kinds of animals. *American Naturalist* 93:145-159.

HUTCHINSON, G. E. and MACARTHUR, R. H. 1959. A theoretical ecological model of size distributions among species of animals. *American Naturalist* 93:117-125.

HUXLEY, J. S. 1932. *Problems in Relative Growth*. London: Methuen.

JERISON, H. J. 1973. *Evolution of the Brain and Intelligence*. New York: Academic Press.

JOHNSTON, I. A. and GOLDSPINK, G. 1973. A study of the swimming performance of the Crucian carp *Carassius carassius (L.)* in relation to the effects of exercise and recovery on biochemical changes in the myotomal muscles and liver. *J. Fish. Biol.* 5:249-260.

KLEIBER, M. 1932. Body size and metabolism. *Hilgardia* 6:315-353.

KUMMER, B. 1951. Zur Entstehung der menschlichen Schädelform (ein Beitrag zum Fetalisationsproblem). Verh. anat. Ges. 49 (suppl. to Anat. Anz. vol. 98):140-145.

LACK, D. 1947. *Darwin's Finches*. Cambridge University Press.

LANG, T. G. 1975. Speed, power, and drag measurements of dolphins and porpoises. In: *Swimming and Flying in Nature,* Vol. 2 (Wu, Brokaw, and Brennen, eds.), New York: Plenum Press, pp. 553-572.

LIETZKE, M. H. 1956. Relation between weight-lifting totals and body weight. *Science* 124:486-487.

LIGHTHILL, J. 1977. Introduction to the scaling of aerial locomotion. In: *Scale effects in animal locomotion* (T. J. Pedley, ed.), New York/London: Academic Press, pp. 365-404.

LINDSTEDT, S. L. and CALDER, W. A. 1981. Body size, physiological time, and longevity of homeothermic animals. *Quart. Rev. Biol.* 56:1-16.

LOCHHEAD, J. H. 1977. Unsolved problems of interest in the locomotion of crustacea. In: *Scale effects in animal locomotion* (T. J. Pedley, ed.), New York/London: Academic Press, pp. 257-268.

MACARTHUR, R. H. 1972. *Geographical Ecology.* New York: Harper & Row.

MACARTHUR, R. H. and WILSON, E. O. 1967. *The Theory of Island Biogeography.* Princeton University Press.

MAY, R. M. 1978. The dynamics and diversity of insect faunas. In: *Diversity of Insect Faunas,* L. A. Mound and N. Waloff, eds. Symposia of the Royal Entomological Society in London, No. 9. Oxford: Blackwell Scientific Publications, Inc., pp. 188-204.

MCCUTCHEN, C. W. 1976. Flow visualization with stereo shadowgraphs of stratified fluid. *J. Exp. Biol.* 65:11-20.

MCMAHON, T. A. 1971. Rowing: a similarity analysis. *Science* 173:349-351.

MCMAHON, T. A. 1973. Size and shape in biology. *Science* 179:1201-1204.

MCMAHON, T. A. 1975a. The mechanical design of trees. *Scientific American* 233:92-102.

MCMAHON, T. A. 1975b. Using body size to understand the structural design of animals: quadrupedal locomotion. *J. Appl. Physiol.* 39:619-627.

MCMAHON, T. A. 1975c. Allometry and biomechanics: limb bones in adult ungulates. *American Naturalist* 109:547-563.

MCMAHON, T. A. 1983. *Muscles, Reflexes, and Locomotion*. Princeton University Press, Princeton, New Jersey.

MEDAWAR, P. B. 1945. Size, shape, and age. In: Essays on growth and form presented to D'Arcy Thompson (W. E. Le Gros Clark and P. B. Medawar, eds.) pp. 157-187. Oxford: Clarendon Press.

MEDAWAR, P. B. 1958. Postscript: D'Arcy Thompson and *Growth and Form*. In: *D'Arcy Wentworth Thompson*, by Ruth D'Arcy Thompson. London: Oxford University Press. pp. 219-233.

MUYBRIDGE, E. 1887. *Animal locomotion. An electro-photographic investigation of consecutive phases of animal movements.* Philadelphia: Lippincott.

NEWELL, N. D. 1949. Pyletic size increase, an important trend illustrated by fossil invertebrates. *Evolution* 3:103-124.

NEWMAN, B. G.; SAVAGE, S. B.; and SCHOUELLA, D. 1977. Model tests on a wing section of an Aeschna dragonfly. In: *Scale effects in animal locomotion* (T. J. Pedley, ed.) New York/London: Academic Press, pp. 445-477.

NOORDERGRAFF, A.; LI, J. K.; and CAMPBELL, K. B. 1979. Mammalian hemodynamics: a new similarity principle. J. Theor. Biol. 79:485-489.

PENNAK, R. W. 1953. *Fresh-water Invertebrates of the United States.* New York: John Wiley and Sons.

PRANGE, H. D. 1977. The scaling and mechanics of arthropod exoskeletons. In: T. J. Pedley (ed.) *Scale effects in animal locomotion.* New York/London: Academic Press, pp. 169-181.

PROSSER, C. L. 1973. *Comparative Animal Physiology.* Philadelphia: Saunders, p. 827.

PURCELL, E. M. 1977. Life at low Reynolds number. *American Journal of Physics* 45:3-11.

RAPER, K. B. 1941. Developmental patterns in simple slime molds. Third Growth Symposium. *Growth* 5:41-76.

RAYLEIGH, J. W. S. 1878. *Theory of Sound.* New York: Dover Publications.

RENSCH, B. 1956. Increase of learning capacity by increase in brain size. *American Naturalist* 90:81-95.

RUBNER, M. 1883. Ueber den Einfluss der Korpergroesse auf Stoff- und Kraftwechsel. *Z. Biol. Munich* 19:535-562.

SACHER, G. A. and STAFFELDT, E. F. 1974. Relation of gestation time to brain weight for placental mammals: Implications for the theory of vertebrate growth. *Am. Nat.* 108:593-615.

SCHMIDT-NIELSEN, K. 1970. Energy metabolism, body size, and problems of scaling. *Fed. Proc.* 29:1524-1532.

SCHMIDT-NIELSEN, K. 1975. Scaling in biology: the consequences of size. *J. Exp. Zoology* 194:287-307.

SCHMIDT-NIELSEN, K. 1977. Problems of scaling: Locomotion and physiological correlates. In: *Scale effects in animal locomotion* (T. J. Pedley, ed.), New York/London: Academic Press, pp. 1-21.

SEDOV, K. I. 1959. *Similarity and Dimensional Methods in Mechanics.* Academic Press, New York, N.Y.

SCHULTZ, ADOLPH H. 1969. *The Life of Primates.* Universe Books, New York, N.Y.

SILVERMAN, M. and SIMON, M. 1974. Flagellar rotation and the mechanism of bacterial motility. *Nature* 249:73-74.

SIMPSON, G. G. 1953. *The Major Features of Evolution.* New York: Columbia University Press.

SLEIGH, M. A. and BLAKE, J. R. 1977. Methods of ciliary propulsion and their size limitations. In: *Scale effects in animal locomotion* (T. J. Pedley, ed.), New York/London: Academic Press, pp. 243-256.

STAHL, W. R. 1962. Similarity and dimensional methods in biology. *Science* 137:205-212.

STAHL, W. R. and GUMMERSON, J. Y. 1967. Systematic allometry in five species of adult primates. *Growth* 31:21-34.

STEINBERG, M. S. 1978. Specific cell ligands and the differential adhesion hypothesis: How do they fit together. In: *Specificity of Embryological Interactions* (D. R. Garrod, ed.) London: Chapman and Hall, pp. 97-130.

STEVENS, P. S. 1974. *Patterns in nature.* Boston: Atlantic Monthly Press.

TAYLOR, C. R. 1977. Why big animals? *The Cornell Veterinarian* 67:155-175.

TAYLOR, C. R.; CALDWELL, S. L.; and ROWNTREE, V. J. 1972. Running up and down hills: some consequences of size. *Science* 178:1096-1097.

TAYLOR, C. R.; MALOIY, G. M. O.; WEIBEL, E. R.; LANGMAN, V. A.; KAMAU, J. M. Z.; SEEHERMAN, H. J.; and HEGLUND, N. C. 1981. Design of the mammalian respiratory system. III. Scaling maximum aerobic capacity to body mass: wild and domestic animals. *Respir. Physiol.* 44:25-37.

TAYLOR, G. I. 1950. The formation of a blast wave by a very intense explosion. I. Theoretical discussion. *Proc. Roy. Soc., A,* 201:159-186.

TESSIER, G. 1931. Recherches morphologiques et physiologiques sur la croissance des insectes. Travaux de la station biologique de Roscoff 9:27-238.

THOMPSON, D'A. W. 1917 (and later editions) *On Growth and Form.* Cambridge University Press, Cambridge.

ULTSCH, G. R. 1973. A theoretical and experimental investigation of the relationships between metabolic rate, body size, and oxygen exchange capacity. *Resp. Phys.* 18:143-160.

VAN VALEN, L. 1973. Body size and numbers of plants and animals. *Evolution* 27:27-35.

WEBB, P. W. 1977. Effects of size on performance and energetics of fish. In: *Scale effects in animal locomotion* (T. J. Pedley, ed.), New York/London: Academic Press, pp. 315-331.

WELLS, H. G.; HUXLEY, J. S.; and WELLS, G. P. 1931. *The Science of Life.* London: Doubleday, Doran and Company.

WENT, F. W. 1968. The size of man. *American Scientist* 56:400-413.

WILKIE, D. R. 1977. Metabolism and body size. In: *Scale effects in animal locomotion* (T. J. Pedley, ed.), New York/London: Academic Press, pp. 23-36.

WILLIAMS, E. E. and PETERSON, J. A. 1982. Convergent and alternative designs in the digital adhesive pads of scincid lizards. *Science* 215:1509-1511.

WU, T. Y. 1977. Introduction to the scaling of aquatic animal locomotion. In: *Scale effects in animal locomotion* (T. J. Pedley, ed.), New York/London: Academic Press, pp. 203-232.

# Index

**References to illustrations are printed in boldface type. The lower case "t" indicates reference to a table.**